Health Care Waste Management and COVID 19 Pandemic

Sadhan Kumar Ghosh • Pariatamby Agamuthu
Editors

Health Care Waste Management and COVID 19 Pandemic

Policy, Implementation Status and Vaccine Management

 Springer

Editors
Sadhan Kumar Ghosh
Mechanical Engineering
Jadavpur University
Kolkata, India

Pariatamby Agamuthu
Center on Sustainable Development
Sunway University
Banndar Sunway, Malaysia

ISBN 978-981-16-9338-0   ISBN 978-981-16-9336-6   (eBook)
https://doi.org/10.1007/978-981-16-9336-6

This Springer imprint is published by the registered company Springer Nature Singapore Pte Ltd.
The registered company address is: 152 Beach Road, #21-01/04 Gateway East, Singapore 189721, Singapore

Contents

Part VII   Third Wave of COVID Pandemic

Part I
COVID 19 Waste Management and Cultural Perspectives

Waste Management During Pandemic of COVID-19 in India, Italy, and the USA: The Influence of Cultural Perspectives

Sadhan Kumar Ghosh, Ronald L. Mersky, Sannidhya Kumar Ghosh, and Francesco Di Maria

1 Introduction

Since its first detection occurred in the city of Wuhan (China) (CDC, 2020) the diffusion of the severe acute respiratory syndrome (SARS) disease due to the novel coronavirus (SARS-CoV-2) has been characterized on March 11th by the World Health Organization (WHO, 2020a) as a pandemic COVID-19. The world was facing a novel pathogen, the scope and spread of which was unknown in the beginning of 2020. WHO declared a Public Health Emergency of International Concern on 30 January and a Global Pandemic on 11 March. The UN and public health partners took early action to ensure access to life-saving supplies, notably PPE and test kits in February–March, notably WHO and UNICEF (WHO, 2021). Total of infected people across the world as on May 10th 2020 was greater than 3,900,000 with a number of total deaths of about 275,000 (WHO, 2020b) spread over more than 210 countries. According to the WHO COVID-19 dashboard, the number of affected confirmed cases are 249,399,713, number of deaths reached 5,046,344 and recovered cases 225,874,632 in the world as on fifth November 2021 (https://www.worldometers.info/coronavirus/). SARS together with middle east respiratory syndrome (MERS) represents the third highly pathogenic human coronavirus that

S. K. Ghosh (✉)
Mechanical Engineering Department, Jadavpur University and ISWMAW, Kolkata, West Bengal, India

R. L. Mersky
Department of Civil Engineering, Widener University, Chester, PA, USA

S. K. Ghosh
Department of Civil, Environmental and Architectural Engineering, University of Colorado Boulder, Boulder, CO, USA

F. Di Maria
Dipartimento di Ingegneria – University of Perugia, Perugia, Italy

3

emerged in the last 20 years (de Wit et al., 2016). As already detected for other respiratory infections the main transmission pathway of SARS-CoV-2 viruses is based on airborne droplets, with average diameter ≥ 5 mm, generated by sneeze, cough, breaths, or during normal speaking of infected subjects (Lewis, 2020; National Research Council, 2020; Yu et al., 2018; WHO, 2020c). Liu et al. (2020) reported the presence of SARS-CoV-2 in the airborne aerosols sampled in the air inside two hospitals of Wuhan during pandemic peaks. Similar results were also reported by Santarpia et al. (2020). In this study they analysed 13 rooms of the Nebraska University Hospital in which were recovered patients infected by SARS-CoV-2. Both authors indicated also frequent ventilation of rooms and staying in open spaces as two effective ways for reducing the spread of the infection.

Together with nose and mouth also eyes have been indicated as another possible access of the SARS-CoV-2 into human body contributing to the pandemic spreading. For this reason, together with social distances and use of protective mask and gloves also particular care in hand cleansing has been recommended. In fact, the persistence of SARS-CoV-2 on different surfaces and materials has been already detected by several authors. Van Doremalen et al. (2020) reported that SARS-CoV-2 remained viable on different surfaces such as cardboard, copper, stainless steel, and plastic from 15 h up to 72 h at laboratory conditions indicating high similarity with SARS-CoV-1 virus. Kampf et al. (2020) detected the persistence of the virus on inanimate surfaces up to 9 days indicating also the effectiveness of its inactivation by the use of different biocidal agents. Of course, ambient and climatic conditions (e.g., temperature, humidity, solar radiation) can influence this period. All this highlights the importance of proper management and disposal of waste materials potentially affected by the SARS-CoV-2. Health care waste generated by hospital or assimilated activities has already their specific regulations, in particular concerning those characterized by risk of infection (DPR, 2003; WHO, 2020d). A different question is presented by household waste generated by patients treated at home or in other areas different from hospitals and medical centers because of positive tests or for quarantine. Even if there is no definitive evidence of the persistence of the virus in such waste there is anyway a diffused perception of the potential risk due to their improper management that could represent a threat for both operators and citizens (Gomes Mol & Caldas, 2020).

Furthermore, improper management of such waste could represent a serious threat for workers of the informal sector, that is a very diffused activity in given areas (Cruvinel et al., 2019). About this topic the European Commission states that there is no evidence of the transmission of the virus via household waste management, recommending the continuity of proper municipal waste management services including separate collection in the EU area (EC, 2020). However, this indication appears quite far from the realistic perception of the problems daily faced by workers and citizens and from the opinion of experts, leading some member states to develop specific recommendations and guidelines for waste management during the coronavirus crisis (ISS, 2020). USA and Italy are among the top ten most affected countries, whereas India, the biggest democracy with 1.38 billion people is a moderately affected country. However, it is good that country like India has reached a target of 1 billion people vaccinated by early October 2021, while in the USA and Italy more than 90% people are vaccinated. Medical organizations and general citizens applauded the rollout of pediatric vaccines in India, Italy, and the USA. Based on these facts the present paper presents the state of the art of indications and guidelines concerning these particular wastes already implemented in the three different countries considered: India, Italy, and the USA.

2 Materials and Methods

The data reported in this study were based on the most recent documents and guidelines, technical reports, and research papers released by recognized entities, legal bodies, central and local governments as of 18th May 2020. Given the relevance of the topic and the continuous evolution of the crisis, documents and guidelines may be modified and updated in the near future. The comparison of the different schemes implemented in the considered countries was performed on the basis of the possible association between the prescription and recommendations reported in the regulations and the different socio-cultural perspectives (Thompson et al., 1990) from which they seem inspired.

Societal reference systems influence the risk judgment and management style. Even if there is a general agreement on the major causes of environmental and health impact different preconditions (e.g., population density, climate, GDP) but also cultural differences generate different comprehensive coverage among the different countries.

According to grid-group cultural theory (Douglas & Wildavsky, 1982) the following four main ways of life can be identified: individualist, egalitarian, hierarchical, and fatalist. The first group is characterized by a social context in which individuals are bound by neither group incorporation nor prescribed roles generating an individualist social context. The second group is characterized by individuals with strong group boundaries and minimal prescriptoins generating an egaliatrian social context. In hierarchical social context individuals have both strong group boundaries and social prescription generating a social environment where people are controlled both by the groups to which they belong but also by the demands of roles imposed by the society. Finally, fatalist social contexts are characterized by low or absence of boundaries of the group (i.e., exclusion from the group) but high binding from social prescriptions. Fatalists are excluded from membership in the groups responsible for making their life rules. In practice these four ways of life represent only archetypes to which a limited number of individuals are fully committed adherents. For this reason, Schwarz and Thompson (1990) reported a list of world visions, management styles, attitudes, and characteristics of the four archetypes with respect to several aspects concerning social life. Starting from this, the main aspects more adherent to the problem of waste management under coronavirus crisis were identified as reported in Table 1. Successively per each country a single score was assigned to the way of life more representative for the management style or attitude

Table 1 Identification of the management style, attitude, and characteristics of the four ways of life archetypes concerning the main aspects of regulation and approaches for waste management under coronavirus

Aspects	Way of life			
	Hierarchist	Egalitarian	Individualist	Fatalist
Procedures applied	Rules	Ethical standard	Skill	–
Management style	Control	Preventive	Adaptive	–
Pollution solution	Change nature to conform society	Change society to conform nature	Market incentives (transferable rights to pollute)	
Attitude toward risk	Risk-accepting	Risk-aversive	Risk-seeking	–
Risk handling style	Rejection and absorption	Rejection and deflection	Acceptance and deflection	Acceptance and absorption

adopted toward each single aspect according to this scheme: 1 = representative; 0 = not representative. Of course, for specific aspects more than one way of life could be representative. In this way the contribution of each socio-cultural perspective to the final system adopted for waste management under coronavirus crisis was ranked.

2.1   The Italian Scenario

Among the occidental countries Italy was the first one where the SARS-CoV-2 pandemic started in a very severe manner making a very huge pressure on the sanitary system. Possibly first cases started since the first half of January 2020. At the end of February, the number of total cases detected were about 888, reaching more than 7300 by the 9th of March 2020. Subsequently, the central government imposed the lockdown of the whole nation with consequent restriction of the mobility of people which had positive effects on the decrease of the epidemiological curve (Fig. 1).

The big efforts put in practice for limiting both the negative consequence of the infection but also its further diffusion (e.g., social distances, personal protective equipment, hand washing, sanitation care) lead also to take in serious consideration the risk represented by the waste generated by private houses and areas in which are treated patients affected by SARS-Co-V-2 or in quarantine. Waste management was indicated by the government as an essential activity that must be operated also during the coronavirus crisis.

One of the first official documents concerning this topic was the Report n.3/2020 of the Italian Institute of Health (ISS) of the Ministry of Health (ISS, 2020) dated

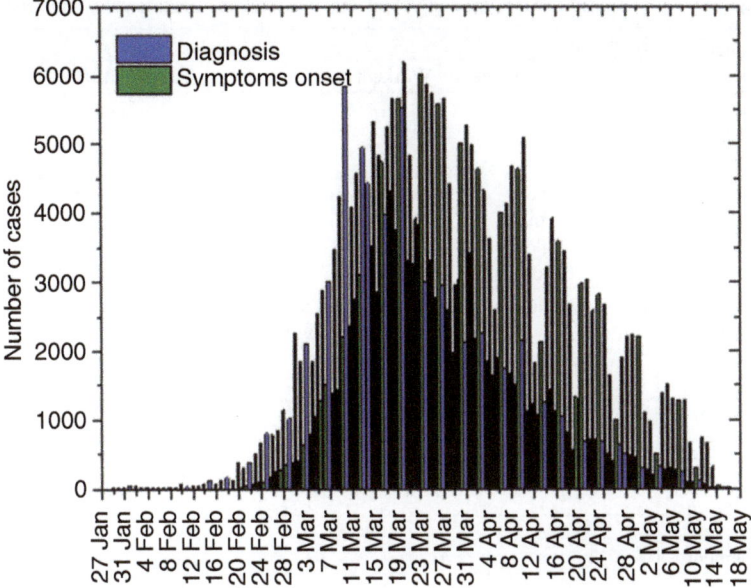

Fig. 1 Number of Italian tests and cases detected from 27th January to 18st May 2020

14th March 2020 and successively updated on 31st March 2020. This gives some instructions to the stakeholders, workers, and trained volunteers aimed to avoid the risk of contagion during waste collection. On 23rd of March the Italian Institute for the Environmental Research and Protection (ISPRA, 2020) published some guidelines including the treatment and the storage of these wastes. On 30th March the Italian Ministry of Environment (2020) published main guidelines concerning the storage, the incineration, and the landfilling of waste potentially affected by SARS-CoV-2 different from those from medical centers. Meanwhile different local governments started to promote specific orders concerning the management of these specific wastes.

In Table 2 are summarized the main indications reported in these official documents including the users, the workers, and the final treatment. A first general recommendation was the interruption of separated collection and to put all the waste from possibly contagious persons in a double plastic bag avoiding the presence of sharp materials. Indication on minimal individual protection systems and how to manage them after their use were also reported with particular care for workers that are the ones characterized by higher risk of becoming in contact with potentially infected household waste.

Specific indications on the adoption of single use individual protection systems as masks and gloves were included for both users, trained volunteers, and workers. Of course the latter have also to wear the other protection systems imposed by the current specific safety regulations. Furthermore, specific indication on how to proceed to the sanitization of professional suits was also supplied. After collection

Table 2 Main guidelines for users, trained volunteers, workers, and treatments at final destination

Guideline/ activities	Recipients			
	Users	Volunteers	Workers	Treatment facilities
Interruption of separated collection	Yes	–	–	–
Single use individual protection system	Yes (gloves)	Yes (mask, gloves, coverall)	Yes (gloves, coverall)	–
Use of durable protection systems	No	No	Yes (filtering mask, other according to worker safety regulation)	–
Sanitization instruction	Yes	Yes	Yes	–
Waste recycling	–	–	–	No
Waste pre-treatment	–	–	–	No
Composting	–	–	–	No
Anaerobic digestion	–	–	–	No
Incineration	–	–	–	Yes (prior)
MBT	–	–	–	Yes but only if able to ensure sanitization of the whole waste before sorting
Landfill	–	–	–	Yes but in dedicated cells properly managed

these wastes must be moved directly to the disposal facilities without any pre-treatment. Direct incineration is the preferable option. Mechanical biological treatment (MBT) can be used only if able to avoid any manual operation and to ensure the sanitization of the whole mass of waste before any sorting operation. Landfilling can also be used but the adoption of dedicated cells clearly identified and to be properly covered every day is strongly recommended.

Concerning the SARS-CoV-2 infected waste generated by hospital and other medical centers, the current legislation (DPR, 2003) indicates that, after being separately collected and sealed in plastic bins clearly identified, these must be disposed, without any pre-treatment, by:

- Co-incineration with municipal waste in facilities equipped with direct loading systems in the combustion chamber or in the furnace hopper;
- Incineration in dedicated facilities;
- Sterilization after which used for the production of solid recovery fuel or incinerated;

– Landfilling but only if there is no incineration capacity and as a transient solution until adequate incineration capacity is delivered.

Of course any manual operation is strictly forbidden.

2.2   The USA Scenario

In the USA SARS-CoV-2 policies regarding closure of facilities and activities are primarily promulgated at the state level. This means that there are 51 different policies among the 50 states and District of Columbia. Even within individual states there are sometimes geographic variations of regulations based on local intensity of the outbreak (ISRI, 2020).

Early in the pandemic, at the national level, the USA Department of Homeland Security (DHS) designated solid waste collection and transport workers as belonging to America's Essential Critical Infrastructure Workforce (Taylor, 2020). While recycling activities were not been specifically designated by DHS as essential, workers necessary for manufacturing of critical materials were so designated, which may imply recycling workers (DeAnne, 2020). Overall waste collection, transport, recycling, and disposal continued, but with many local disruptions.

As people spent more time in homes, residential waste collection volumes increased, with some estimates expecting growth of up to 30%. This placed a greater burden on workers and equipment. Some communities responded by suspending the pickup of bulky wastes and yard wastes (which may have reduced composting rates) and changing collection schedules and procedures for other wastes (WPXI, 2020; GCH, 2020; Staub, 2020; MacFarland, 2020). In some places residential collections were delayed and recycling completely suspended as staff were unable to meet the increased generation.

Furthermore, the shift from commercial to residential generation adversely affected the industry by impacting quality of materials and costs. Offices, stores, and other businesses generally produce a recycling stream that is less contaminated than that produced residentially. Furthermore, residential recycling collection is more costly than commercial (Minter, 2020). The shift from commercial to residential generation thus put an added strain on the recycling industry which had already been suffering from reduced markets and less tolerance of contamination among materials buyers. As the USA economy has begun to return closer to normal in 2021, another issue that impacted the waste industry was, and continues to be, a shortage of workers. This has also resulted in collection delays, mostly involving recycling.

In summary, the USA waste industry has functioned throughout the pandemic and that is unlikely to change due to its essential nature. However, the shift from commercial to residential generation and current labor shortage has resulted in temporary procedural and economic challenges. Most significantly, recycling has been adversely affected by increased costs and contamination potential as well as reduced collection services.

2.3 The Indian Scenario

2.3.1 Present Situation

The rate of affected persons was going up slowly in the country at first but in the prior 2 weeks the rate was increasing with 52,952 recorded confirmed active cases, 1783 deaths, 15,267 cured and discharged cases, and 1 migrated case according to the update on seventh May 2020, at GMT+5:30 released on the COVID-19 Dashboard, Government of India. India is a federal union comprising 28 states and 8 union territories (UT), for a total of 36 entities (MyGov, 2020). Citizens in most out of these 36 entities have been affected and many efforts have been taken by respective state governments in the country. Table 3 shows state-wise data as of May 2020. The current population of India is 1,377,937,289 as of May 7, 2020, based on Worldometer elaboration of the latest United Nations data which is equivalent to 17.7% of the total world population (Table 3). At the initial stage the number of affected people in India was very low. From the very beginning of the insurgence of coronavirus, the Government of India as well as most of the states and UT started taking actions to arrest and prevent the effect of SARS-CoV-2. The pandemic has changed the way people live and work in India as with other countries throughout the world. In the case of communities affected by COVID 19, the importance of provisioning safe water, sanitation, and hygiene (WASH), Solid Waste Management (SWM), and other cities' services is very important. Planning, waste management, sewage management, and how densely packed urban areas will deal with some of the important issues.

On 22nd March 2020, with the lockdown of the whole nation with consequent restriction of the mobility of people, there were positive effects on the decrease of the epidemiological curve.

2.3.2 Challenges

There are a few challenges identified for the provisioning of Solid Waste Management (SWM), safe water, sanitation, and hygiene (WASH) during the pandemic outbreak of COVID-19 virus as follows:

- hand hygiene as per WHO directives,
- inactivation of surfaces that can spread the COVID-19,
- risk of transmission of COVID-19 virus from faces of an infected person,
- movement of people and necessary protection,
- drinking-water and sanitation services applied to the COVID-19 outbreak as per WHO guidelines,
- handling and disposal of infectious waste from confirmed COVID-19 patients,
- WASH and SWM preparedness and response plans during normal and emergency situation,
- additional precautions for formal and informal waste management workers,

Table 3 State-wise data of confirmed, active, recovered, and deceased cases as of 7th May 2020

State/union territories (UT)s	Confirmed	Active	Recovered	Deceased
Maharashtra	16,758	13,013	3094	651
Gujarat	6625	4729	1500	396
Delhi	5532	3925	1542	65
Tamil Nadu	4829	3278	1516	35
Rajasthan	3317	1629	1596	92
Madhya Pradesh	3138	1854	1099	185
Uttar Pradesh	2998	1808	1130	60
Andhra Pradesh	1777	1012	729	36
Punjab	1516	1354	135	27
West Bengal	1456	948	364	144
Telangana	1107	450	628	29
Jammu and Kashmir	775	445	322	8
Karnataka	693	310	354	29
Haryana	594	327	260	7
Bihar	542	350	188	4
Kerala	503	30	469	4
Odisha	185	122	61	2
Jharkhand	127	87	37	3
Chandigarh	120	98	21	1
Uttarakhand	61	21	39	1
Chhattisgarh	59	23	36	0
Assam	45	12	32	1
Himachal Pradesh	45	5	38	2
Tripura	43	41	2	0
Ladakh	41	24	17	0
Andaman and Nicobar	33	1	32	0
Meghalaya	12	1	10	1
Puducherry	9	3	6	0
Goa	7	0	7	0
Manipur	2	0	2	0
Arunachal Pradesh	1	0	1	0
Dadra and Nagar Haveli	1	1	0	0
Mizoram	1	1	0	0

- adjusting recycling activities to avoid cross-contamination and infections, and,
- safe treatment and disposal of the increased quantities of healthcare and medical waste.

All these are very challenging areas of intervention to reduce and eliminate the impact of the virus. Key points that are to be addressed very carefully are as follows:

1. Likelihood of SARS-CoV-2 virus contaminating the MSW at household level should be ensured by following all protocols.

2. Impacts of this virus on waste generation have to be assessed, understood, and reduced or zeroed down by following Biomedical Waste Management Rules 2016 and the protocols issued by the Central Pollution Control Board (CPCB).
3. The remedial measures to deal with the current situation need to be known and practiced. Contact details of specific agencies of help should be made available including emergency hospitalization and scientific and safe waste handling and disposal system.
4. Readiness of the municipal bodies and the governments to reduce/eliminate impacts anticipated by SARS-CoV-2 on existing waste management practices and the interventions desired from technology, policy and financing point of view. The implementation strategies of evolved system to combat the SARS-CoV-2 virus have to be followed up for effective implementation.
5. The work environment of waste workers including the PPEs (personal protective equipment) should be improved during and post-pandemic.

2.3.3   Actions Taken

The Government of India as well as all the states and union territories initiated preventive and reactive measure as soon as the outbreak was witnessed in the state of Kerala on 30th January 2020 for the first time. Taking rapid actions to limit travel by suspending visas and quarantining all incoming travelers has helped. All international passengers entering India undergo Universal Health Screening. According to health officials, more than one million passengers have been screened at airports, limiting the entry of coronavirus. Another quick recommendation and enforcement are for residents to avoid or postpone mass gatherings until the virus is contained. Country-wide mass awareness campaign through electronic and print media started in mid-February 2020 giving a number of precautionary measures to be practiced, namely use of face mask as per the guidelines issued by WHO. It was slowly realized that the virus cannot be protected by only mask. Interactions with other affected countries started and the central government imposed several actions, namely "Janta Curfew" (People's Curfew), country-wide lockdown from March 25th that continues till August 2020 after which partially lockdown was withdrawn in different states based on the affected cases. Several government and private agencies issue different circulars on various issues including waste management. Since the month of March 2020 all the educational institutes including schools all over the country were declared closed. During September 2020 when the number of affected cases came down, the higher educational institutes started opening. Again, in March 2021 those institutes were closed because of the surge of second wave of the COVID-19. June 2021 onwards the higher educational institutes started operating with 30% manpower with partial activities. However, with the successful vaccination program in India, the government in different states started opening the schools from September 2021. Thus, the COVID waste generation also fluctuates with time. The municipalities in India by now have become smarter in handling the health care waste from COVID units. The Biomedical Waste Treatment facilities handle

COVID waste effectively. However, with challenges in some of the cities and villages there is no major failure in COVID waste management reported in India.

As identified by the compendium of Medical Education and Drugs Department (COMP, 2020), Government of Maharashtra, India, the High-Risk Contact SARS-CoV-2 includes: (a) Touched body fluids of the patient (respiratory tract secretions, blood, vomit, saliva, urine, faces), (b) Had direct physical contact with the body of the patient including physical examination without PPE, (c) Touched or cleaned the linens, clothes, or dishes of the patient, (d) Lives in the same household as the patient, (e) Anyone in close proximity (within 3 ft) of the confirmed case without precautions, (f) Passenger in close proximity (within 3 ft) of a conveyance with a symptomatic person who later tested positive for SARS-CoV-2 for more than 6 h and the low Risk Contacts include: (a) Shared the same space (same class for school/ worked in same room/similar and not having a high-risk exposure to confirmed or suspect case of SARS-CoV-2) and (b) Traveled in same environment (bus/train/ flight/any mode of transit) but not having a high-risk exposure. Among these risks a few of the points are highly necessary to have effective waste management.

2.3.4 Legislative Guidelines in India for SARS-CoV-2 Generated Wastes

In India main concern in waste management is the waste pickers in the formal and informal sector as in most of the cases the activities are performed manually. They are involved in collecting, sorting, recycling, and selling materials manually which create virus vulnerability. They contribute to local economies, public health, and environmental sustainability. A major portion of waste pickers in India are women who are often family members.

Preventing SARS-CoV-2 infection is a critical public health priority globally at present. In India there are several rules related to the management and handling of different types of wastes. Solid Waste Management (SWM, 2016) Rules 2016 and Bio-Medical Waste Management (BMWM, 2016) Rules, 2016 are related to the activities associated with SARS-CoV-2. The Government of India including all the states has initiated various steps to combat the pandemic, which include setting up of quarantine centers/camps, isolation wards, sample collection centers, and laboratories.

The Central Pollution Board, Government of India issued specific guidelines (GHTDW, 2020) for management of waste generated during diagnostics and treatment of SARS-CoV-2 suspected/confirmed patients on 19th March and subsequently revised on 25th March 2020. Personal protective equipment and hand washing are the major issues to deal with the virus. Guideline on rational use of personal protective equipment (PPE) has also been issued by Ministry of Health and Family Welfare Directorate General of Health Services. Guidelines for Handling, Treatment and Disposal of Waste Generated during Treatment/Diagnosis/Quarantine of COVID-19 Patients were revised on 17th July 2020.

In case of generation of large volume of yellow color coded (incinerable) SARS-CoV-2 waste, permit HW incinerators at existing Treatment Storage and Disposal Facilities (TSDFs) to incinerate the same by ensuring separate arrangement for handling and waste feeding. Dedicated carts/trolleys/vehicles are used for transport of biomedical waste ensuring sanitization of vehicles with 1% hypochlorite after each trip. Waste collectors arriving at quarantine center or at home care spray the disinfectant (1% hypochlorite solution) on the bin used for yellow bag.

2.3.5 COVID-19 Second Wave

Most of the countries are facing a real threat of COVID-19 resurgence, or already fighting it during the third and fourth quarters of 2021. The current pace of transmission across the 53 countries of the WHO European Region is of grave concern. COVID-19 cases are once again approaching record levels, with the more transmissible Delta variant continuing to dominate transmission across the world.

2.4 Discussion and Analysis

The countries are gaining experience with time, as this situation is an unprecedented one. Large exchange of information and data of course affects the medical assistance for patients infected by TSDF and the principal routes of coronavirus outbreak. Furthermore, there are many other aspects where sharing of knowledge could be necessary and useful for decrease in the TSDF spreading along with ensuring a safe operation of other economic activities during and after the lockdown. The most important aspects are the collaborative efforts and exchange of cases and experience among the affected countries. As is seen in India guidelines with respect to handling of MSW where there is a possibility of mixing of general wastes and waste generated out of TSDF affected areas has to be developed. Based on the data generated from the implementation of the guidelines, the documents need to be amended at a regular interval. USA approach is more oriented to ensuring as much as possible the continuance of waste management and recycling mainly oriented to supply to the recyclable materials industry. Some modifications were introduced for the household waste pickup due to their significant increase, up to 30%, due to the implementation of the "stay at home" program. In Italy the central government is operating for giving adequate guidelines for avoiding the contagious spreading for all the different working activities both those classified as essential and strategic and for those that are going to restart during the next second phase. Several stakeholder, public, private, citizens are putting queries on how to deal with their activities during this coronavirus crisis. In particular the probable risk of further virus spreading by objects and materials as well as the lack of knowledge on the ability of the SARS-CoV-2 to survive on waste caused some uncertainties in both the users and waste management operators. Particular threat for their health was perceived for the handling of household waste produced by patients affected by SARS-CoV-2 or in quarantine treated at home. For this reason, specific multidisciplinary working groups for giving adequate responses and viable guidelines, able to consider the different socio-economic contexts, to these queries were activated at the Italian Institute of Health (ISS). Along with an objective hindrance of virus spreading these guidelines contribute also to more safety and comfortable working conditions for the workers, the trained volunteers as well as the users.

These differences highlight that above what is already confirmed or not by scientific knowledge, there is a not uniform perception of the risk from waste generated from private houses and/or areas, different from hospital and medical centers, where are treated patients affected by SARS-CoV-2 or quarantine. Of course there is no direct scientific evidence that waste management could represent another route for coronavirus spreading but the contrary cannot be stated since the permanence from 24 h up 9 days of SARS-CoV-2 on different materials surface has been already detected.

Despite near-record COVID-19 cases, new deaths are at approximately half the peak levels. This reflects the life-saving effects of vaccines and the herculean task of health authorities, the health workforce and communities to develop, administer, and accept vaccines. One billion doses have now been administered in Europe and Central *Asia. As of 28th October*, one billion COVID-19 vaccine doses were administered in the WHO European Region—but risks for the unvaccinated leave no room for complacency. *Very recently India reached the mark of one billion people getting vaccinated and the start of vaccine for children of age between 2 and 18 very soon.*

The different approaches detected in dealing with waste management under coronavirus crisis show some significant differences related to the prevalence of some socio-cultural groups prevailing in the different countries. The Italian approach is mainly addressable to a hierarchal way of life (Table 4, Fig. 2) concerning the procedures applied and also somewhat the management style even if for this latter also egalitarian way of life plays an important role. Egalitarian perspective seems to be prevalent concerning pollution solutions, attitude toward risk and risk handling style. For India similar approaches were detected for procedure applied and management style and pollution solutions, whereas different results were detected for attitude toward risk and risk handling style. USA shows a more distributed contribution of the all ways of life in the approach implemented.

2.5   Conclusion

Waste management is a big issue in most of the developing and all underdeveloped countries. Since the early 2020, an additional pressure emerged on the waste management systems with the insurgence of waste generated from the units treating thousands of COVID affected patients at hospitals and small health care units, quarantine centers, including the crematorium/graveyards. Handling COVID waste

Table 4 Score assigned to the management style, attitude, and characteristics of the four way of life archetypes concerning the main aspects of regulation and approaches for waste management under coronavirus crisis for Italy, India, and the USA

Aspects	Way of life			
	Hierarchist	Egalitarian	Individualist	Fatalist
Italy				
Procedures applied	Rules—1	Ethical standard—0	Skill—0	–
Management style	Control—1	Preventive—1	Adaptive—0	–
Pollution solution	Change nature to confirm society—0	Change society to confirm nature—1	Market incentives (transferable rights to pollute)—0	
Attitude toward risk	Risk-accepting—0	Risk-aversive—1	Risk-seeking—0	–
Risk handling style	Rejection and absorption—0	Rejection and deflection—1	Acceptance and deflection—0	Acceptance and absorption—0
Total score	2	4	0	
India				
Procedures applied	Rules—1	Ethical standard—0	Skill—0	–
Management style	Control—1	Preventive—1	Adaptive—0	–
Pollution solution	Change nature to confirm society	Change society to confirm nature—1	Market incentives (transferable rights to pollute) —0	
Attitude toward risk	Risk-accepting—1	Risk-aversive—1	Risk-seeking—0	–
Risk handling style	Rejection and absorption—0	Rejection and deflection—0	Acceptance and deflection—1	Acceptance and absorption—0
Total score	3	3	1	
USA				
Procedures applied	Rules—1	Ethical standard—0	Skill—1	–
Management style	Control—0	Preventive—1	Adaptive—1	–
Pollution solution	Change nature to confirm society	Change society to confirm nature	Market incentives (transferable rights to pollute) —0	
Attitude toward risk	Risk-accepting—1	Risk-aversive—0	Risk-seeking—0	–
Risk handling style	Rejection and absorption—0	Rejection and deflection—0	Acceptance and deflection—0	Acceptance and absorption—1
Total score	2	1	2	1

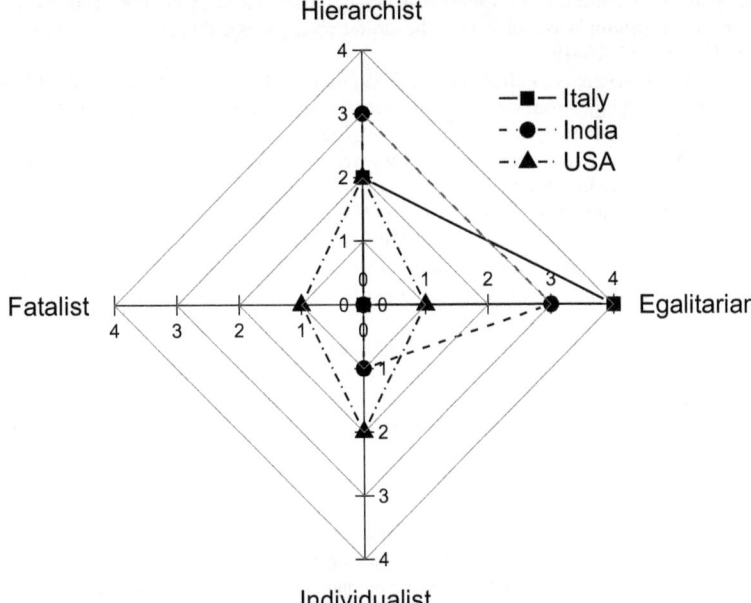

Fig. 2 Kiviat diagram representing the main socio-cultural perspective affecting the regulation of waste management under coronavirus crisis in Italy, India, and the USA

is very dangerous but the work must be performed with utmost care as per the protocols, guidelines, and expertise. However, all efforts should be made to reduce the generation of COVID waste to zero. This is only possible with the help of COVID-19 vaccines which has reached billions of people worldwide, the evidence is overwhelming that no matter which one you take, the vaccines offer life-saving protection against a disease that has killed millions. The pandemic is far from over, and they are our best bet of staying safe and reduce waste generation making the environment cleaner. Let us wish that our sincere efforts eradicate the coronavirus from the world, destroy the source of this dangerous virus, and develop immunity to fight in need.

References

BMWM. (2016) *Bio medical waste management rules, 2016*. Retrieved May 18, 2020, from https:// dhr.gov.in/document/guidelines/bio-medical-waste-management-rules-2016

CDC (Centers of Disease Control and Prevention). (2020). *Coronavirus disease 2019 (COVID-2019)*. Retrieved May 3, 2020, from https://www.cdc.gov/coronavirus/2019-ncov/cases-updates/summary.html?CDC_AA_refVal=https%3A%2F%2Fwww.cdc.gov%2Fcoronavirus%2F2019-ncov%2Fsummary.html

COMP. (2020). *Compendium of guidelines, instruction and standard operative procedures for COVID-19* (1st ed.). Medical Education and Drugs Department Government of Maharashtra.

Cruvinel, V. R. N., Marques, C. P., Cardoso, V., et al. (2019). Health conditions and occupational risks in a novel group: Waste pickers in the largest open garbage dump in Latin America. *BMC Public Health, 581,* 16–19.

de Wit, E., van Doremalen, N., Falzarano, D., & Munster, V. J. (2016). SARS and MERS: Recent insights into emerging coronaviruses. *Nature Reviews. Microbiology,* 14, 523–534.

DeAnne, T. (2020). ISRI: Recyclers included among DHS essential industries. *Recycling Today.* Retrieved May 8, 2020, from https://www.recyclingtoday.com/article/recyclers-deemed-essential-industry-covid-19-outbreak/

Douglas, M., & Wildavsky, A. (1982). *Risk and culture: An essay on the selection of technological and environmental dangers.* University of California Press.

DPR. (2003). Decreto del Presidente della Repubblica 15 luglio 2003, n.254. Regolamento recante disciplina della gestione dei rifiuti sanitari a norma dell'articolo 24 della legge 31 luglio 2002, n. 179. (Regulation for the management of medical waste according to art. 24 of the legislation n. 179 of 31st July 2002). *Official Journal of Italian Republic* n. 211.

EC. (2020). *Waste management in the coronavirus crisis.* Retrieved May 3, 2020, from https://ec.europa.eu/info/sites/info/files/waste_management_guidance_dg-env.pdf

GCH. (2020). Goodhue County Household hazardous waste collection changes to reflect COVID-19 policies. *Rivertowns.net.* Retrieved May 8, 2020, from https://www.rivertowns.net/news/government-and-politics/6262947-Goodhue-County-Household-hazardous-waste-collection-changes-to-reflect-COVID-19-policies

GHTDW. (2020). *Guidelines for handling, treatment and disposal of waste generated during treatment/diagnosis/quarantine of COVID-19 patients.* CPCB, Govt of India.

Gomes Mol, M. P., & Caldas, S. (2020). Can the human coronavirus epidemic also spread through solid waste? *Waste Management and Research, 38*(5), 485–486.

ISPRA. (2020). *Prime indicazioni generali per la Gestione dei Rifiuti – Emergenza COVID-19.* Retrieved May 3, 2020, from https://www.snpambiente.it/2020/03/24/emergenza-covid-19-indicazioni-snpa-sulla-gestione-dei-rifiuti/

ISRI. (2020). *Institute of scrap recycling industries COVID-19 state and local policy dashboard.* Retrieved from https://docs.google.com/spreadsheets/d/e/2PACX-1vRlJWZJ7OkGUW57_rdA2n3xBJ3qjW6u4Z9N6K9Y5L4bM_6H7-S308qdKmJfpVstYWf300nyujvZPFSy/pubhtml?urp=gmail_link

ISS. (2020). *Rapporto ISS COVID-19 n.3/2020 Rev. Indicazioni ad interim per la Gestione dei Rifiuti urbani in relazione alla trasmissione dell'infezione da virus SARS-CoV-2.* Retrieved May 3, 2020, from https://www.iss.it/rapporti-covid-19/-/asset_publisher/btw1J82wtYzH/content/id/5299930?_com_liferay_asset_publisher_web_portlet_AssetPublisherPortlet_INSTANCE_btw1J82wtYzH_redirect=https%3A%2F%2Fwww.iss.it%2Frapporti-covid-19%3Fp_p_id%3Dcom_liferay_asset_publisher_web_portlet_AssetPublisherPortlet_INSTANCE_btw1J82wtYzH%26p_p_lifecycle%3D0%26p_p_state%3Dnormal%26p_p_mode%3Dview%26_com_liferay_asset_publisher_web_portlet_AssetPublisherPortlet_INSTANCE_btw1J82wtYzH_cur%3D0%26p_r_p_resetCur%3Dfalse%26_com_liferay_asset_publisher_web_portlet_AssetPublisherPortlet_INSTANCE_btw1J82wtYzH_assetEntryId%3D5299930

Italian Ministry of Environment. (2020). *Circolare ministeriale recante "Criticità nella gestione dei rifiuti per effetto dell'Emergenza COVID 19" – indicazioni.* Retrieved May 3, 2020, from https://www.minambiente.it/pagina/circolari-0

G., Kampf D., Todt S., Pfaender E., Steinmann (2020) Persistence of coronaviruses on inanimate surfaces and their inactivation with biocidal agents. Journal of Hospital Infection 104(3) 246-251 10.1016/j.jhin.2020.01.022

Lewis, D. (2020). Is the coronavirus airborne? Experts can't agree. *Nature.* https://doi.org/10.1038/d41586-020-00974-w

Liu, Y., Ning, Z., Chen, Y., Guo, M., Liu, Y., Gali, N. K., Sun, L., Duan, Y., Cai, J., Westerdahl, D., Env, D., Liu, X., Ho, K.-F., Kan, H., Fu, Q., & Lan, K. (2020). Aerodynamic characteristics and RNA concentration of SARS-CoV-2 aerosol in Wuhan hospitals during COVID-19 outbreak. https://doi.org/10.1101/2020.03.08.982637

MacFarland, M. (2020, April 9). Republic: 'It's just a nightmare.' Trash collectors overwhelmed by rising amount of household waste. *CNN Business*.

Minter, A. (2020, May 4). As recycling dwindles, supply chains are buckling. *Bloomberg Opinion*.

MyGov. (2020). *India fights corona COVID-19*. Retrieved May 10, 2020, from https://www.mygov.in/covid-19

National Research Council. (2020). *Rapid expert consultation on the possibility of bioaerosol spread of SARS-CoV-2 for the COVID-19 pandemic*. National Academies Press. https://doi.org/10.17226/25769

Santarpia, J. L., Rivera, D. N., Herrera, V., Morwitzer, M. J., Creager, H., Santarpia, G. W., Crown, K. K., Brett-Major, D., Schnaubelt, E., Broadhurst, M. J., Lawler, J. V., Reid, S. P., & Lowe, J. J. (2020). Transmission potential of SARS-CoV-2 in viral shedding observed at the University of Nebraska Medical Center. *Scientific Reports, 10*, 12732. https://doi.org/10.1101/2020.03.23.20039446

Schwarz, M., Thompson, M. (1990). Divided We Stand: Redefining Politics, Technology and Social Choice. University of Pensylvania press.

Staub, C. (2020, March 31). Republic: COVID-19 brings 'uncontrollable challenges'. *Resource Recycling*.

SWM. (2016). *Solid waste management rules, 2016*.

Taylor, B. (2020). DHS designates waste collection as essential service. *Waste Today*. Retrieved May 8, 2020, from https://www.wastetodaymagazine.com/article/nwra-dhs-waste-collection-essential-covid-19/

Thompson, M., Ellis, R., & Wildavski, A. (1990). *Cultural theory*. Political culture series. ISBN 978-0813378640.

Van Doremalen, N., Bushmaker, T., Morris, D. H., Holbrook, M. G., Gamble, A., Williamson, B. N., Tamin, A., Harcourt, J. L., Thornburg, N. J., Gerber, S. I., et al. (2020). Aerosol and surface stability of SARS-CoV-2 as compared with SARS-CoV-1. *The New England Journal of Medicine, 382*, 1564–1567.

WHO. (2020a). *COVID-19 2020 situation summary*. Updated 15 March 2020. Retrieved May 3, 2020, from https://www.cdc.gov/coronavirus/2019-ncov/summary.html

WHO. (2020b). *Coronavirus disease (COVID-19)*. Situation report 103. Retrieved May 11, 2020, from https://www.who.int/docs/default-source/coronaviruse/situation-reports/20200510covid-19-sitrep-111.pdf?sfvrsn=1896976f_2

WHO. (2020c). *Modes of transmission of virus causing COVID-19: Implications for IPC precaution recommendations*. Retrieved May 3, 2020, from https://www.who.int/publications-detail/modes-of-transmission-of-virus-causing-covid-19-implications-for-ipc-precaution-recommendations

WHO. (2020d). *Laboratory biosafety guidance related to coronavirus disease 2019 (COVID-19)*. Retrieved May 3, 2020, from https://apps.who.int/iris/bitstream/handle/10665/331138/WHO-WPE-GIH-2020.1-eng.pdf

WHO. (2021). Assessment of the COVID-19 Supply Chain System. Available at: file:///C:/Users/Francesco/Downloads/21_02_26-CSCS_Assessment_Summary-Written-Report-.pdf. (accessed 28.02.2022)

WPXI. (2020). *New rules in effect for curbside trash pickups during COVID-19 pandemic*. Retrieved May 8, 2020, from https://www.wpxi.com/news/top-stories/new-rules-effect-curbside-trash-pickups-during-covid-19-pandemic/VWAASZ5POJEHRNDL4QIPYFCDPM/

Yu, H., Afshar-Mohajer, N., Theodore, A. D., Lednicky, J. A., Fan, Z. H., & Wu, C. Y. (2018). An efficient virus aerosol sampler enabled by adiabatic expansion. *Journal of Aerosol Science, 117*, 74–84. https://doi.org/10.1016/j.jaerosci.2018.01.001

Part II
Health Care Waste Management and Case Studies

Clinical Waste Management in Malaysia and COVID-19 Waste Management Case Study

P. Agamuthu, S. B. Mehran, and B. Jayanthi

1 Introduction and Definition

Any waste that is generated from hospitals and healthcare facilities is classified as clinical waste. The composition of clinical waste can be heterogeneous including infectious waste, radioactive waste, hazardous waste, pharmaceuticals waste, and general waste. Clinical waste can also be known as healthcare waste, medical waste, or scheduled waste (DOE, 2009). Ministry of Health (MoH) Malaysia defines clinical waste which may consist wholly or partly of human or animal tissues, blood or other bodily fluids, excretions, drugs or other pharmaceutical products, swabs or dressings, syringes, needles or other sharp instruments, and waste which, unless rendered safe, may prove hazardous to any person coming into contact with waste.

World Health Organization (WHO) defines clinical waste as waste and by-products that cover a diverse range of materials such as infectious waste, pathological waste, sharps waste, chemical waste, pharmaceutical waste, cytotoxic waste, radioactive waste, and also non-hazardous or general waste. According to WHO clinical or biomedical waste contains potentially harmful microorganisms that may easily infect other patients, healthcare workers, or the general public if not handled and disposed in the proper way (WHO, 2018). Besides that, it may also potentially spread drug-resistant microorganisms to the environment. Moreover,

P. Agamuthu (✉)
Jeffrey Sachs Center on Sustainable Development, Sunway University, Sunway, Malaysia
e-mail: agamutup@sunway.edu.my

S. B. Mehran
Faculty of Science, Institute of Biological Sciences, University of Malaya, Kuala Lumpur, Malaysia

B. Jayanthi
Faculty of Health and Life Sciences, INTI International University, Nilai, Malaysia

© The Author(s), under exclusive license to Springer Nature Singapore Pte Ltd. 2022
S. K. Ghosh, P. Agamuthu (eds.), *Health Care Waste Management and COVID 19 Pandemic*, https://doi.org/10.1007/978-981-16-9336-6_2

clinical waste may also pose health risks in the form of sharps-inflicted injuries, toxic exposure due to pharmaceutical products such as antibiotics and cytotoxic drugs or hazardous chemicals like mercury, and the release of dioxins and particulate matter from incineration of clinical wastes. In addition to that, handling of clinical waste may cause radiation burns, chemical burns from disinfection, sterilization or waste treatment activities, and thermal injuries occurring from open burning and the operation of clinical waste incinerators (WHO, 2018, Agamuthu & Jayanthi, 2020).

Globally, an average of 0.5 kg of clinical waste per hospital bed is generated in high-income countries every day, while low-income countries generate an average of 0.2 kg of clinical waste per bed per day. However, clinical waste is often not separated into hazardous or non-hazardous wastes in low-income countries, hence making the real quantity of hazardous waste much higher. Due to its composition, clinical waste requires proper management and disposal method; otherwise, it could potentially bring harm to human health and the environment and other living organisms. The major sources of clinical waste are from hospitals and other health facilities, laboratories and research centers, mortuary and autopsy centers, animal research and testing laboratories, blood banks and collection services, nursing homes for the elderly.

Generally, increase in healthcare facilities results in increased clinical waste generation. Due to increase in the demand for medical services, clinical waste from healthcare centers, hospitals, or clinics is growing proportionally in Malaysia too (Ambali et al., 2013). However, epidemics or pandemics like COVID-19 also result in higher generation of clinical waste within a shorter span (Agamuthu & Jayanthi, 2020). Coronavirus disease (COVID-19) first began in December 2019 at Wuhan, China which is transmitted from human to human (Yu et al., 2020). Since then, the number of people infected with corona virus is increasing rapidly and has caused global emergency to the most parts of the world. WHO announced COVID-19 outbreak a pandemic in March 2020, and it is rapidly spreading to all over the world. The total number of cases reported globally as of second August 2021 were 199,022,838 cases and the total death toll has reached 4,240,374. With the increase in the number of infections of COVID-19, the increase in clinical waste from healthcare centers such as hospitals, clinics, laboratories, temporary quarantine centers, and research laboratories is observed in almost every part of the world. Similar increase in clinical waste generation is recorded in Malaysia. This chapter elucidates the policies and regulations for the management of clinical waste, reports the generation and management of clinical waste and illuminates on the management of COVID-19 related clinical waste in Malaysia. Finally, the management of COVID-19 related clinical waste is reinforced with a case study of a Hospital in Malaysia.

2 Policy/Regulations/Guidelines in Malaysia on Clinical Waste

2.1 Policy and Regulations

Management of clinical waste in Malaysia follows the guidelines prepared by MoH and Department of Environment (DOE) Malaysia. The preliminary guidelines were first introduced in Malaysia in the year 1998 by the MoH for the management of hospital waste, together with guidelines on the biomedical waste management at state level. Besides that, "Hospital Waste Management Manual," which is a detailed guideline for handling and disposing wastes (Saw, 1994) was also introduced. DOE has formulated the Waste Pollution Prevention and Control Law and/or the Regulations on management of biomedical hazardous wastes. Environmental Quality Act (EQA) of 1974 is a provision for prevention, abatement, control, and protection of the environment, which is administered by DOE. The EQA was enforced in 1975 and had been amended in 1976, 1985, and 1996 in accordance with international standards (Noor Artika et al., 2019).

Before the introduction of EQA 1974 regulation, waste generated from hospitals was disposed of together with other domestic waste according to the provided guidelines at that time. Currently biomedical, clinical waste, or any hazardous waste is listed as scheduled waste and is categorized according to Environmental Quality (Scheduled Wastes) Regulations 2005 ("Regulations") which serves as a key legal framework for handling of hazardous waste. There are subsidiary legislations for schedule waste that fall under Environmental Quality (Scheduled Wastes) Regulations 2005 which include;

1. The Environmental Quality (Prescribed Premises) (Scheduled Wastes Treatment and Disposal Facilities) Order 1989.
2. The Environmental Quality (Prescribed Premises) (Scheduled Wastes Treatment and Disposal Facilities) Regulations 1989.

The guidelines consist of a comprehensive framework of federal, state, regional, and local laws, whereas licenses and permits govern virtually all aspects of management activities of clinical waste including labeling, identification, on-site storage and management, transportation, treatment, and disposal which are clearly addressed in guidelines on the handling and management of clinical waste in Malaysia (2009) (Ambali & Bakar, 2012). The clinical waste generated in Malaysia is basically classified into four different categories as stated in Environmental Quality (Scheduled Wastes) Regulations 2005. However, in Environmental Quality Reports, published by DOE, a fifth type of schedule waste "discarded drugs" is also reported. Table 1 tabulates the classification of clinical waste according to schedule waste regulation.

During the COVID-19 pandemic, no new policies or regulations were introduced for the management of COVID-19 related clinical waste. The management of clinical waste generated from COVID-19 related cases has been managed in

Table 1 Clinical waste classification under the Environmental Quality (Scheduled Wastes) Regulations, 2005

Scheduled waste codes	Types of clinical waste
SW403	Expired drugs containing psychotropic substances or containing substances that are toxic, harmful, carcinogenic, mutagenic, or teratogenic
SW404	Pathogenic and clinical wastes and quarantined materials
SW 405	Discarded drugs
SW421	A mixture of scheduled wastes
SW422	A mixture of scheduled and non-scheduled wastes

compliance with existing Environmental Quality Act 1974 Act (DOE, 2009). According to DOE, the responsibility of management of clinical waste falls on the Ministry of Environment and Water and national authorities, under the supervision of Federal Government, and, as well as, on the local municipal council. Similarly, Federal Government and DOE administer the responsible authorities for daily management of COVID-19 related clinical waste.

3   Generation of Clinical Waste

The generation of clinical waste in several countries is shown in Table 2. But clinical waste generation data may be approached with some circumspection as one of the issues highlighted by academia is the lack of distinction between the generation of hazardous and non-hazardous clinical waste in the reported figures (Minoglou et al., 2017; Windfeld & Brooks, 2015). As reported by Windfeld and Brooks (2015), some countries such as the USA, Brazil, Taiwan reported clinical waste generation numbers of both infectious (hazardous) waste and general (non-hazardous) waste. Table 2 tabulates the total clinical waste generation according to income level and clinical waste generation rate is based on both hazardous and non-hazardous clinical waste. The analysis conducted by Minoglou et al. (2017) shows that countries with high GDP (gross domestic product) and higher health expenditure generate more clinical waste due to higher treatment capacity based on the availability of higher number of medical consumables and equipment. Malaysia, being upper-middle income country, generates 1.9 kg of clinical waste (both hazardous and non-hazardous) per hospital bed per day.

Clinical waste constituted approximately 1.44% of total schedule waste generated in 2017 in Malaysia. On average, approximately 1.28% of schedule waste is comprised of clinical waste in Malaysia every year, where pathogenic clinical waste, expired and discarded drugs make up the total hazardous clinical waste. The total generation of hazardous clinical waste in Malaysia for the last 5 years is shown in Table 3.

Table 2 Generation rate of clinical waste in selected countries (Adapted from Minoglou et al., 2017)

Income Level	Country	Clinical Waste Generation (kg/bed/day)
Low-Income	Ethiopia	1.10
	Nepal	0.50
	Tanzania	0.75
Lower-Middle Income	Bangladesh	1.24
	Cameroon	0.55
	Egypt	1.03
	El Salvador	1.85
	India	1.55
	Indonesia	0.75
	Jordan	2.69
	Lao	0.51
	Morocco	0.53
	Pakistan	2.07
	Palestine	2.02
	Sudan	0.87
	Vietnam	1.57
Upper-Middle Income	Algeria	0.96
	Argentina	3.00
	Brazil	2.94
	Bulgaria	2.00
	China	4.03
	Ecuador	2.09
	Iran	3.04
	Kazakhstan	5.34
	Lebanon	5.70
	Malaysia	1.90
	Mauritius	0.44
	Thailand	2.05
	Turkey	4.55
High-Income	Canada	8.20
	France	3.30
	Germany	3.60
	Greece	3.60
	Italy	4.00
	Japan	2.15
	Korea	2.40
	Latvia	1.18
	Netherlands	1.70
	Norway	3.90
	Spain	4.40
	United Kingdom	3.30
	United States	8.40

Table 3 Generation of clinical waste in Malaysia (Environmental Quality Reports 2013–2017)

Types of Waste	2017	2016	2015	2014	2013
Pathogenic Clinical Waste—SW 404	28,375.24	23,844.91	25,523.32	21,976.12	18,152.95
Expired Drug—SW 403	458.97	14,250.60	282.31	447.97	1470.14
Discarded Drug—SW 405	298.53	337.77	112.01	110.59	120.36
TOTAL (tonnes)	29,132.74	38,433.28	25,917.64	22,534.68	19,743.45

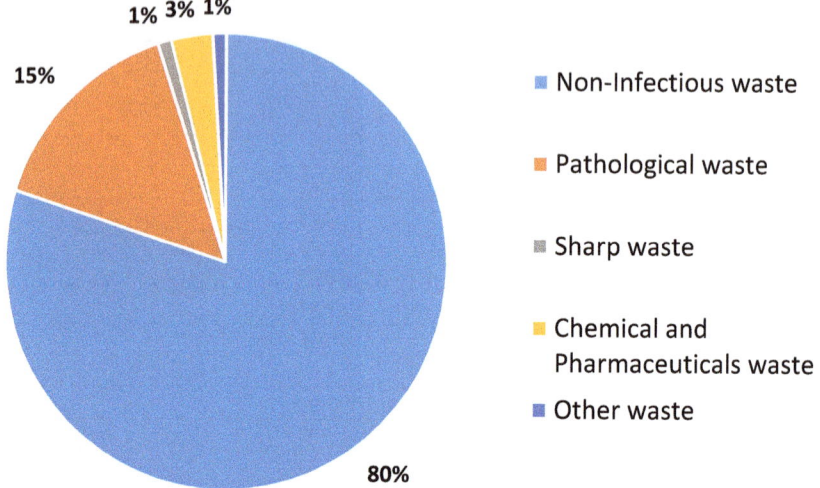

Fig. 1 Composition of clinical waste in Malaysia

4   Composition of Clinical Waste in Malaysia

Majority of clinical waste generated in Malaysia is composed of general or non-hazardous waste, as 80% of clinical waste is non-infectious. The composition of clinical waste in Malaysia is shown in Fig. 1. Approximately 15% of clinical waste is pathological waste, which is the main hazardous clinical waste that is generated in Malaysia. About 28,375 tonnes of pathological waste was generated in 2017 (see Table 3). Third main component of clinical waste in Malaysia is chemical and pharmaceutical waste which is also considered as scheduled waste.

5   Management of Clinical Waste

The management of clinical waste in Malaysia is carried out by the concessionaire companies. There are currently seven concession companies that are permitted for the management of clinical waste in Malaysia. The concession company is responsible for handling the clinical waste from point of waste generation to the final treatment and/or disposal level. There are currently 144 government hospitals with 42,000 beds and 240 private hospitals with 16,000 beds that generate clinical waste on a daily basis in Malaysia. The management of clinical waste follows the guidelines provided by DOE, comprising of the regulations and specific requirements that require proper management of generated clinical waste before disposal, as far as possible, since it could be harmful to human health and environment. The generators of schedule waste are required to notify DOE of any scheduled waste (including hazardous clinical waste) that is generated and must keep an up-to-date inventory of scheduled waste that is generated, treated, and disposed as per regulation. Scheduled waste may be stored, recovered, and treated within the premises of a waste generator. Whereas if the management of schedule waste such as storage, incineration, and disposal needs to be performed offsite, waste generators are obliged to the requirements of the consignment note system for transportation as it ensures that schedule waste reaches the approved destination and is carried out by licensed transporter. Similarly, the treatment must also be performed at prescribed premises that are licensed by the DOE. These protocols for the management of schedule waste are also applicable to hazardous clinical waste. Moreover, clinical waste must be stored in durable waste containers with clear labels and transporters must be informed, by waste generators, about the nature of the wastes that is being transported and action to be taken in case of accidents (DOE, 2009). The hazardous clinical waste is basically classified into five different categories as abovementioned in Table 1 of Sect. 2.

5.1 Privatization and Companies Involved

Seven concessionaire companies are permitted for the collection of clinical waste from both public and private sectors in Malaysia. Table 4 shows the list of concession companies that are registered with DOE Malaysia. The licensed contractors for the management of clinical waste in Malaysia are given license for managing clinical waste under Section 18 (Environmental Quality Act 1974) regulation. Five out of seven companies are the concession companies such as Radicare Sdn Bhd, Edgenta

Table 4 List of concessionaire companies that manage clinical wastes in Malaysia

List of companies	Service for government/private
Radicare Sdn Bhd	Government hospital, clinic, medical institute
Edgenta Medisure Sdn Bhd	Government hospital, clinic, medical institute
Medivest Sdn Bhd	Government hospital, clinic, medical institute
Sedafiat Sdn Bhd	Government hospital, clinic, medical institute
One Medicare Sdn Bhd	Government hospital, clinic, medical institute
Future Nrg Sdn Bhd	Private hospital and clinics
Kualiti Alam Sdn Bhd	Private hospital and clinics

Medisure Sdn Bhd, Medivest Sdn Bhd, Sedafiat Sdn Bhd, and One Medicare Sdn Bhd. These concession companies are appointed by the Malaysian Government under MoH Malaysia to manage generated clinical waste from hospitals, government clinics, or medical institute under the service of Clinical Waste Management System (CWMS). Kualiti Alam Sdn Bhd and Future Nrg Sdn Bhd are the companies responsible for management of clinical waste from private hospitals and clinics. The companies that were originally managing the clinical waste are now also managing COVID-19 related clinical waste as well.

DOE, who is responsible for all aspects of clinical management, from the collection, transportation, treatment, and disposal, carries out monitoring of clinic waste management. DOE uses online system called electronic scheduled waste management system (eSWIS) to oversee the compliance of companies with the environmental regulations.

5.2   Collection of Clinical Waste

The first step in management of clinical waste involves segregation of clinical waste based on the composition and hazardousness. Generated clinical waste is labeled and disposed into proper containers and bags by the waste generator, which are then collected by the licensed contractors for disposal and treatment. Figure 2 shows the different bins used in hospital or healthcare facilities in Malaysia for disposal of clinical waste. The clinical waste or any type of waste generated at the healthcare facilities is disposed of according to color coded containers or plastic bags (Zaimastura, 2005).

Blue plastic bag/container is used for waste that requires autoclaving, yellow bag/container is designated for waste that needs to be incinerated and black bag/container is assigned for general municipal waste (DOE, 2009). Figure 3 shows the yellow and blue plastic bag used for disposal of clinical waste. All bags and drum containers used for clinic waste disposal must be identified at the point of generation and should be indelibly and clearly marked with biohazard symbol. The other details that must be displayed on the container include date of generation of scheduled waste, as well as name, address, and telephone number of the waste generator.

Some hospitals store the generated clinical waste in cold rooms temporarily until the dedicated lorry arrives. Collection of clinical waste can be daily or 3 times a week, depending on the quantity. Transportation of clinical waste can only be carried out by the special lorry licensed to transport schedule waste which belong to the waste companies.

Sharp waste bins   Pedal operated clinical waste bins

Mobile Pedal operated waste bins   Two-wheel clinical waste bin

Fig. 2 Different types of bins used in hospital or healthcare facilities for disposal of clinical waste

Fig. 3 Bags used for disposal of clinical waste

5.3   Treatment of Clinical Waste

Almost all clinical wastes in Malaysia are incinerated at 12 incineration facilities nationwide. Figure 4 shows the amount of total generated clinical waste in respective years and the amount of clinical waste incinerated. Approximately all pathological clinical waste is incinerated in Malaysia, whereas discarded or expired drugs are only incinerated if they are of cytotoxic characteristic or are hazardous. Otherwise, discarded or expired drugs may be disposed of in landfills (secure or sanitary depending on the hazardousness of drugs). The Bukit Nanas Integrated Waste Treatment Facility (Kualiti Alam) is the country's first comprehensive integrated treatment plant possessing various facilities with high temperature incinerator, physical, and chemical treatment. The incineration plant is suitable for organic wastes such as hazardous, toxic, clinical, and pathological wastes in all forms, which requires thermal treatment for destruction efficiency. The rotary kiln incinerator operates at temperatures above 1000 °C to ensure the highest possible destruction efficiency, followed by heat recovery system and finally an extensive multi-stage flue-gas treatment system.

Moreover, there are several on-site clinical waste incinerators or regional incinerators in Malaysia due to privatization of Malaysian clinical waste management and hospital support services in 1995 (Ambali & Bakar, 2012). In some states, private hospitals have installed an on-site incinerator for the treatment of clinical wastes (DOE, 2009). Most of the collected clinical waste is incinerated; however, some waste may be pre-treated by autoclaving, if necessary. Only one state (Penang) has implemented recycling of non-pathogenic waste after sterilization but this is not practiced for COVID-19 related clinical waste.

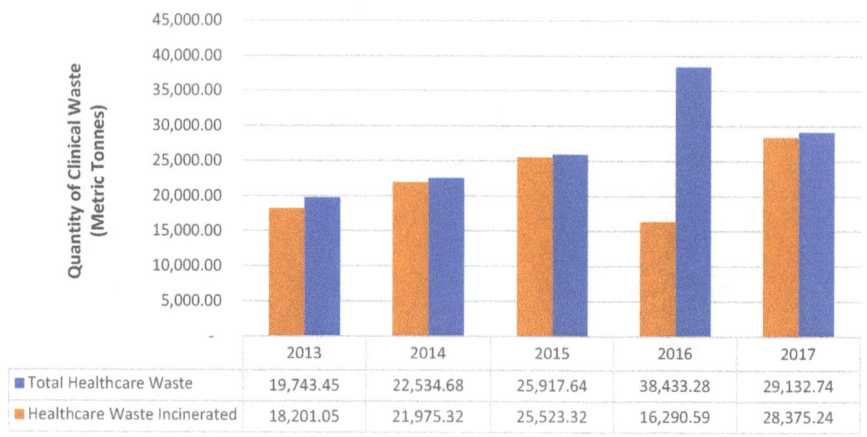

Fig. 4 Amount of clinical waste incinerated (Environmental Quality Reports 2013–2017)

5.4   Ash Management by Kualiti Alam

Secure landfill is the final destination of stabilized or reduced schedule waste. Incineration by-products like slags, fly ash, and flue-gas cleaning products with other residues undergo solidification and are then deposited in the secure landfill. All the ash from incineration of clinical waste from the incineration plants are taken to the Integrated Hazardous Waste Treatment Center and is solidified with cement before being disposed into the secured landfill. Kualiti Alam produces about 14,000 metric tonnes of bottom ash from incineration process and disposes of the ash into their secured landfill (Noor Artika et al., 2019).

5.5   Additional Regulation

DOE Malaysia has an additional regulation, where public is not allowed to handle and dispose any type of scheduled waste including clinical waste without proper license. Those found guilty can be sentenced to a mandatory prison and maximum fine of RM500,000 (115,000 USD) (DOE, 2009).

6   COVID-19 and Clinical Waste

6.1   COVID-19 Variants

Since the start of COVID-19 pandemic, several variants of SARS-CoV-2 have emerged which have been classified as variants of interest (VOI) and variants of concern (VOC) by WHO depending on its "disease severity, transmissibility, therapeutic and diagnostic escape, immune escape" and others (WHO, n.d.). These classifications from VOI to VOC may change depending on the research available. COVID-19 variants that are reported in Malaysia are listed in Table 5. COVID-19 cases are not reported according to variants in Malaysia yet so the table below gives only qualitative report of COVID-19 variants in Malaysia.

Table 5 COVID-19 variants that are reported in Malaysia

Variant	Originated from	Date of designation	Presence in Malaysia
Alpha (VOC)	United Kingdom	18 Dec 2020	Yes
Beta (VOC)	South Africa	18 Dec 2020	Yes
Gamma (VOC)	Brazil	11 Jan 2021	No
Delta (VOC)	India	11 May 2021	Yes
Lambda (VOI)	Peru	14 June 2021	No

Source: Daily Express (2020) and WHO (n.d.)

6.2   COVID-19 Vaccine

The administration of COVID-19 vaccine started on 24 February 2021 in Malaysia which was kick-started by Malaysian Prime Minister receiving the first dose Pfizer/ BioNTech vaccine (Hamid, 2021). On the first day of vaccination campaign, 60 inhabitants received the first doses of vaccine. Since then, the vaccine administration rate has been roughly increasing. As of 3 August 2021, a total of 22,152,367 doses have been given to Malaysians at a rate of 69.16 doses per 100 people. Approximately 45.45% of Malaysians have been vaccinated with first dose, whereas 22.99% of Malaysians are completely vaccinated as of 3 August 2021. Three vaccines are being administered in Malaysia including Oxford/AstraZeneca, Pfizer/BioNTech, and Sinovac until now. It is expected that 70% of Malaysians will be inoculated with both doses of vaccine by 17 September 2021 if current rate of 69.2 doses per 100 people is continued (Covidvax, 2021).

6.3   Generation of COVID-19 Related Clinical Waste

The daily generation of clinical waste in year 2013 was 50 tonnes per day and 18,000 tonnes annually, whereas currently it is estimated that the generation of clinical waste is 90 tonnes per day or 33,000 tonnes annually. Clinical waste contained about 75% to 80% of non-pathogenic waste. The novel coronavirus outbreak has become a global issue because it poses a serious risk to human health due to its rate of infection (Khan et al., 2020; Al-Qaness et al., 2020). Since the declaration of COVID-19 epidemic to pandemic by WHO in March 2020, the number of infections is increasing tremendously. Owing to the increase in number of infections, the increase in generation of clinical waste also can be observed. In addition to the normal protocols for clinical/infectious waste management, the management of COVID-19 clinical waste may require specific precautions, operations, and handling. Most countries reported an increase for waste generated and Table 6 presents the amount of COVID-19 related clinical waste generated in several countries, including Malaysia. The amount of clinical waste generated in Malaysia increased by 111.94% in June 2021 as compared to the previous year. Conversely, in South Korea 295 tonnes of COVID-19 related clinical waste was generated in 1 month. It is noticeable that countries such as Jordan and Wuhan, China showed the highest generation of COVID-19 clinical waste as compared to other countries. Approximately, 600% increase in infectious clinical waste was recorded during COVID-19 outbreak in Hubei Province, China (ADB, 2020).

During this COVID-19 pandemic, a 111.94% increase in the generation of clinical waste was reported in Malaysia by the ministry. The increase in the number of COVID-19 infections parallels the increase in the clinical waste generation from healthcare centers such as hospitals, clinics, laboratories, temporary quarantine centers, and research laboratories in almost every part of the world. The increase

Table 6 Generation of COVID-19 clinical waste by country

Country	Amount of waste generated during COVID-19 (kg/bed/day)	Percentage increase in clinical waste generation during COVID-19 pandemic	Reference
Taiwan	0.9–2.7	No data	Chiang et al. (2006)
Jordan	3.95	1000%	Abu-Qdais et al. (2020)
Wuhan, China	0.6–2.5	213%	Yu et al. (2020)
Bandung, Indonesia	2.2	17.1%	Damanhuri (2020)
Penang, Malaysia	0.4–1.0	27%	Astro Awani (2020)
Thailand	2.9	No data	IGES (2020)
Mexico	2.0–2.2	No data	IGES (2020)

in COVID-19 related clinical waste is mainly attributed to the increased usage of disposable gloves, face masks, and personal protective equipment (PPE) by the medical staffs (Astro Awani, 2020). Additionally, the usage of face masks and disposable gloves by public for protection from the infection is also well noticed. Incidentally improper disposal of the waste along shop lots, rubbish bins, and public areas is observed even though there is no published article on this matter.

Asian Development Bank (ADB) estimated the generation of COVID-19 related infectious clinical waste (in about 60 days) in major cities of Southeast Asian countries including Manila, Philippines (16,800 tonnes), Jakarta, Indonesia (12,750 tonnes), Hanoi, Vietnam (9600 tonnes), Bangkok, Thailand (12,600 tonnes), and Kuala Lumpur, Malaysia (9240 tonnes). According to ADB, infectious clinical waste could be estimated by following equation (ADB, 2020):

$$\text{Estimated Infected Persons} \times 3.4 \text{ kilograms} = \text{Increase in Infectious Clinical Waste per day of Outbreak} \quad (1)$$

Based on the equation above, the generation of infectious clinical waste from the start of COVID-19 outbreak to 30 June 2020 in respective countries was estimated, which is tabulated in Table 7. The data on number of positive cases for each country was extracted from Worldometer (2021) website.

It is also estimated that the increase in the amount of clinical waste could increase the cost of clinical waste management and this will increase the overall cost since clinical waste quantity has increased by 111.94% in Malaysia in June 2021. Training and awareness program related to COVID-19 waste management in Malaysia was carried out mainly on the proper use of PPE. Whereas public was given awareness

Table 7 Estimated generation of infectious clinical waste during COVID-19 pandemic in respective countries

Countries	Total no. of COVID-19 cases[a]	Total estimated generation of infectious clinical waste (tonnes)	Total no. of days of COVID-19 outbreak[b]	Daily generation of infectious clinical waste (tonnes)
United States	2,727,357	9273	170	54.5
Brazil	1,408,485	4789	126	38
Russia	647,849	2203	152	14.5
India	585,792	1992	153	13
United Kingdom	282,432	960	152	6.3
Italy	240,578	818	152	5.4
Malaysia	8639	29	158	0.18

[a]Total number of positive COVID-19 cases as of 30 June 2020
[b]Total number of days of COVID-19 outbreak from the first patient to 30 June 2020 (including 30 June 2020)

through all social media on proper sanitation and washing hands regularly. Regular periodic training for healthcare staff is provided by DOE, MoH, and the hospitals.

6.4   COVID-19 and Composition of Clinical Waste

During COVID-19 pandemic, the increase in clinical waste generation could be from swab, syringes, needles, sharps, blood or body fluid, excretions, mixed waste, laboratory waste, material or equipment contaminated with the virus, mask or disposable gloves, and PPE that are used for screening and treatment for COVID-19 infected patients.

6.5   COVID-19 and Household Clinical Waste Management

While clinical waste management in the hospital and clinical facilities is well managed during COVID-19, it is evident that the abundance of masks is not properly disposed by the public. Penang city council observed a sharp increase in the amount of clinical waste generated daily and warned that it could be an environmental disaster if no action is taken for managing the waste. Some of the COVID-19 related clinical waste has ended up in the solid waste landfills which includes PPE and face masks. It also reported that about 9000 face masks are disposed of daily in Penang. This also poses health risk to the municipal workers who collect waste from households, shop lots, and industrial area (The Sunday Mail, 2020) (Fig. 5).

Fig. 5 Improper disposal of face masks

7   COVID-19 Clinical Waste Management: A Case Study

A study on the current practices in handling the clinical waste during COVID-19 pandemic was carried at a northern state Hospital, Malaysia. This case study discusses the results of the questionnaire directed to personnel who are responsible for COVID-19 clinical waste management at the hospital. Edgenta Mediserve is the national agency responsible for clinical waste management. The workers of Edgenta Mediserve are dealing with COVID-19 clinical waste daily in all government hospitals throughout Malaysia. The hospital is located on Jalan Tun Hussein Onn, Seberang Jaya, 13,700 Permatang Pauh, Pulau Pinang and it is a government hospital. It has a total of 12 wards and 314 beds for the admission of various cases. At this hospital in Seberang Jaya, selected places were identified and reserved as areas for COVID-19 clinical waste, which included SARI (severe acute respiratory infection) ward, the emergency department, the pediatric ward, and CSSD (central sterile services department).

7.1   Management of COVID-19 Clinical Waste at the Hospital

The cleaning and collection of COVID-19 related clinical waste at the hospital is carried out by seven appointed cleaners in four different areas. Three cleaners are in charge of SARI ward working in three shifts, whereas one cleaner is in charge of the emergency department (observation ward, COVID-19 screening booth and decontamination room) and three more cleaners are in charge of the pediatric ward and CSSD unit. The COVID-19 clinical waste is collected in the waste bags (yellow bags) and then labeled as COVID-19 clinical waste. These yellow bags are placed into the properly labeled COVID-19 collection bins, which are well locked. Only the designated cleaners who are dealing with the COVID-19 clinical waste have the access key. The collection bins are changed daily.

7.1.1 Amount of Clinical Waste Generated

All the types of clinical waste generated from dealing with COVID-19 cases are treated as clinical waste. The daily generation of clinical waste at this hospital is in the range of 800 to 1000 kg/day of clinical wastes, whereas COVID-19 related clinical waste is about 120 kg/day. In Wuhan, China, the generation of clinical waste for treating COVID-19 patients ranged from 0.6 to 2.5 kg/per bed in a day for each patient (Yu et al., 2020). The clinical waste generated at Seberang Jaya Hospital mostly consists of sharps needles, scalpels, knives, blades.

7.1.2 Source Separation of Clinical waste

Cleaners at Seberang Jaya hospital segregate clinical waste. The clinical waste is segregated according to the guidelines provided by Government of Malaysia. Clinical waste is segregated into following types of waste:

1. Sharps—needles, scalpels, knives, and blades
2. Infectious waste which may transmit infection to human being
3. Waste suspected to contain pathogen—lab culture, swabs, gauzes, and bandages
4. Pathological waste, human tissue or fluid, body parts, blood, and body fluids
5. Chemical waste—lab reagents and disinfectants
6. Pharmaceutical waste—expired drugs

The clinical waste is further classified according to Scheduled Waste Regulations 2005 as SW403 Chemical and Pharmaceutical waste, SW404 Pathological waste, SW421 Mixture of Hazardous waste, and SW 424 Mixture of Hazardous and Non-Hazardous Waste.

7.1.3 Storage of Clinical Waste

The collection bins with clinical waste, as well as, COVID-19 clinical waste is stored in cold storage before transportation to the incinerator outside the hospital on daily basis.

7.1.4 Transport and Treatment of Clinical Waste

The transportation and treatment of clinical waste is important due to its negative impacts to human health and environment (WHO, 2018). The treatment of clinical wastes could be autoclave, incineration, and chemical disinfection. All the clinical waste storage bins including the COVID-19 waste are transported on daily basis by dedicated lorry of the concession contractor to a specific site in Kamunting for incineration. All the clinical waste including the COVID-19 waste is incinerated and the ash is disposed after incineration. The incinerator decontaminates the clinical waste by subjecting it to thermal destruction process at high temperature (1100–1600 °C) in controlled operational conditions. The products of combustion are ash residue, water, and carbon-dioxide (Olanrewaju & Fasinmirin, 2019).

7.1.5   Training Related to Management of Clinical Waste Related to COVID-19

Training is regularly conducted to equip the staff with adequate knowledge and information for protection from contracting the infection while dealing with COVID-19 clinical waste. The healthcare staff are trained on the correct technique of donning and doffing of PPE and as well as on the correct technique to dispose the general clinical waste and the COVID-19 waste.

7.1.6   Personal Protective Equipment (PPE)

Any personnel handling the COVID-19 related cases or handling clinical waste, including the cleaner, are well equipped with PPE with a three-ply surgical mask, disposable glove, disposable gown. Moreover, some additional PPE also include face shield, cap, and shoe cover.

8   Conclusion

In Malaysia, clinical waste management is under the Federal Government and Department of Environment (DOE). Environmental Quality Act 1974 with recent amendments is currently being implemented in Malaysia for the management of clinical waste. The guidelines on the management of clinical waste provide comprehensive information on appropriate handling and disposal of clinical wastes generated from hospitals and other health care facilities. The estimated generation of clinical waste is 90 tonnes/day or 33,000 tonnes/annually, where approximately 75–80% is non-pathogenic waste and remaining clinical waste is hazardous. Only licensed companies are allowed to transport, store, treat, and dispose clinical waste in Malaysia. Almost all of the hazardous clinical waste is incinerated, whereas non-hazardous clinical waste is disposed of along with municipal waste. From the case study, it is well established that the management of clinical and COVID-19 related waste is properly managed according to the Malaysian Schedule Waste Regulation (2005) under Environmental Quality Act 1974.

Specific practices currently being implemented for COVID-19 clinical waste management:

1. All the COVID-19 waste collection bags are properly labeled.
2. All the COVID-19 waste collection bags are placed into the collection bin, which is properly labeled and locked.
3. Only the designated staffs for handling COVID-19 waste have the access key to open the COVID-19 waste collection bin.
4. On top of the usual PPE that staff wears; three ply mask, disposable glove, disposable gown, additional PPE including cap, face shield, and shoes cover are also advised to be put on.

References

Abu-Qdais, H. A., Al-Ghazo, M. A., & Al-Ghazo, E. M. (2020). Statistical analysis and characteristics of hospital medical waste under novel Coronavirus outbreak. *Global Journal of Environmental Science and Management, 6*(Special Issue (COVID-19)), 21–30.

Agamuthu, P., & Jayanthi, B. (2020). Clinical waste management under COVID-19 scenario in Malaysia. *Waste Management & Research*. https://doi.org/10.1177/0734242X20959701

Al-Qaness, M. A., Ewees, A. A., Fan, H., & Abd El Aziz, M. (2020). Optimization method for forecasting confirmed cases of COVID-19 in China. *Journal of Clinical Medicine, 9*(3), 674.

Ambali, A. R., & Bakar, A. N. (2012). Medical waste management in Malaysia: Policies, strategies and issues. In *2012 IEEE colloquium on humanities, science and engineering (CHUSER)* (pp. 672–677). IEEE.

Ambali, A. R., Bakar, A. N., & Merican, F. M. (2013). Environmental policy in Malaysia: Biomedical waste, strategies and issues. *Journal of Administrative Science, 10*(1), 1–17.

Asian Development Bank. (2020). *Managing infectious medical waste during the COVID-19 pandemic*. Retrieved July 29, 2020, from https://www.adb.org/sites/default/files/publication/578771/managing-medical-waste-COVID19.pdf

Astro Awani. (2020). *COVID-19: Clinical waste management still under control - Dr Noor Hisham*. Retrieved May 25, 2020, from http://english.astroawani.com/malaysia-news/COVID-19-clinical-waste-management-still-under-control-dr-noor-hisham-244319

Chiang, C. F., Sung, F. C., Chang, F. H., et al. (2006). Hospital waste generation during an outbreak of severe acute respiratory syndrome in Taiwan. *Infection Control Hospital Epidemioly., 27*(5), 519–522.

Covidvax. (2021). *Live Covid-19 vaccination tracker*. Retrieved August 4, 2021, from https://covidvax.live/location/mys

Daily Express. (2020). *Breaking: 119 new delta, alpha, beta Covid-19 cases detected in Malaysia*. Retrieved August 4, 2021, from https://www.dailyexpress.com.my/news/175022/breaking-119-new-delta-alpha-beta-covid-19-cases-detected-in-malaysia/

Damanhuri, E. (2020). *Medical waste management for COVID-19: Current practices/responses in managing medical waste under the COVID 19 situation in Indonesia* (Personal Communication).

Department of Environment (DOE). (2009). *Guideline on the handling and management of clinical wastes in Malaysia* (3rd ed.). Government Printers.

Environmental Quality (Scheduled wastes) Regulation 2005. (2005). Retrieved from https://www.env.go.jp/en/recycle/asian_net/Country_Information/Law_N_Regulation/Malaysia/Malaysia%20EQA%20Scheduled%20Waste%202005.pdf

Hamid, A. A. (2021). *Program Imunisasi Covid-19 Kebangsaan bermula [METROTV]*. Retrieved August 4, 2021, from https://www.hmetro.com.my/utama/2021/02/677761/program-imunisasi-covid-19-kebangsaan-bermula-metrotv

Institute for Global Environmental Strategies (IGES). (2020). *Implications of COVID-19 for the environment and sustainability*. Retrieved from https://www.iges.or.jp/en/publication_docu ments/pub/policysubmission/en/10841/%231+COVID-19+Position+Paper+Final+%28EN%2 9.pdf

Khan, S., Ali, A., Siddique, R., & Nabi, G. (2020). Novel coronavirus is putting the whole world on alert. *Journal of Hospital Infection, 104*(3), 252–253.

Minoglou, M., Gerassimidou, S., & Komilis, D. (2017). Healthcare waste generation worldwide and its dependence on socio-economic and environmental factors. *Sustainability, 9*, 220. https://doi.org/10.3390/su9020220

Noor Artika, H., Yusof, M. Z., & Nor Faiza, M. T. (2019). An overview of scheduled wastes management in Malaysia. *Journal of Wastes and Biomass Management (JWBM), 1*(2), 01–04.

Olanrewaju, O. O., & Fasinmirin, R. J. (2019). Design of medical wastes incinerator for health care facilities in Akure. *Journal of Engineering Research and Reports*, 1–13.

Saw, C. B. (1994). Introduction to the Malaysian guidelines for clinical waste management. In *Proceedings of the workshop on clinical waste management held on 28th–2nd December*.

The Sunday Mail. (2020). *Govt urged to act on 'COVID refuse' disposal*. Retrieved June 23, 2020, from https://www.thesundaily.my/local/govt-urged-to-act-on-COVID-refuse-disposal-IX261 7578

WHO. (n.d.). *Tracking SARS-CoV-2 variants*. Retrieved August 4, 2021, from https://www.who.int/en/activities/tracking-SARS-CoV-2-variants/

Windfeld, E. S., & Brooks, M. S.-L. (2015). Medical waste management – A review. *Journal of Environmental Management, 163*, 98–108.

World Health Organization (WHO). (2018). *Healthcare waste*. Retrieved May 23, 2020, from https://www.who.int/news-room/fact-sheets/detail/health-care-waste

Worldometer. (2021). *Coronavirus update*. Retrieved August 2, 2021, from https://www.worldometers.info/coronavirus/

Yu, H., Sun, X., Solvang, W. D., & Zhao, X. (2020). Reverse logistics network design for effective management of medical waste in epidemic outbreaks: Insights from the coronavirus disease 2019 (COVID-19) outbreak in Wuhan (China). *International Journal of Environmental Research and Public Health, 17*(5), 1770.

Zaimastura, I. (2005). *Management and disposal of medical waste: Case study: Hospital Universiti Kebangsaan Malaysia*. Universiti Teknologi Malaysia. Retrieved May 20, 2020, from www.efka.utm.my

Waste Management Related to COVID-19 in the Noncontact Era: Case Study of the Republic of Korea

Seung-Whee Rhee and Dal-Ki Min

1 Introduction

The human history against specific micro-organisms and viruses such as plague in Europe and smallpox in a new continent has been recorded. After these infectious diseases, it was recognized the existence of micro-organisms which were invisible and caused to produce symptoms such as fever, weakness, and headache (CDC, 2020). Medical technologies were dramatically improved to manage several infectious diseases during several decades and the World Health Organization (WHO) activated in 1948 to attain the highest possible level of health, to promote health, to keep the world safe, and to serve the vulnerable social groups for global people at a minimal safe country level (WHO, 1948). However, it was faced with more strong and diverse micro-organisms and viruses after the later half of twentieth century.

At the beginning of 2020, COVID-19 has been surprisingly spread to the human society in all over the world. Even though COVID-19 was not the first virus disease spreading to the human society, its effect was incredibly great to global human society not likely the adverse effect from SARS, MERS, and Ebola virus. The global trend of COVID-19 is still increasing and total confirmed case on COVID-19 in the world is more than 198 million and 4.2 million deaths globally at 2 August 2021. All people, all locations, and all nations have been struggling against COVID-19 to overcome the current disease situation (WHO, 2021).

S.-W. Rhee (✉)
Department of Environmental Energy Engineering, Kyonggi University, Suwon, Republic of Korea
e-mail: swrhee@kyonggi.ac.kr

D.-K. Min
Department of Civil and Environmental Engineering, Gachon University, Seongnam, Republic of Korea

© The Author(s), under exclusive license to Springer Nature Singapore Pte Ltd. 2022
S. K. Ghosh, P. Agamuthu (eds.), *Health Care Waste Management and COVID 19 Pandemic*, https://doi.org/10.1007/978-981-16-9336-6_3

In Korea as well as most countries, COVID-19 has been urgently concerned from January 2020. Since Korea has some experiences of infectious diseases such as SARS, MERS, and Ebola virus, Korean government has managed strongly for COVID-19 at the beginning stage. The number of confirmed COVID-19 cases was more than 1000 at 1st of March and then stable to be less than 100 after the middle of March. However, the confirmed cases of COVID-19 sharply increased in winter season because COVID-19 was still rampant in Korea.

Lifestyle in COVID-19 pandemic situation is totally changed for public people to wear a mask, to use a hand sanitizer, and to keep a social distance with noncontact mode in order to protect from COVID-19. The traditional market has been converted to delivery service through online and mobile channels in the noncontact era (GRI, 2020). In addition, consumers prefer to use disposable products such as food containers and coffee cup to prevent indirective infection of COVID-19. Hence, the use of packaging and disposable product has increased significantly.

In order to manage COVID-19 sanitarily and safely, vaccine against COVID-19 must be developed but it still takes some time for most people to vaccinate even though some biopharmacy companies such as Modena and Pfizer developed vaccines. In addition, several used products from the confirmed patient on COVID-19 also should be managed safely. Accordingly, the Korean government took an additional measures to utilize combustible waste from households (e.g. plastic packaging) as an energy source in cement industries to cope with the increasing those wastes. In the COVID-19 era, medical waste management strategies should be strengthened in terms of sustainability, transparency, and sanitation to prepare for the increase in the generation of medical wastes, waste personal protective equipment (PPE), and household wastes.

The current status of COVID-19 and a social distance control system to protect from COVID-19 in Korea is described initially in this manuscript. The medical waste management including classification system and treatment facilities is described. And the management of waste PPE related to COVID-19 is expressed from sources to final treatment. Also household waste due to the change of lifestyle caused by COVID-19 is examined to manage properly and safely. Finally issues and perspectives of waste related to COVID-19 are suggested to solve the abnormal situation on waste generation and to sustain the waste management sanitarily.

2 Current Status of COVID-19 Pandemic in Korea

2.1 Status on the Spread of COVID-19

In the Republic of Korea (ROK), confirmed cases on COVID-19 were seriously increased from about 100 people daily in mid-February 2020 to about 1000 people daily in early April 2020. In order to minimize the confirmed cases, the government of Korea established a social distance control system with 3 levels to prevent the spread of COVID-19 (KDCA, 2020a).

As the number of confirmed cases on COVID-19 increased, the demand for personal hygiene products such as masks and hand sanitizers increased significantly. In February, personal hygiene products were very short so that most people could not buy them to protect from COVID-19 and the government of Korea tried to stabilize the distribution of masks through the 5 division system for purchasing masks. The 5 division system allowed consumers to purchase only 2 or 3 masks per person in a certain day among five weekdays from Monday to Friday using the last digit of their birth year (The Government of the Republic of Korea, 2020a).

As a result of the prevention of COVID-19, the number of confirmed cases on COVID-19 was reduced to about 30 people per day from April 19 to August 15, 2020. As the spread of COVID-19 decreased, the level of the social distance control system was changed from level 4 division to level 3 division to sustain a daily life normally. As the number of outdoor activities for most people increases in daily life, however, the local spread of COVID-19 was increased to about 200 people per day from August 16 to October 11, 2020.

As the spread of COVID-19 increased again seriously, the Korean government made a level change for restaurants not to available after 21:00 in order to limit people's gatherings, to recommend working at home, and to mandate wearing masks in outdoor life. As outdoor activities and gatherings of people were decreased, the spread of COVID-19 also decreased after October 11, and the system was converted to level down for the normal life system daily (KDCA, 2020a).

Due to low level of the government's strengthened social distance control system, the number of COVID-19 confirmed cases significantly increased more than 1000 confirmed cases per day from December 2020 to early January 2021. In this situation, the Korean government restricted gatherings of people as part of stronger quarantine measures. As a result, from mid-January to June 2021, confirmed cases of COVID-19 have been reduced to the confirmed case of 300–600 per day.

In Korea, AstraZeneca, Pfizer, Moderna, and Janssen vaccines are being introduced to prevent the spread of COVID-19, and vaccination started on February 26, 2021. In August of 2021, about seven million people (14% of the total population) have been fully vaccinated and about 19.94 million people who have been vaccinated at least once (38.8% of the total population) (MOHW, 2021b). As South Korea has fewer people fully vaccinated than other developed countries, challenges still remain for the Korean government to introduce a COVID-19 vaccine.

As the spread of COVID-19 has stabilized to some extent and started the vaccination of COVID-19, the Korean government attempted to come down the level of the social distance control system in consideration of the people's activities in accordance with the strengthened quarantine measures. During summer period from mid-July 2021, 1000–1900 confirmed cases of COVID-19 occurred daily, and the Korean society entered a new period of pandemic again. Since the spread of COVID-19 is rapidly progressing to a very serious situation, it is currently trying to prevent the spread of COVID-19 by restricting gatherings of people with the highest level of the system such as limiting gatherings to no more than 3 people after 6 pm.

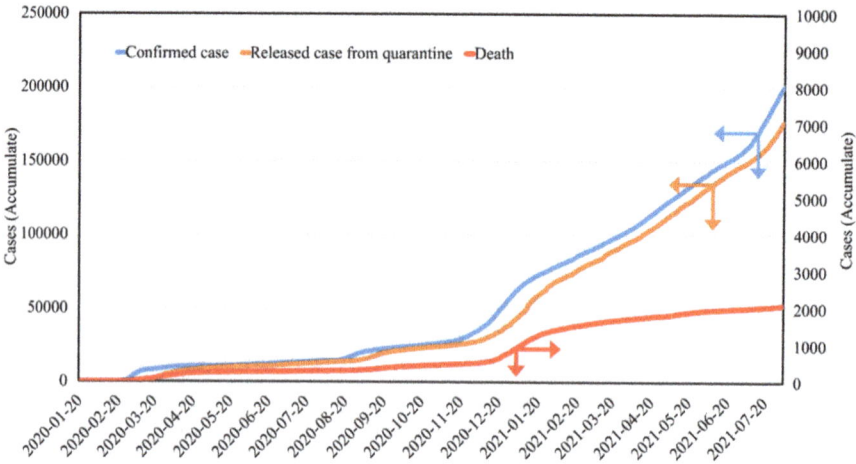

Fig. 1 Status on spread of COVID-19 in Korea

In Korea, the confirmed cases on COVID-19 were 202,203 people, where 177,909 people recovered, and 2104 people died on August 3, 2021. The status on spread of COVID-19 with time in Korea is shown in Fig. 1 (MOHW, 2021a).

2.2 Lifestyle Change in Noncontact Society

Due to the rapid spread of COVID-19 in Korea from February to March 2020, lifestyle has been significantly changed by the introduction of the concept of social distancing. The Korean government established a social distance control system to prevent the infection of COVID-19 by minimizing contact people with social distance levels 1–3 depending on the situation of the spread of COVID-19 (KDCA, 2020a). The WHO also strongly recommends that social distancing is important to prevent the infection of COVID-19 by expressing physical distancing between persons.

In Korea, the social distance control system is commonly used to limit the number of persons in multi-use facilities and people usually follow quarantine rules in the social distance control system with the decided certain level. In the case of level 2 of the social distance control system, for example, in-store dining after 21:00 is prohibited in restaurants and only takeout or delivery service is allowed. The quarantine rules according to the level of the social distance control system in Korea are shown in Table 1 (KDCA, 2020a).

The level in the social distance control system according to the period of spread and the confirmed case of COVID-19 in Korea is shown in Fig. 2. With the introduction of the social distance control system in March 2020, level 1 of social distancing was sustained from April to August 15 because the spread of COVID-19

Table 1 Quarantine rules for each life facility in the social distance control system (KDCA, 2020a)

Level of social distancing	Entertainment facilities	Door-to-Door sales	Karaoke	Indoor concert hall	Restaurant, cafe, bakery, etc.
1	– Limit of number of user (1 person/ 4 m² of facility area)	– Limit of number of user (1 person/4 m² of facility area) – Prohibition of offering songs and food	– Disinfection and do not use for 30 min after use	– Limit of number of user (1 person/4 m² of facility area)	– Installation of sneeze guard on the table – Leave 1 m interval between table
1.5	– Limit of number of user (1 person/ 4 m² of facility area) – No dancing and moving between seats	– Limit of number of user (1 person/4 m² of facility area) – Shutdown after 21:00 – Prohibition of offering songs and food	– Limit of number of user (1 person/4 m² of facility area) – Prohibition of eating food – Disinfection and do not use for 30 min after use	– Limit of number of user (1 person/4 m² of facility area) – Prohibition of eating food	– Installation of sneeze guard on the table – Leave 1 m interval between table
2	– Ban gatherings	– Limit of number of user (1 person/8 m² of facility area) – Shutdown after 21:00 – Prohibition of offering songs and food	– Limit of number of user (1 person/4 m² of facility area) – Shutdown after 21:00 – Prohibition of eating food – Disinfection and do not use for 30 min after use	– Shutdown after 21:00 – Leave 1 m interval between seat – Prohibition of eating food	– Restaurants can only be takeout or delivery service after 21:00 – Restaurants can only be takeout or delivery service
2.5	– Ban gatherings				– Restaurants can only be takeout or delivery service after 21:00 – Restaurants can only be takeout or delivery service
3	– Ban gatherings – Facilities essential to life such as selling daily necessities can be operated exceptionally				

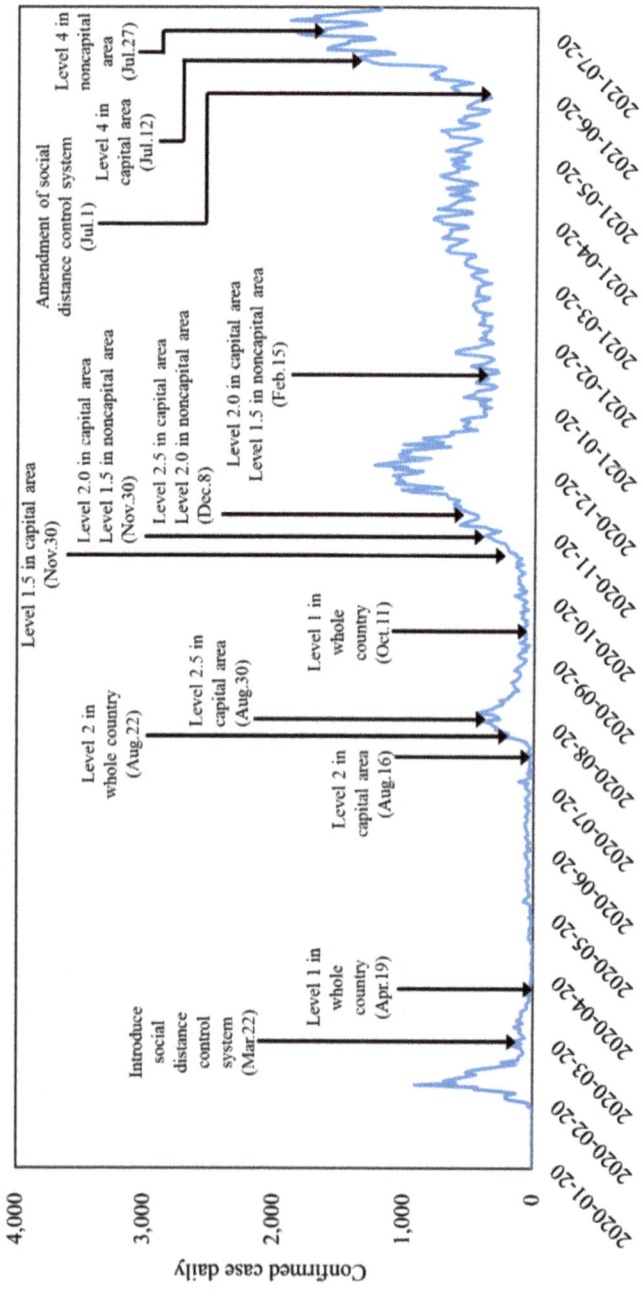

Fig. 2 Level in the social distance control system by confirmed cases of COVID-19

was relatively stable. After the period of summer vacation, it was upgraded to level 2 of social distancing in the capital area from August 16 and nationwide from August 22. Since the spread of COVID-19 was severe in the Seoul metropolitan area from August 30, the social distancing level was raised to 2.5. When the spread was relatively stabilized, the social distancing level was reduced to level 1 from October 11, 2020. And from the end of November, the spread of COVID-19 became serious again and the level of social distancing was sharply raised up to 2.5 (The Government of the Republic of Korea, 2020c). In June 2021, the Korean government amended the social distance control system into four stages in consideration of the local economy of Korea. In the amended social distance control system, most of the contents are similar to the previous ones, and it was intended to slightly recover the profits of small business owners by extending the time limit for use of stores until 10 pm.

Due to the spread of COVID-19, lifestyles have totally changed in Korea as well as most countries in the world. All people wear masks outside home and wash hands frequently. Most schools and universities turned into online lectures and many companies have used platforms of teleconference such as ZOOM, Webex, and others to transform the working form with online working at home. In addition, many self-employment businesses have discontinued by decreasing social gatherings but the use of online shopping and food delivery services has increased. Marketing scale of online shopping including household goods, food delivery service, and others was increased from about 307 billion U$ in the third quarter of 2019 to about 382 billion U$ in third quarter of 2020 (Statistics Korea, 2020a). Therefore, the spread of COVID-19 has accelerated the conversion to a noncontact society and an online society with enhancing the delivering system.

2.3   Changes in Household Waste Generation in COVID-19 Pandemic

The total waste generation has increased from 374,642 ton/day in 2010 to 446,102 ton/day in 2018 with a Compounded Annual Growth Rate (CAGR) of 2.2% for the period of 8 years. In 2018, total waste is generally composed of municipal solid waste (12.6%), industrial general waste (37.6%), construction waste (46.4%), and designated waste (3.4%) (MoE, 2019a).

In Korea, there is concern that specific wastes such as plastic waste and packaging waste will increase due to the increase in online shopping and food delivery service as the transition to a noncontact society due to COVID-19 pandemic. As shown in Table 2, the packaging wastes such as waste paper and plastic bag separated and discharged from household and residence before and after COVID-19 pandemic were compared from January to March (Statistics Korea, 2020b).

The amount of waste paper, plastic bag, and plastic waste is increased mainly due to the use of paper boxes and packaging materials for the delivery of goods or food.

Table 2 Major recyclable wastes among household waste in Korea (2020) (Statistics Korea, 2020b)

Type	January		February		March	
	Generation[a] (ton/day)	Variation[b] (%)	Generation[a] (ton/day)	Variation[b] (%)	Generation[a] (ton/day)	Variation[b] (%)
Total	5349	10.8	5355	10.2	5521	9.1
Paper	816	12.2	888	19.7	830	14.3
Plastic bag	937	7.6	905	6.6	926	10.0
Plastic	809	16.7	839	23.4	868	18.1
Resins	112	19.6	114	−0.9	112	−0.9
Others	2665	9.4	2609	5.5	2785	5.4

[a]This value is the sorting of generation of wastes separated and discharged among household wastes
[b]The variation represents the variation compared to the same month last year

The total amount of waste separated and discharged from household and residence is 16,225 tons/day from January to March 2020 with an increase of about 10% compared to the same period last year. In particular, the amount of plastic waste was increased about 20% compared to the same period last year. Since the amount of waste separated and discharged from household and residence in 2018 was 15,985 ton/day, it is found that the amount of waste separated and discharged from household between January and March 2020 exceeded the total amount of waste separated and discharged from household in 2018 as shown in Table 2. Therefore, it can be estimated that the changes in lifestyle caused by COVID-19 have a significant effect on the generation of packaging waste and plastic waste.

Because the amount of packaging waste has been increased, the storage capacity of recycled plastic and waste plastic bag in the separation companies should be necessary to increase. Since it is difficult to expand the storage capacity, most separation companies refused to collect the plastic waste and packaging waste from household and residence. Also, the price of plastic raw materials went down and the profit for recycling companies of plastic waste was almost nothing. Therefore, the Ministry of Environment (MoE) intends to maintain smoothly the plastic waste flow in Korea by controlling the storage amount of recycled plastic materials through public purchase. The market scale of public purchase is about 3.9 million US$, and the amount of recycled plastic materials by public purchase is about 10,000 tons (MoE, 2020a). In order to consume domestic plastic waste (PET, PP, PS, and PE) preferentially in Korea, furthermore, measures to ban the import of plastic waste from foreign countries are being promoted just like China ban on imported wastes (MoE, 2020b).

3 Medical Waste Management in Korea

3.1 Classification of Medical Waste

In Korea, waste is generally managed under the Waste Control Act. The classification of waste is divided into household waste and industrial waste according to the generation source (NIER, 2017). Industrial waste can be classified into general industrial waste and designated waste in consideration of the generation source and hazardous characteristics. Medical waste is classified and managed as a type of designated waste.

Medical waste under the Waste Control Act is defined as waste requiring special consideration for protection of human health and environment, such as waste that may harm the human health discharged from health care facility, testing institution, animal carcasses, and other wastes (KLRI, 2019a). And the medical waste is classified as an isolated medical waste, hazardous medical waste, and general medical waste as shown in Table 3.

Table 3 Classification of medical waste under the Waste Control Act in Korea

Type and code		Description
Isolated medical waste (10-11-00)		All wastes generated from medical measures conducted on persons who were quarantined in order to protect other persons from infectious diseases specified in Article 2, Clause 1 of the 「Infectious Disease Control and Prevention Act」
Hazardous medical waste (10-12-00)	Biopsic waste (10-12-01)	Part or whole of human or animal tissues, organs, system, animal carcass, blood, pus, and blood-generated substances (serum, plasma, blood derivatives)
	Pathological waste (10-12-02)	Culture medium used in tests, exams, etc., medium containers, stored culture collection, used test tubes, slides, cover glass, used culture medium, used gloves
	Damage-causing wastes (10-12-03)	Syringe needle, suture needle, surgical blade, acupuncture needle, dental needle, damaged glass-made testing devices
	Organic and chemical waste (10-12-04)	Used vaccine, used anti-cancer compound, used chemical treatment substances
	Blood-contaminated waste (10-12-05)	Used blood bags, wastes used during blood dialysis, and other wastes that contain blood that may leak and thus require special management
	Placenta (10-12-06)	Placenta among biopsic waste. It is managed for recycling purposes under the Waste Control Act
General medical waste (10-13-00)		Cotton clothes, bandage, gauze, single-use diaper, menstrual pad, single-use syringe, and infusion sets containing blood·bodily fluid·secretion·excrement

3.2 Medical Waste Generation

In Korea, medical waste is managed in 2 main categories according to the generation source. For the case of animal bodies, they are transferred to animal burial facilities. Therefore, the corpses of animals are not recorded in the amount of medical waste generated and processed. All other medical wastes are actively managed from the generation source (KLRI, 2019b).

With respect to storage, medical waste is stored in a special container as shown in Table 4 (MoE, 2019b). Special container for medical waste is used as plastic bag and box with cardboard or plastics. Each medical waste should be stored in a special container with different symbol colors. Red symbol color used for isolated medical waste and green symbol color used for placenta. General and hazardous medical wastes except placenta are stored in black bag or yellow box. Most liquid wastes should be stored in plastic box and other medical waste should be stored in plastic bag or cardboard box. Placenta is packaged one by one in a transparent inner bag, then stored in plastic box. And all containers for medical waste are controlled below 4 °C.

In Korea, facilities for discharging medical waste are as shown in Table 5 (MoE, 2019b). As of 2018, the number of medical waste discharging facilities is 66,142, which are managed by 7 basin and regional environmental offices. All facilities for discharging medical waste are classified into 364 general hospitals, 3061 hospitals, 50,245 clinics, 2518 public health centers, 11 midwifery centers, 2307 veterinary hospitals, 700 testing institutions, 592 funeral homes, 11 correctional facilities, 3106 welfare institutions for senior citizens, and 3227 other facilities (NHIS, 2020).

The type of medical waste according to generation source is shown in Table 6. Damage-causing waste and general medical waste are commonly generated from all generation sources and placenta for recycling purposes is managed specifically. However, placenta generated from maternity hospital is managed as a biopsic waste because it is not easy to store for recycling.

The total amount of medical waste generated is increased from 115 thousand tons in 2010 to 238 thousand tons in 2018 and the Compound Annual Growth Rate (CAGR) for same period was about 10% (Table 7). In 2018, the total amount of medical waste generated was 238 thousand tons and personal generation amount is about 4.7 kg/cap./year. The highest portion of medical waste is general medical waste, which accounts for 73% of total medical waste, followed by blood-contaminated waste and pathological waste. Among the medical waste in Korea, isolated medical waste has the most prominent trends in generation (MoE, 2019c).

From 2010 to 2018, the CAGR of isolated medical waste was the highest rate of 42%. It is because the residue of the test in the process of confirming the diagnosis of infectious diseases such as MERS and Ebola virus in those period and waste generated from the care of the confirmed patients has increased with the strengthening of the quarantine system to prevent the spread of infectious diseases. In addition, the number of confirmed cases of infectious diseases including MERS, Ebola virus, Varicella, Scrub typhus, and other diseases sharply increased from

Table 4 Standards for storage of medical waste

Type		Special container			Infectious symbol	Symbol color	Storage period
		Plastic bag	Box				
			Cardboard	Plastic			
Isolated medical waste		✓				Red	7 days
General medical waste		✓	✓			Black (bag), yellow (box)	30 days in refrigeration
Hazardous medical waste	Biopsic waste			✓			15 days (60 days for teeth)
	Pathological waste	✓	✓				15 days
	Damage-causing waste			✓			30 days
	Organic and chemical waste	✓	✓				15 days
	Blood-contaminated waste	✓	✓				15 days
	Placenta			✓		Green	–

Table 5 Facilities for discharging medical waste under the Waste Control Act

Medical waste generating institution	Related law and act
1. Medical institution	Article 3 of Medical Service Act
2. Public health center	Articles 3 and 7 of Regional Public Health Act
3. Public health clinic	Act on Special Measures for Health and Medical Service in Agricultural and Fishing Villages, etc.
4. Blood center	Article 2, subparagraph 3 of Blood Management Act
5. Quarantine station and animal quarantine agency	Article 30, subparagraph 1 of Quarantine Act and Article 30 of Act on the Prevention of Contagious Animal Diseases
6. Veterinary hospital	Article 2, subparagraph 4 of Veterinarians Act
7. Testing and research institution of national or local government (institutions related to medical science, Korean medicine, veterinary science, others)	
8. University, college and its affiliated testing and research institutes (institutions related to medical science, Korean medicine, veterinary science, others)	
9. Research institutes that conduct academic research or test and research on product manufacturing and invention (research institute related to medical science, Korean medicine, veterinary science, others)	
10. Funeral home	Article 29 of Act on Funeral Service, etc.
11. Medical facility installed in correctional facility, juvenile correctional facility, detention centers	Article 11 of Administration and Treatment of Correctional Institution Inmates Act
12. Medical facility with an area of 100 m^2 or more as an affiliated institution of a corporation	Article 35 of Medical Service Act
13. Medical facility installed in military units of division level or higher	Decree on the Armed Forces Medical Command
14. Welfare institution for senior citizens	Article 34 of Welfare of Senior Citizens Act
15. Company that has obtained permission for waste recycling business for placenta	Article 25 of Waste Control Act
16. Tissue bank	Article 13, subparagraph 1 of Safety, Management, etc. of Human Tissue Act
17. Other institutions notified by the Minister of Environment	

39,624 in 2010 to 168,219 in 2018 so that the amount of isolated medical waste was increased accordingly (KDCA, 2020b).

Specifically, the generation of placenta decreases from 2010 and then increased for 2016. In Korea, total fertility rate has decreased from 1.244 in 2011 to 0.977 in 2018 (MOIS, 2020), whereas the increase in the generation of placenta was

Table 6 Type of medical waste generated by generation source

Waste type		Generation source										
		General hospital	Hospital	Clinic	Public health center	Midwifery clinic	Veterinary hospital	Testing institution	Funeral home	Correctional facility	Welfare institution for senior citizens	Others
Isolated medical waste		✓	✓	✓	✓						✓	✓
Hazardous medical waste	Biopsic waste	✓	✓	✓	✓	✓	✓	✓			✓	✓
	Pathological waste	✓	✓	✓	✓		✓	✓			✓	✓
	Damage-causing waste	✓	✓	✓	✓	✓	✓	✓	✓	✓	✓	✓
	Organic and chemical waste	✓	✓	✓	✓		✓	✓			✓	✓
	Blood-contaminated waste	✓	✓	✓	✓		✓	✓	✓	✓	✓	✓
	Placenta	✓	✓	✓					✓	✓		
General medical waste		✓	✓	✓	✓	✓	✓	✓	✓	✓	✓	✓

Table 7 Generation of medical waste in Korea (unit: ton)

Type		2010	2012	2014	2016	2018	CAGR[a] (%)
Isolated medical waste		223	414	621	1764	3972	42
Hazardous medical waste	Subtotal	22,804	29,665	34,173	55,153	60,379	13
	Biopsic waste	523	5730	4643	7368	8932	8
	Pathological waste	7038	9360	10,024	12,163	12,973	8
	Damage-causing wastes	2308	2882	3247	4380	5202	11
	Organic and chemical waste	1708	2594	2467	3881	6226	17
	Blood-contaminated waste	6494	9087	13,783	27,337	27,016	20
	Placenta	33	13	9	24	31	9
General medical waste		92,028	117,577	135,133	164,676	173,922	8
Total		115,054	147,658	169,926	221,592	238,272	10

[a]CAGR is calculated from 2010 to 2018

estimated as a result of the fact that most pregnant woman prefer to give birth in hospitals and that hospitals manage the placenta usefully (Statistics Korea, 2020c).

3.3 Management of Medical Waste

Discharged and stored medical waste is transferred to incineration facilities using the special vehicle for medical waste. In the transportation of medical waste, the special vehicle for medical waste with closed loading box must be used, and the vehicle must be colored in white and marked with infectious symbol. As of 2018, the number of medical waste transportation company is 190, with the special vehicles of 1093 and loading capacity of 1908 tons (MoE, 2019c).

Medical wastes can be treated by incineration. Medical wastes subject to incineration are isolated medical waste, biopsic waste, organic and chemical waste, and waste that can flow down during storage and transport such as blood, body fluid, secretion, and excretion. The incineration residue of medical waste must be disposed of in the sanitary landfill.

In Korea, there are 14 incineration facilities for medical waste with the total capacity of 613.4 ton/day, which have been managed by the Basin and Regional Environmental Office. The location and capacity of incineration facilities for medical waste are shown in Fig. 3 according to the jurisdiction of the Basin and Regional Environmental Office (MoE, 2019c).

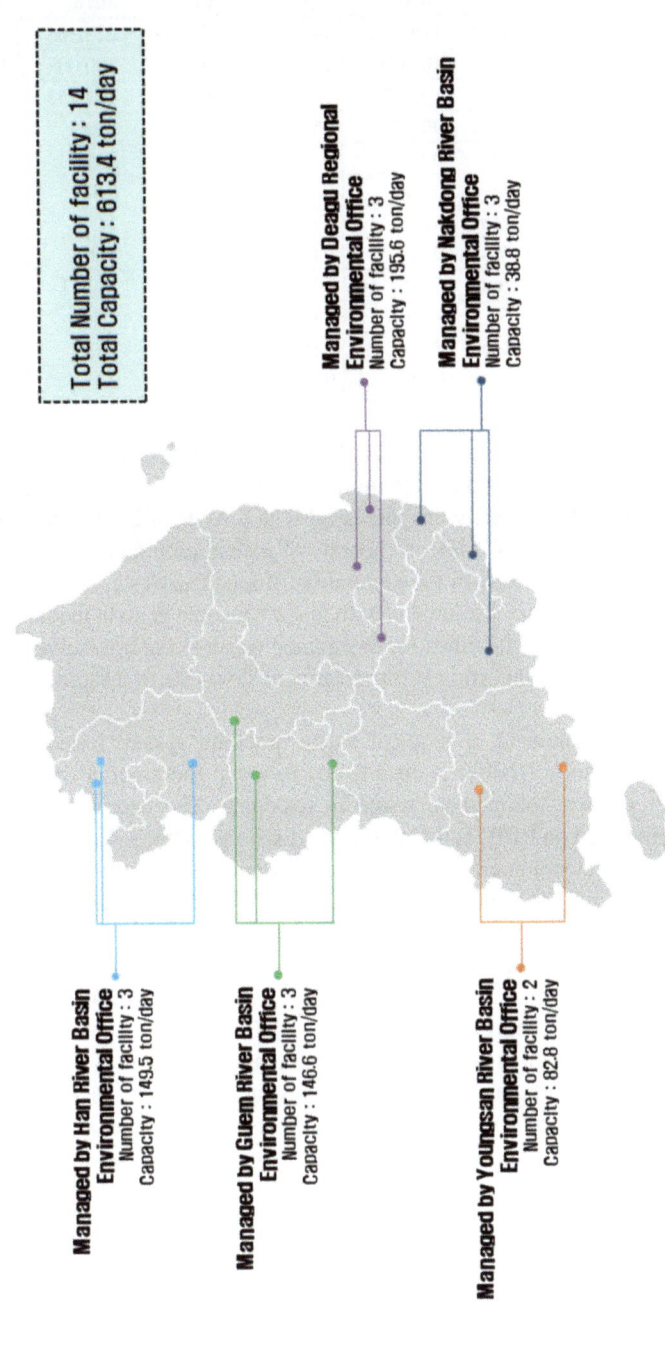

Fig. 3 Location of medical waste incineration facilities in Korea

Table 8 Treatment of medical waste in Korea (unit: ton)

Year		2010	2012	2014	2016	2018
Total	ton	115,396	148,593	170,468	223,303	238,789
	%	100.00	100.00	100.00	100.00	100.00
Incineration	ton	109,078	141,879	163,041	204,753	221,420
	%	94.52	95.48	95.64	91.69	92.73
Sterilization with shredding	ton	0	0	750	935	1171
	%	0.00	0.00	0.44	0.42	0.49
Recycling	ton	31	13	9	24	30
	%	0.03	0.01	0.01	0.01	0.01
Others	ton	5405	5448	6051	16,521	14,820
	%	4.68	3.67	3.55	7.40	6.21
Storage	ton	882	1254	616	1070	1348
	%	0.76	0.84	0.36	0.48	0.56

In 2010, there are 16 incineration facilities for medical waste with the total capacity of 508.9 ton/day. Although the number of incineration facilities for medical waste has decreased, the capacity has increased with CAGR of 2% for the period of 2010 to 2018. However, the CAGR in capacity of incineration facilities for medical waste is lower than the CAGR in the generation of medical waste. Accordingly, in 2019, the Waste Control Act was amended to allow for some general medical wastes to be treated at incineration facilities for designated waste in order to solve the event of a problem such as a congestion in the treatment flow due to a rapid increase in medical waste.

Also, in the management of medical waste, tracking system based on Radio Frequency Identification (RFID) is established to grasp the amount of generation and treatment, pathway of medical waste in real time. The operation method of tracking system based on RFID is that the waste discharger attaches an electronic tag to the medical waste container, and the transporter authenticates the amount of the medical waste through the reader when the waste is delivered. In addition, when the medical waste is unloaded to the treatment facility, the electronic tag is recognized again and the information on the treatment amount of medical waste is transmitted to the central information system. Through those processes, the overall management flow of medical waste is being tracked (KECO, 2020).

The treatment status of medical waste in Korea is shown in Table 8. Of the medical waste treatment status, incineration takes up the highest share with about 91% or more from 2010 to 2018, and recycling is focused on placenta only. And the amount of medical waste treated is slightly higher than the amount generated because the amount of medical waste stored in the previous year was included.

3.4   Management of Isolated Medical Waste Related to COVID-19

The management system of medical waste related to infectious virus in Korea has been developed and focused to treat safely through the experience of infectious diseases such as severe acute respiratory syndrome (SARS), Middle East respiratory syndrome (MERS), and Ebola virus (KDCA, 2015).

In the face of the COVID-19 era, medical waste related to COVID-19 can be generated from hospitals, public health center, quarantine, community treatment center, and even residence. Since the generation source of medical waste related to COVID-19 was diverse, it is necessary to manage the medical waste from the source control initially. The generation amount of medical waste should be considered by establishing a special management strategy to manage the medical waste relate to COVID-19 sanitarily and safely.

The medical waste was initially anticipated to increase a lot because almost all persons thought that the number of confirmed patients on COVID-19 might be sharply increased. The comparison of medical waste generation before and after the COVID-19 in Korea is shown in Fig. 4 from January to August between 2019 and 2020. The total amount of medical waste generated monthly in 2020 was declined to 85% of 2019 because hospital assess for most general mild patients was significantly decreased due to the concern of COVID-19 infection shown in Fig. 4a. However, the amount of isolated medical waste generated in 2020 is much higher than that in 2019 because quarantine and treatment activities increased with the spread of COVID-19 from January 2020. From March to April 2020, the amount of isolated medical waste generated more about 5 times than that of 2019 and the generation amount of isolated medical waste has increased more than double after May 2020 shown in Fig. 4b.

The correlation between COVID-19 patients and isolated medical waste from January to August 2020 was clearly shown in Fig. 5. The reason for the large amount of isolated medical waste generated in March and April 2020 is that all confirmed patients of COVID-19 were managed in the intensive care unit (ICU) because of

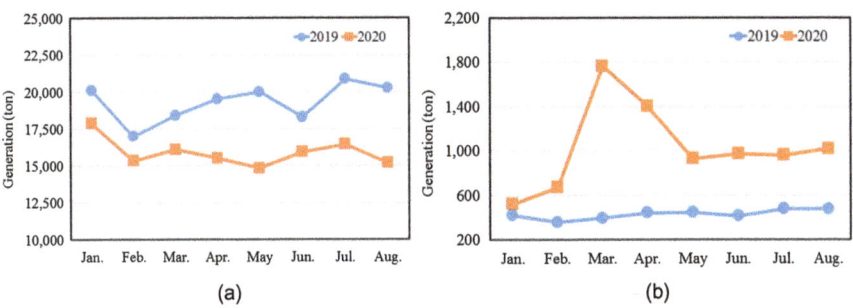

Fig. 4 Changes in pattern of medical waste generation by COVID-19 in Korea. (**a**) Total medical waste. (**b**) Isolated medical waste

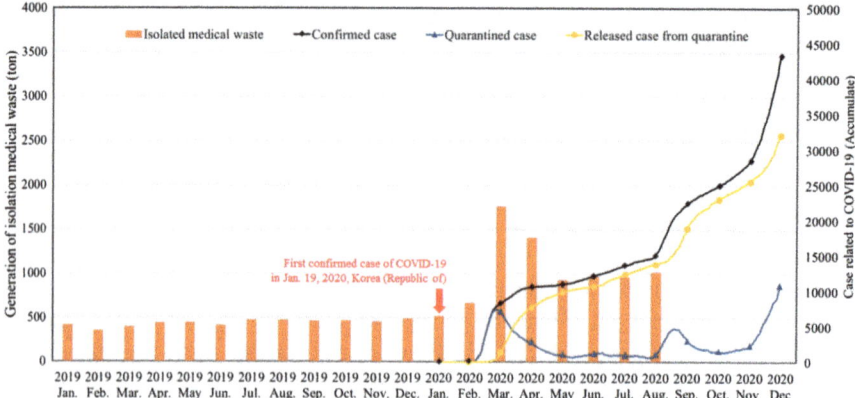

Fig. 5 Generation of isolated medical waste with spread of COVID-19 in Korea

treating separately at the initial response to COVID-19. Afterwards, the amount of isolated medical waste has been stabilized at 900 ton/month because only critical patients of COVID-19 are treated in ICU separately.

In particular, about 50 confirmed cases on COVID-19 occurred per day until the beginning of August 2020 and about 100 confirmed cases occurred daily by the end of October 2020 with a slight increase. From November 2020, the spread of COVID-19 has increased significantly in winter season with about 1000 confirmed cases a day. Accordingly, it is expected that the isolated medical waste will be increased significantly from the quarantine facilities, community treatment center, and hospitals by the treatment of confirmed patients on COVID-19.

The management of medical waste in the period of spreading COVID-19 should follow the flow stage of Classification—Discharge—Storage—Transportation— Treatment and Disposal (MoE, 2020c). The management strategy for medical wastes related to COVID-19 was established by Ministry of Environment (MoE) and medical waste related to COVID-19 was always inspected by government officers from stage to stage. The MoE announced special measures to strengthen the management of isolated medical waste related to COVID-19 compared with the management of medical wastes in normal situation as shown in Table 9.

Medical waste generated by COVID-19 has been basically managing as an isolated medical waste. Unlike conventional isolated medical waste, isolated medical waste related to COVID-19 is carried with a double-packed storage container of plastic bag inside and plastic box on the outside, and is stored at 4 °C. The isolated medical waste related to COVID-19 is collected daily, and temporary storage for transportation is not permitted (MoE, 2020c).

Table 9 Special measures on medical waste related to COVID-19

Type	Special measures on COVID-19	Current regulation		
	Isolated medical waste	Isolated medical waste	Hazardous medical waste	General medical waste
Classification				
Discharge	– Double sealed structure of outer plastic box and inner plastic bag (infectious symbol is red)	– Double sealed structure of outer plastic box and inner plastic bag (infectious symbol is red)	– Biopsic waste: Plastic box (infectious symbol is yellow) – Placenta: Plastic box (infectious symbol is green) – Pathological, organic, chemical and blood-contaminated waste: Using cardboard box (infectious symbol is yellow) or plastic bag (infectious symbol is black) – Damage-causing wastes: Using plastic box (infectious symbol is yellow)	– Using plastic bag (infectious symbol is black) – Using cardboard box (infectious symbol is yellow)
Storage	– Refrigerated storage (4 °C) – Disinfection in storage site – Take out on the day the waste was generated	– Refrigerated storage (4 °C) – Disinfection in storage site. – Storage until 7 days	– Biopsic waste and placenta: Refrigerated storage (4 °C) – Pathological, organic, chemical, blood-contaminated and sharp waste: Closed storage warehouse – Storage until 15~60 days	– Closed storage warehouse – Storage until 30 days
Transportation	– Use the refrigerator car – No temporary storage	– Use the refrigerator car – Temporary storage (2 days)	– Use the refrigerator car – Temporary storage (2~5 days)	– Use the refrigerator car – Temporary storage (2~5 days)
Treatment	– Incineration – Treatment period (on the day)	– Incineration – Treatment period (2 days)	– Incineration – Treatment period (2~5 days)	– Incineration – Treatment period (5 days)

4 The Waste of Personal Protective Equipment

4.1 Source for the Waste Personal Protective Equipment (PPE)

Medical waste generated from general medical institutions not related to COVID-19 is discharged, stored, collected, transported, and incinerated in accordance with the existing medical waste management system. In the case of medical institutions related to COVID-19,, all medical wastes such as masks, residue from test and others must be treated as isolated medical waste and incinerated on the same day. Also, all waste from intensive care units or quarantine treatment centers where COVID-19 patients are staying should be treated as isolated medical waste and incinerated on the same day (MoE, 2020c).

Anyone suspected of being infected with COVID-19 will be quarantined for 14 days by community treatment center or self-quarantine. Those who are quarantined at the community treatment center are given a plastic bag for medical waste, and the plastic bag for medical waste generated from community treatment center is basically collected every 3 days. Plastic bag for medical waste is also provided to people who stays in self-quarantine. All wastes from self-quarantine must be loaded in the plastic bag for medical waste with a double packaging and discharged with standard garbage bag. The waste PPE such as masks, plastic gloves, and containers of sanitizer discharged from general household and workplace is regarded as general waste and is carried in a standard garbage bag for incineration. The generation sources of the waste PPE and types of major wastes are summarized in Table 10.

Table 10 Generation source and type of the waste PPE

Type		Major type of PPE	Management
Patient	Intensive care unit	All medical waste and PPE generated by patients and medical workers	All medical waste and PPE are treated as isolated medical waste and incinerated on the same day
	Quarantine treatment center		
Quarantine station, public health center, hospital, and others		All PPE, residue from test	
People suspected of being infected*	Community treatment centers	All PPE	After disinfection, it is managed as general waste and incinerated.
	Self-quarantine	Mask, wet wipes, toilet paper, and others	
Household and workplace		Mask, wet wipes, sanitizer, and others	Discharged to general waste

*Those who have developed COVID-19 clinical manifestation such as fever, cough, respiratory difficulty, chills and others within 14 days of contact with COVIV-19 patient

4.2   The Waste PPE Generation

The waste PPE can be generated simply and in large quantities in a special day. Despite the concern about the spread of COVID-19, the election for the National Assembly members was held on April 15, 2020 in Korea. Peoples who voted for the election of members in the National Assembly wore disposable sanitary gloves (vinyl gloves) on both hands when entering into the polling place. The number of voters who participated at the election of members for the National Assembly in Korea is about 29.1 million out of 44 million eligible voters, and the voting rate of this election is 66.2% (NEC, 2020). Accordingly, it may be estimated that about 58.3 million disposable sanitary gloves were discarded. Since the mass of a piece of disposable sanitary gloves in Korea is about 1.3 g, the weight of disposable sanitary gloves discarded in the election is about 75.7 tons a day. In principle, these discarded disposable sanitary gloves should be incinerated without any recycling because it may be possible to infect the corona virus from the globes indirectly.

In addition, since the spread of COVID-19 in Korea, the demand for PPE such as face masks and globes for personal hygiene had increased rapidly. In particular, in the early stages of the spread of COVID-19, the supply of face masks was difficult, so the Korean government introduced a publicly distributed face mask (The Government of the Republic of Korea, 2020a). The publicly distributed face mask is a system in which face masks can be purchased only on designated day at designated location such as pharmacies to stabilize the supply and demand of face masks. This system began in March 2020 and was abolished on July 12 2020, when the supply and demand for face masks stabilized (The Government of the Republic of Korea, 2020b). After November 2020, the central and local governments have made it mandatory for people to wear face masks during outdoor activities specifically in bus and subway, and face masks have become a necessity in the lives of people amid the spread of COVID-19 (KDCA, 2020c).

In Korea, about 5.27 billion face masks were produced from February 19 to December 18, 2020 (MFDS, 2020). Face masks distributed to people on the market are about 4.0 g in packaging and about 4.3 g in face masks. Therefore, it is roughly estimated that about 22 thousand tons of face masks and 21 thousand tons of packaging material of face masks will be generated when considering the production of face masks and the weight of the face masks distributed in the market. Thus, the amount of the waste PPE generated can be very large if it includes PPE other than face masks. In addition, the waste PPE generated from general household or workplace is discharged in standard garbage bag for incineration and is managed as household waste, not medical waste. If it is assumed that the half of production of face mask is used by consumers, the amount of waste mask may be estimated to be about 11 thousand ton. Taking into account the other PPE, such as medical gown, gloves, plastic shield for eye protector and sterilized products in hospitals, community treatment center and quarantine facilities, the amount of the waste PPE can be anticipated to be much higher than 11 thousand ton. In order to treat the waste PPE safely and sanitarily, the amount of the waste PPE generated from several sources should be estimated in detail with an available methodology.

4.3   Management of the Waste PPE

Confirmed COVID-19 patients are generally treated in hospital care, community treatment centers, and at residence for self-quarantine in Korea (KDCA, 2020d). Hospital care is applied to serious confirmed cases on COVID-19. When hospital care is not required by the doctor's diagnosis, confirmed COVID-19 patients are admitted at community treatment centers to improve the symptoms of COVID-19. Self-quarantine at residences for 2 weeks is applied to suspected persons of having COVID-19 infection and to persons traveled to foreign countries. The indirect infection of COVID-19, which means the spreading from the used PPE by the confirmed patient without any direct contact with the confirmed patient, has recently occurred.

The Ministry of Environment, in cooperation with the health authorities, has strengthened the management of the waste PPE as a part of medical waste related to COVID-19 in order to prevent the indirect spread of COVID-19 from the waste PPE. The management flow of the waste PPE related to COVID-19 by special measures is shown in the Fig. 6 (MoE, 2020c; Rhee, 2020). The management flow of the waste PPE related to COVID-19 was established from the sources of hospitals, community treatment centers and self-quarantine facilities through collection and transportation to treatment facilities safely and sanitarily. All the waste PPEs related to COVID-19 from hospitals, community treatment center, and quarantine facilities are classified as isolated medical wastes. When discharging of this, it should be used a medical waste bag with a specific color. And the waste PPE should be sealed in special containers made of plastic and disinfected. In addition, the waste PPE related to COVID-19 should be stored in a facility with closed system and maintained at 4 °C. The waste PPE related to COVID-19 from hospital should be taken out every day and incinerated on the same day (MoE, 2020c).

The waste PPE related to COVID-19 from the community treatment center can be divided into isolated medical waste and general medical waste. When the waste PPE from the community treatment center is proved as an isolated medical waste, it is followed by the management flow of medical waste related COVID-19 from hospitals. If the waste PPE in the community treatment center is generated from working operator without contact with confirmed patients of COVID-19, the waste PPE is managed as general medical wastes. The working operator is not related to COVID-19 directly but performs maintenance and administrative work of the community treatment center. General medical waste is also placed in medical waste bag with a specific color. And general medical waste from the community treatment center should be sealed in container made of cardboard box with non-visible structure from the outside and taken out every day to incinerate within a day.

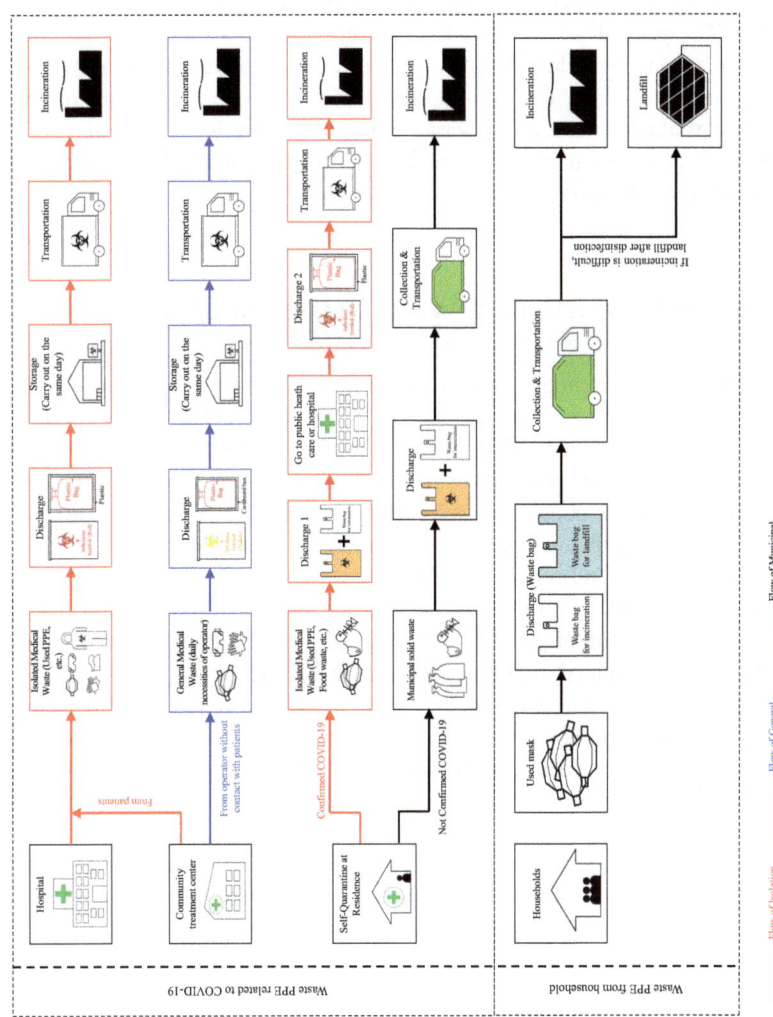

Fig. 6 Management flow of the waste PPE in Korea (Modified from Rhee, 2020)

The wastes from self-quarantine at residence can be divided into isolated medical waste or household waste in accordance of the confirmation of COVID-19. If persons confirmed by COVID-19 during self-quarantine, the waste PPE from confirmed persons in self-quarantine is classified to be isolated medical waste. In the case, it should be sent to public hospitals using double packaging with a medical waste bag and a standard garbage bag. Both the medical waste bag and standard garbage bag are provided free of charge by the Korea Environment Corporation to self-isolator. The packed the waste PPE from self-isolator is transported to public health care or hospital to follow the management flow of isolated medical waste. In the case of not confirmed COVID-19, the waste PPE from self-quarantine at residence is classified as household waste and collected by local governments for incineration. If incineration of the waste PPE is difficult, it can be disposed into landfill site after sterilization.

5   Issues and Perspectives of Waste Related to COVID-19

Because of the first experience of corona virus pandemic, issues on wastes related to COVID-19 may be variety with the situation of each country. One of the hypotheses on waste management during COVID-19 pandemic may be that waste related to COVID-19 should be managed more strengthen than recycling wastes. In reality, almost all people did not consider wastes related to COVID-19 in public area and open business places but did take care the infection of COVID-19 in their life activities. Since wastes related to COVID-19 can indirectly cause the COVID-19 infection, it will be very important to manage them safely and sanitarily. In addition, during the period of post COVID-19 pandemic, the composition of household waste may be changed significantly because lifestyle in major area of businesses and schools will be adopted by the online activities. In household residence, plastic wastes and packaging wastes increased due to the increase in online shopping and food delivery service as the transition to a noncontact society due to COVID-19 pandemic. Hence, it should be considered to manage plastic wastes and packaging wastes efficiently and properly.

The first issue of wastes related to COVID-19 is to set the category of the waste PPE from generation source considering the risk of the infection. The waste PPE should be collected separately by the category to manage efficiently and sanitarily. During the period of COVID-19 pandemic, the PPE such as mask, glove, gown, and eye protection are intended to provide for the protection against the infection of COVID-19. The guideline of the PPE regulation in EU suggested to set the guide for categorization of the PPE based on the risk type (European Commission, 2018; ECDC, 2020). The PPE shall be placed in the category corresponding to the risk of the infection and the category of the waste PPE should be considered to collect them from the sources separately and efficiently.

The second issue of wastes related to COVID-19 is to anticipate the generation amount of the waste PPE from generation source. The capacity of waste treatment facilities should be suitable to receive the collecting amount of the waste PPE. If not, the waste PPE may be one of the infection sources indirectly because the waste PPE can be treated inefficiently and unsanitarily. However, it is difficult to collect and to figure out the generation amount of the waste PPE during the period of COVID-19 pandemic. There are so many sources to generate the waste PPE and the sources are not classified clearly. Hence, the category of the PPE based on the risk type should be used to find out the generation sources. The category corresponding to the risk can be decided by the generation sources. If the generation sources are decided, the methodology of the generation amount is developed to get the generation amount of the waste PPE accurately. The goal of collecting the generation amount of the waste PPE is to bring the better real-time waste management data and facilities.

The third issue of wastes related to COVID-19 is to decide a treatment method of the waste PPE. Since the waste PPE may be one of the infection sources indirectly, the treatment of the waste PPE should be used by incineration sanitarily. Hence, it is necessary to secure the capacity of incineration facilities in each country even though there are some movements by residences or NGOs against building the incinerators. In Korea, some medical wastes such as diapers and plastic materials for liquid protection can be treated in the incinerators for designated wastes. If there is no incineration facilities to treat the waste PPE, it should be sterilized at first and then disposed in landfill site.

The fourth issue is to establish the management guideline of the waste PPE. Eventually, all waste streams of the waste PPE including source category and control, collection, transportation, and treatment should be involved in the management guideline. In each waste stream, the management of the waste PPE is established by the consideration of the difficulties in local area. And the responsibility and role for stakeholder from source through transportation to final treatment should be clarified in the guideline to manage the waste PPE safely and sanitarily.

For the perspective of waste related to COVID-19, the waste PPE will be gradually disappeared as the confirmed patients on COVID-19 will recover with the use of vaccines. In the periods of post COVID-19 pandemic, disposable waste and packaging waste may be still increased by adopting virtual activities through online system because anti-infection and anti-flu are used for most people not to use multiple products. Since it is concerned with human health over environmental health, plastic reduction policies and disposable waste management strategies have recently been revised or temporarily postponed (Prata et al., 2020). Immunity system should be reinforced to reduce and to recycle disposal products and packaging product. Hence, it should be reinforced the management of disposable waste and packing wastes with the consideration of anti-infection and anti-flu because most people do not want to use second hand products.

For medical wastes, monitoring system should be intensified to figure out transportation flow from generation to final treatment. In Korea, all transportations for medical wastes must get the license from the central government and build a global positioning system (GPS) to record the transport route (KECO, 2020; Kim, 2020). However, illegal transportation and treatment may be occurred because the treatment cost is very expensive. Because the capacity of medical waste incineration is not enough, the treatment cost of medical waste will be increased significantly. Hence, transportation of medical waste will just visit treatment facilities but will not unload the transportation at treatment facilities. The objectives of monitoring system of the waste PPE are to provide the information of mitigation and loading measures on real time and to protect illegal transportation and treatment. Using the GPS with gaging the weight measure, the weight of transportation is also recorded whenever medical waste is unloaded at treatment facilities in order to prevent illegal transportation and treatment.

Finally education program based on the management guideline for the management of medical waste including the waste PPE should be prepared to inform for source control, collection method, transportation, and treatment technique. The education program should be developed in collaboration with all stakeholder including public health authorities and/or hospitals. In order to aware the education program, it should be established communication channels to provide the information of the minimum requirement for all stakeholder and residents to understand the reason of the management of the waste PPE safely and sanitarily. Also most people and residents should understand the reason why does the capacity of the treatment facilities expand and build to manage them sanitarily and safely.

References

CDC (Centers for Disease Control and Prevention). (2020). *Smallpox, history of smallpox.* Retrieved December 3, 2020, from https://www.cdc.gov/smallpox/history/history.html

ECDC (European Centre for Disease Prevention and Control). (2020). *Guidance for wearing and removing personal protective equipment in healthcare setting for the care of patients with suspected or confirmed COVID-19.*

European Commission. (2018). *Guide to application of Regulation (EU) 2016/425 of the European Parliament and of the Council of 9 March 2016 on Personal Protective Equipment and repealing Council Directive 89/686/EEC.*

GRI (Gyeonggi Research Institute). (2020). *Makes a better future for Korea.* Retrieved May 1, 2020, from https://www.gri.re.kr/%ec%9d%b4%ec%8a%88-%ec%a7%84%eb%8b%a8/?pageno=2&ptype2=&sc=&sv=%EC%BD%94%EB%A1%9C%EB%82%98&limit=10&searchcode=&pcode=&brno=14491&prno=20200233

KDCA (Korea Disease Control and Prevention Agency). (2015). *Ebola virus response guidelines.* Retrieved April 6, 2020, from http://www.cdc.go.kr/board.es?mid=a20504000000&bid=0014&tag=&act=view&list_no=128408

KDCA (Korea Disease Control and Prevention Agency). (2020a). *COVID-19, Social distancing,* http://ncov.mohw.go.kr/socdisBoardView.do?brdId=6&brdGubun=1 (accessed 16 August 2020).

KDCA (Korea Disease Control and Prevention Agency). (2020b). *Infectious disease portal.* Retrieved December 16, 2020, from http://www.kdca.go.kr/npt/biz/npp/ist/bass/bassDissStatsMain.do

KDCA (Korea Disease Control and Prevention Agency). (2020c). *COVID-19, wearing mask.* Retrieved November 10, 2020, from http://ncov.mohw.go.kr/guidelineView.do?brdId=7&brdGubun=71&dataGubun=&ncvContSeq=4056&contSeq=4056&board_id=&gubun=

KDCA (Korea Disease Control and Prevention Agency). (2020d). *COVID-19, Quarantine system in Korea*. Retrieved November 10, 2020, from http://ncov.mohw.go.kr/baroView2.do?brdId=4&brdGubun=42

KECO (Korea Environment Corporation). (2020). *Waste management, RFID-based medical waste management system*. Retrieved December 16, 2020, from https://www.keco.or.kr/kr/business/resource/contentsid/3247/index.do

Kim, J. H. (2020). *Medical waste management in Korea's COVID-19 response*. UNDP Webinar series: Building Back better in Korea and Elsewhere. Retrieved May 20, 2020, from https://www.undp.org/content/seoul_policy_center/en/home/presscenter/articles/2019/building-back-better%2D%2Dgreen-new-deal-in-korea-and-elsewhere.html

KLRI (Korea Legislation Research Institute). (2019a). *Korea law translation center, waste control act*. Retrieved October 14, 2020, from https://elaw.klri.re.kr/kor_service/lawView.do?hseq=51523&lang=ENG

KLRI (Korea Legislation Research Institute). (2019b). *Korea law translation center, enforcement decree of waste control act*. Retrieved October 14, 2020, from https://elaw.klri.re.kr/kor_service/lawView.do?hseq=51207&lang=ENG

MFDS (Ministry of Food and Drug Safety, Korea). (2020). *Mask production/supply trend*. Retrieved December 22, 2020, from https://www.mfds.go.kr/brd/m_99/view.do?seq=44882&srchFr=&srchTo=&srchWord=%EB%A7%88%EC%8A%A4%ED%81%AC&srchTp=0&itm_seq_1=0&itm_seq_2=0&multi_itm_seq=0&company_cd=&company_nm=&Data_stts_gubun=C9999&page=1

MoE (Ministry of Environment, Korea). (2019a). *National waste generation and treatment in 2018*.

MoE (Ministry of Environment, Korea). (2019b). *Guidance on separate discharge of medical waste*.

MoE (Ministry of Environment, Korea). (2019c). *National designated waste generation and treatment in 2018*.

MoE (Ministry of Environment, Korea). (2020a). *Press Release, Public purchase of recycled PET respond to COVID-19*. Retrieved May 7, 2020, from http://me.go.kr/home/web/board/read.do?pagerOffset=0&maxPageItems=10&maxIndexPages=10&searchKey=title&searchValue=%EB%B9%84%EC%B6%95&menuId=286&orgCd=&boardId=1369720&boardMasterId=1&boardCategoryId=&decorator=

MoE (Ministry of Environment, Korea). (2020b). *Press Release, Ban on import of unnecessary waste*. Retrieved March 24, 2020, from http://me.go.kr/home/web/board/read.do?pagerOffset=0&maxPageItems=10&maxIndexPages=10&searchKey=title&searchValue=%EC%88%98%EC%9E%85&menuId=286&orgCd=&boardId=1358830&boardMasterId=1&boardCategoryId=&decorator=

MoE (Ministry of Environment, Korea). (2020c). *3rd special measures for safety management of wastes related to COVID-19*. Retrieved March 3, 2020, from http://ncov.mohw.go.kr/duBoardList.do?brdId=2&brdGubun=28

MOHW (Ministry of Health and Welfare, Korea). (2021a). *Coronavirus disease-19, Cases in Korea*. Retrieved August 3, 2021, from http://ncov.mohw.go.kr/en/bdBoardList.do?brdId=16&brdGubun=161&dataGubun=&ncvContSeq=&contSeq=&board_id=

MOHW (Ministry of Health and Welfare, Korea). (2021b). *Coronavirus Disease-19, Press release, Status of vaccination in Korea*. Retrieved August 3, 2020, from http://ncov.mohw.go.kr/tcmBoardView.do?brdId=3&brdGubun=31&dataGubun=&ncvContSeq=5747&contSeq=5747&board_id=312&gubun=ALL

MOIS (Ministry of the Interior and Safety, Korea). (2020). *Korean city statistics*. Retrieved December 16, 2020, from https://www.mois.go.kr/frt/bbs/type001/commonSelectBoardArticle.dobbsId=BBSMSTR_000000000014&nttId=81704

NEC (National Election Commission, Korea). (2020). *Election statistics system*. Retrieved October 3, 2020, from http://info.nec.go.kr/main/main_previous_load.xhtml

NHIS (National Health Insurance Service, Korea). (2020). *Health insurance statistics*. Retrieved October 26, 2020, from https://kosis.kr/statHtml/statHtml.do?orgId=354&tblId=DT_HIRA44

NIER (National Institute of Environmental Research). (2017). *Manual of waste classification system and classification method.*

Prata, J. C., Patricio Silva, A. L., Walker, T. R., Duarte, A. C., & Rocha Santos, T. (2020). COVID-19 pandemic repercussions on the use and management of plastics. *Environmental Science & Technology, 54*(13), 7760–7765.

Rhee, S. W. (2020). Management of used personal protective equipment and wastes related to COVID-19 in South Korea. *Waste Management & Research, 38*(8), 820–824.

Statistics Korea. (2020a). *Press Release, Online shopping in October 2020.* Retrieved December 3, 2020, from https://kostat.go.kr/portal/korea/kor_nw/1/12/3/index.board?bmode=read& bSeq=&aSeq=386445&pageNo=1&rowNum=10&navCount=10&currPg=&searchInfo=& sTarget=title&sTxt=

Statistics Korea. (2020b). *Press Release, Korean social trends.* Retrieved December 11, 2020, from http://kostat.go.kr/portal/korea/kor_nw/1/1/index.board?bmode=read&aSeq=386561

Statistics Korea. (2020c). *Birth statics in 2019.* Retrieved December 16, 2020, from http://kostat.go.kr/portal/korea/kor_nw/1/1/index.board?bmode=read&aSeq=384631

The Government of the Republic of Korea. (2020a). *Mask supply and demand stabilization measures by COVID-19.* Retrieved March 6, 2020, from https://www.gov.kr/portal/ gvrnPolicy/view/H2004000000530859?policyType=G00301&srchTxt=%EB%A7%88%EC %8A%A4%ED%81%AC%20%EC%88%98%EA%B8%89%20%EC%95%88%EC%A0% 95%ED%99%94%20%EB%8C%80%EC%B1%85

The Government of the Republic of Korea. (2020b). *Policy briefing.* Retrieved July 7, 2020, from https://www.korea.kr/news/policyNewsView.do?newsId=148874399

The Government of the Republic of Korea. (2020c). *Policy briefing.* Retrieved October 6, 2020, from https://www.korea.kr/news/policyNewsView.do?newsId=148880568

WHO (World Health Organization). (1948). *Constitution of the World Health Organization.*

WHO (World Health Organization). (2021). *WHO coronavirus disease (COVID-19) dashboard.* Retrieved August 3, 2021, from https://covid19.who.int/

Healthcare Waste Management and Post-Pandemic Countermeasures: The Case of the Philippines

Vella Atienza, Lynlei Pintor, and Arlen Ancheta

1 Introduction

The COVID-19 pandemic has created various challenges such as the sudden increase of healthcare waste and the call for urgent actions to properly manage it to avoid environment and health hazards. As one of the countries with high number of COVID-19 transmissions, the Philippines has continued to take actions to address this healthcare waste related concern in collaboration with state and non-state actors. As defined in the Department of Health-Health Facility Development Bureau (DOH-HFDB)'s Health Care Waste Management Manual, health care waste (HCW) includes "all the solid and liquid waste generated as a result of any of the following: diagnosis, treatment, or immunization of human beings; research pertaining to the above activities; research using laboratory animals for the improvement of human health; production or testing of biological products; and other activities performed by an health care facility defined as an institution that has health care as its core service, function, or business." It also includes "same types of waste originating from minor and scattered sources, such as waste produced in the course of health care undertaken in the home" (DOH-HFDB, 2020, p. 7).

The importance of healthcare waste management has gained significant attention due to the disease burdens related to improper handling and disposal of infectious

V. Atienza (✉)
Institute for Governance and Rural Development, College of Public Affairs and Development, University of the Philippines Los Baños, College, Laguna, Philippines
e-mail: vaatienza@up.edu.ph

L. Pintor
Ecosystems Research and Development Bureau, College, Laguna, Philippines

A. Ancheta
Research Center for Social Sciences and Education, University of Santo Tomas, Sampaloc, Metro Manila, Philippines

and toxic substances (i.e. pathological, pharmaceutical, genotoxic, chemical, and radioactive waste, sharps, wastes with high heavy metal content, pressurized containers, etc.) (WHO, 2014; Cruz et al., 2014). These healthcare wastes, if not properly handled, put health-care workers, waste handlers, and the community at risk to infections, toxic effects, and injuries (WHO, 2015). Unfortunately, studies showed that approximately 5.2 million deaths worldwide were related to healthcare waste exposure. Among the diseases caused by exposure to healthcare waste include "hormonally triggered cancers, mutagenicity, dermatitis, asthma, and neurological disorders in children; typhoid, cholera, hepatitis, AIDS, and other viral infections through sharps contaminated with blood" (Malekahmadi et al., 2014). Thus, an effective and efficient management of health care waste is a critical element for environmental health protection (WHO, 2015). Malekahmadi et al. (2014) further stressed that healthcare waste management must pay special attention to "waste generation, segregation, transportation, storage, treatment, and final disposal of all types of waste generated in the healthcare facilities."

The prevalence of healthcare waste management problems is high in developing countries including the Philippines. In fact, it is one of the countries whose healthcare waste management system practices, capacities, and policies are insufficient (Ananth et al., 2009). According to Kenworthy (2017), the Philippine healthcare system does not meet the standards set by the World Health Organization (WHO) and is labeled as "fragmented." With the occurrence of the COVID-19 pandemic, the problem of managing healthcare waste has become more challenging in the country.

Proper healthcare waste management is crucial to effectively control or limit the spread of the COVID-19 virus to protect the healthcare workers and the general public. On March 8, 2020, the State of Public Health Emergency was declared throughout the Philippines due to the threats posed by the Coronavirus Disease 2019 (COVID-19). After more than a week, the State of Calamity was declared throughout the country on March 16, 2020 and the Enhanced Community Quarantine (ECQ) was imposed throughout Luzon. Given this condition, there has been an increasing generation of COVID-19 related waste not only from healthcare facilities, community quarantine units, and temporary treatment and monitoring facilities but also from those within the ECQ areas. To combat the spread of the virus, the mandatory wearing of face masks was issued in the country; however, these usually are disposed of as general waste and mixed with the municipal waste stream. Hence, the Philippine government has issued several guidelines to properly manage health care waste including the COVID-19 related waste, such as the National Solid Waste Management Commission (NSWMC) Resolution No. 1364, Series of 2020: Adopting the Interim Guidelines on the Management of Covid-19 Related Health Care Waste. It provides "guidelines on proper handling and management of all COVID-19 related healthcare waste generated from all households, offices, schools, churches, and other facilities deemed as safe shelters which temporarily house frontliners, public markets, commercial and industrial sources, and the like, including guidelines on ensuring the occupational health and safety of waste workers" (NSWMC, 2020).

2   Objectives

This chapter provides the status of healthcare waste management and the post-pandemic measures in the country with the following main sections: discussion of the definition and categories of health care waste; the policies and regulations related to healthcare waste; the health care waste management systems including the generation, collection, storage, treatment, and disposal of healthcare waste; and the post-pandemic counter measures and innovations with selected case studies highlighting the effective approaches in managing healthcare waste in the country. Further, it provides recommendations toward achieving more effective and sustainable management of healthcare waste in the country.

3   Brief Background of the Philippines

The Philippines, an archipelago of more than 7000 islands, is gifted with beautiful landscapes and seascapes. It has irregular land surface, mountainous with long coastlines. Located at the Pacific ring of fire, the country has numerous volcanoes and often experiences earthquakes. The country is composed of three major island groups: Luzon, Visayas, and Mindanao. As shown in Fig. 1, Luzon in the North is where Metropolitan Manila other urban centers are located; Mindanao in the South, a rich agricultural and forest area; and in between Luzon and Mindanao are small groups of Visayan islands with beautiful beaches and abundant marine resources.

However, the Philippines is vulnerable to environmental problems. It has depleting upland forests, flooding lowland areas, and congested urban cities. In the study of Jambec (2015), the Philippines after China and Indonesia ranks as the world's third biggest polluter, with 2.7 million metric tonnes of plastic waste generated each year. According to a 2018 study of Food Industry Asia, waste disposal is a major problem in the Philippines though it has a high garbage collection rate among Southeast Asian countries. The COVID-19 pandemic aggravates garbage problem especially in the urban centers as more infectious wastes were generated but carelessly disposed of.

4   Definition and Categories of Healthcare Wastes in the Philippines

As defined in the Department of Health-Health Facility Development Bureau (DOH-HFDB)'s Health Care Waste Management Manual, health care waste (HCW) includes "all the solid and liquid waste generated as a result of any of the following: diagnosis, treatment, or immunization of human beings; research pertaining to the above activities; research using laboratory animals for the

Fig. 1 Philippine General Map. (googgle.com)

improvement of human health; production or testing of biological products; and other activities performed by an health care facility defined as an institution that has health care as its core service, function, or business." It also includes "same types of waste originating from minor and scattered sources, such as waste produced in the course of health care undertaken in the home" (DOH-HFDB, 2020, p. 7).

Healthcare wastes in the country fall under two broad categories: hazardous and non-hazardous. Hazardous wastes are those that pose environmental and health risks, while non-hazardous or general wastes are those that do not pose handling problems or hazards to the environment or health of humans. Non-hazardous or general wastes have not been in contact with any infectious or radioactive agents and harmful chemicals.

Each umbrella term may further be characterized into more specific types of HCWs. These categories are available in the fourth edition of the Department of Health's (DOH) Healthcare Waste Management Manual (2020) and are presented in Table 1.

- *Sharps* are the most hazardous HCW that may cause accidental physical injuries like pricks, cuts, or punctures or cause infection through these injuries. Some common sharps wastes are needles, syringes, scalpels, saws, blades, broken glass, infusion sets, knives, and nails.

Table 1 Categories of HCWs in the Philippines

Hazardous	Non-hazardous (general)
Sharps	Recyclable
Infectious	Biodegradable
Pathological	Residual
Anatomical	
Pharmaceutical	
Genotoxic	
Chemical	
Radioactive	
Pressurized containers	

- *Infectious* wastes contain pathogens in sufficient concentration that may cause disease to susceptible hosts when exposed to it. Some examples of infectious wastes are cultures from the laboratory, surgery and autopsy wastes, wastes from isolation wards, and infected animals from research laboratories.
- *Pathological* and *anatomical* wastes consist of tissues, organs, body parts, blood, body fluids, and other wastes from surgeries and autopsies. As such, these wastes are often considered as a subcategory of infectious waste but their handling, treatment, and disposal warrant the separate classification of these wastes.
- *Pharmaceuticals* include pharmaceutical products, drugs, vaccines, and sera that have expired, spilt, contaminated and are no longer required.
- *Genotoxic* wastes may have mutagenic, teratogenic, or carcinogenic properties. Certain cytostatic drugs, vomit, urine, or feces from patients treated with cytostatic drugs, chemicals, and radioactive material are examples of genotoxic wastes.
- *Chemical* wastes are considered as hazardous waste if it has at least one of the following properties: Toxic, reactive, flammable, corrosive, and oxidizing. These wastes are usually used in diagnostic and experimental work or in housekeeping and disinfecting procedures. Some of the most common chemical HCWs are formaldehyde, photographic fixing and developing solution used in X-ray, and waste organic chemicals among others.
- *Radioactive* wastes are those contaminated with radionuclides as a consequence of procedures such as in vitro analysis of body tissue and fluid, in vivo organ imaging and tumor localization, and other investigative and therapeutic practices.
- *Pressurized containers* such as aerosol cans that require disposal and must be handled properly to avoid explosion when incinerated or accidentally punctured.

5 Policies and Regulations on Managing Healthcare Waste in the Philippines

This section presents some of the related policies and regulations on managing healthcare waste in the Philippines. As cited and described in the DOH-HFDB's fourth Edition of the Health Care Waste Management Manual (DOH-HFDB, 2020), there are several laws related to health care waste management that have been implemented in the country since 1960s (Table 2).

Aside from these laws, the Department of Health also developed three (3) editions of the Health Care Waste Management Manual. The first edition entitled "Hospital Waste Management Manual" was issued in 1997 through the Environment Health Service. The second edition was issued in 2004 with the Environment and Occupational Health Office and renamed it "Health Care Waste Management Manual" to provide guidance and information not only for hospitals but for other health facilities in the country. The third edition was published in 2011 in collaboration with various stakeholders in reviewing, enhancing, and updating its policies and regulation. Aside from being more user-friendly, this edition includes new trends and universally accepted technologies (DOH-HFDB, 2020).

The Philippines is also a signatory to various international agreements/conventions such as the Basel Convention on the Control of Trans-boundary Movements of Hazardous Wastes and their Disposal; the Stockholm Convention on Persistent Organic Pollutants; the Minamata Convention on Mercury; the World Health Assembly Resolution on Water, Sanitation and Hygiene (WASH) in Health Care Facilities.

The management of health care waste has become more challenging with the occurrence of the COVID-19 pandemic. This situation contributes to the increasing volume of healthcare waste generation such as the masks, PPEs, and other related COVID-19 wastes. Although the country has existing policies and regulations on hazardous waste including healthcare waste, the COVID-19 related waste is new and highly contagious. Hence, the Philippine government has crafted several guidelines in response to the pandemic (Table 3).

In addition to these Memorandum Circulars, the DOH through the Health Facility Development Bureau (HFDB) published the fourth Edition of the Health Care Waste Management Manual (2020) in partnership with the World Health Organization. Incorporating the requirements of all the laws and regulations in the country on HCWM, this manual serves as the "most comprehensive set of guidelines on the safe management of waste generated from health care activities in the country" (DOH-HFDB, 2020, p. 2). This will be discussed in detail later in the next section, "Status of the Health Care Waste Management System."

5.1   Issues and Challenges in the Implementation of Health Care Waste Management Policies and Regulations

As presented above, the policies and regulations on managing health care waste are already in place even before the COVID-19 pandemic. Hence, interim guidelines were issued to include the COVID-19 related waste and to provide specific guidelines on managing and handling this type of waste. However, the challenge is on their effective implementation. The common issues and challenges include the

Table 2 Policies and regulations on healthcare waste enacted prior to COVID-19 pandemic

Policies and regulations	Description
Republic Act (RA) No. 4226: Hospital Licensure Act (1965)	Requires the registration and licensure of all hospitals in the country and mandates the DOH-Bureau of Medical Services (presently the Health Facilities and Services Regulatory Board) as the acting licensing agency in setting standards in the construction and operation of hospitals One of the DOH Administrative Orders relevant to this Act is the No. 2005-0029 (2005) which requires hospitals and other health facilities applying for initial license to operate to accomplish/submit a waste management plan in addition to other documents
Presidential Decree (PD) No. 856: The Code on Sanitation of the Philippines (1975)	Mandates the DOH to promote and preserve public health and upgrade the standards of medical practice, among other functions and provides the other legal basis for the DOH to issue and require compliance with the Health Care Waste Management Manual
PD No. 984: Providing for the Revision of RA No. 3931, Commonly Known as the Pollution Control Law, and other Purposes (1976)	The Pollution Control Law governs discharges of potentially polluting substances to air and water
PD No. 1586: Environmental Impact Statement (EIS) System (1978)	Requires development projects, including HCFs, to undergo Environmental Impact Assessment (EIA) and secure an Environmental Compliance Certificate (ECC) from the DENR-EMB prior to construction and operation
RA No. 6969: Toxic Substances and Hazardous and Nuclear Wastes Control Act of 1990	Requires the registration of waste generators, waste transporters, and operators of toxic and hazardous waste treatment facilities with the EMB One of the Administrative Orders relevant to this Act is the Joint DENR-DOH Administrative Order No. 02, Series of 2005, Policies and Guidelines on effective and proper handling, collection, transport, treatment, storage, and disposal of health care waste
RA No. 8749: The Philippine Clean Air Act of 1999	Promotes the use of state-of-the-art, environmentally sound, and safe thermal and non-burn technologies for the handling, treatment, thermal destruction, utilization, and disposal of sorted biomedical and hazardous wastes and prohibits incineration
RA 9003: Ecological Solid Waste Management Act of 2000	Seeks to ensure the protection of public health and the environment by utilizing sound strategies for treating, handling, and disposing of solid wastes. It also mandates waste segregation at source including households and institutions such as hospitals

(continued)

Table 2 (continued)

Policies and regulations	Description
RA No. 9275: The Philippine Clean Water Act of 2004	Pursues policy of economic growth which is consistent with the protection, preservation, and revival of the quality of fresh, brackish, and marine waters in the country
RA No. 11223: Universal Health Care Act	Aims the progressive realization of universal health care in the country by applying systematic approach and clear delineation of roles of key agencies and other stakeholders for better performance in the health system. Further, PhilHealth is tasked to provide an incentive scheme to reward health facilities that deliver better service quality, efficiency, and equity which shall include proper health care waste management

following: the lack of capacity of the LGUs in terms of technical and financial constraints, limited number of TSD facilities, and the lack or limited awareness of the general public on the proper management of health care waste, particularly the COVID-19 related waste.

In terms of the capacity of the LGUs, this COVID-19 related waste is new and highly contagious; hence, it complicates all the more the waste management and handling. Even the drafting of resolutions is challenging especially in the early part of the pandemic wherein the enhanced community quarantine is strictly enforced and most are working from home. Aside from the technical difficulties such as the weak internet connection, many are not familiar yet on virtual meetings; hence, deliberations also take time. As cited in the NSWMC Resolution No. 1364 s. 2020, LGUs are also mandated to develop a COVID-19 waste management plan. To comply with this, they need support both technically and financially. Hence, the EMB-DENR will conduct a virtual training to all LGUs nationwide this March 2021 to assist LGUs in the preparation of the said plan. However, since the COVID-19 pandemic happened around March 2020 many LGUs have already allocated their budget for the year. Therefore, they have to find other sources of funds to implement the activities cited in the guidelines to properly manage and handle COVID-19 related waste. In addition, the lack of human resources who are technically equipped to handle the management of this type of waste. To cite, each health care facility (HCF) should have one pollution control officer (PCO); however, many public HCFs cannot afford to employ PCOs.

Another major concern in the management of health care waste is the limited number of TSDs even before the pandemic. Most of the TSDs are located in the National Capital Region and Luzon areas, but still many regions have no TSDs. Hence, even there are clear guidelines on managing and handling health care waste, the availability of the TSDs particularly those with separate cells for infectious waste remains a challenge. Many LGUs have difficulty establishing its own TSDs primarily due to high investment requirement. Hence, it is recommended to promote inter-

Table 3 Memorandum Circulars related to health care waste management in response to the COVID-19 pandemic

Memorandum circulars	Description
EMB Memorandum Circular No. 2020-14 Interim Guidelines on Issuance of Special Permit to transport (SPTT) for the transport of hazardous waste within the quarantine period	The guidelines aim for the unhampered transportation of hazardous waste specifically waste from the health care facility (HCF) to treatment, storage, and disposal (TSD) facilities despite the implementation of the Community Quarantine over the entire Luzon. This covers the registered transporters and registered TSD facilities in the Luzon and they shall be exempted from travel ban upon presentation of the required documents and the special permit to transport
EMB Memorandum Circular No. 2020-15 Addendum to the Interim Guidelines on Issuance of SPTT for the transportation of hazardous waste within the community quarantine period	This further cites that all TSD registration certificates handling M501 which shall expire during the ECQ period shall be automatically extended for 60 days and that renewal shall be done within five (5) days after lifting the ECQ
EMB Memorandum Circular No. 2020-16 Amendment of the Interim Guidelines on issuance of SPTT for the transportation of hazardous wastes within the community quarantine period	This expands the coverage of the EMB MC 2020–14 and EMB MC 2020–15 which allows registered transporters and TSD facilities to haul, treat, and/or dispose health care waste nationwide
EMB Memorandum Circular No. 2020-20 Provisional Guidelines on the hazardous waste management during the extended enhanced community quarantine	This states that the EMB shall simplify existing requirements and procedures for the issuance of PTT to registered transporters which haul to and from healthcare facilities and generators of other hazardous waste from essential industries and TSD facilities
DOH Circulars No. 2020–0152 Guidelines to All Public and Private Healthcare Facilities and other Concerned Establishments to Assist Surveillance Officers in the Epidemiologic Investigation on the Coronavirus Disease 2019 (COVID-19)	It reiterates AO 2020–2012 "Guidelines for the inclusion of the COVID-19 in the list of notifiable disease for mandatory reporting to the DOH." The DOH deploys teams to conduct epidemiologic investigation and contact tracing to expedite gathering of information on new COVID-19 cases
DOH Circular No. 2020–0170 Interim Guidelines on the Management of Health Care Waste in Health Facilities, Community Quarantine Units, and Temporary Treatment and Monitoring Facilities with Cases of Coronavirus Disease 2019 (COVID-19)	This provides guidance on the proper management of all COVID-19 related health care waste in all health facilities, community quarantine units, and temporary treatment and monitoring facilities with suspect, probable, and confirmed COVID-19 patients. This includes guidelines on the proper segregation and collection, storage, transport, and disposal of COVID-19 related health care waste
DOH Memorandum No. 2020-0208 Interim Guidelines on Enhancing the Infection Prevention and Control Measures through Engineering and Environmental Controls in all Health Facilities and Temporary Treatment and	This provides guidance on the engineering and environmental controls to enhance the prevention and control of infection in all health facilities, and temporary treatment and monitoring facilities

(continued)

Table 3 (continued)

Memorandum circulars	Description
Monitoring Facilities during the COVID-19 pandemic	
DOH Circular No. 2021-0031 Interim Guidelines on the Management of Health Care Wastes Generated from COVID-19 vaccination	This provides guidelines on proper management of health care wastes generated from the COVID-19 vaccination activities in all health care facilities, Centers for Health Development, and local government units involved in the COVID-19 vaccination
NSWMC Resolution No. 1364, Series of 2020: Adopting the Interim Guidelines on the Management of Covid-19 related health care waste	Provides guidelines on the proper handling and management (waste generation, segregation and storage at source, collection and transport, treatment and disposal, occupational health and safety) of all COVID-19 related waste generated from all households, institutions, commercial and industrial sources. It also mandates local government unities (LGUs) to develop a COVID-19 waste management plan and that a team maybe created composed of City/Municipal Environment and Natural Officer or his designate and the City/Municipal Health Officer to develop, implement, and monitor these guidelines
DILG Memorandum Circular No. 2020-147: Guidelines on the Management of COVID-19 Related Health Care Wastes	In support to the implementation of the NSWMC Resolution No. 1364 s. 2020, this MC provides LGUs with comprehensive information on their roles, duties, and responsibilities on the proper handling and management of all COVID-19 related health care wastes

local collaboration or clustering among LGUs and to establish partnerships with other non-state actors. "Recognizing the limitations of local government in terms of capacity and resources to provide more efficient and effective public service delivery; and the advantages of collaborative approaches between and amongst local governments or through public-private-partnerships (PPPs) provide strong justification to promote local collaboration on waste management" (Atienza, 2020, p. 132).

6   Healthcare Waste Management System in the Philippines

Under the Joint Administrative Order 2005–02, the Department of Environment and Natural Resources (DENR) and the DOH provided the guidelines on the management of health care wastes. Furthermore, hospitals in the country are subject to the provisions of the Health Care Waste Management Manual of the DOH as

Table 4 Healthcare waste in the Philippines

City	Population	Medical waste generated pre-COVID-19 (tons/day)	Medical waste generated (tons/day)	Total possible production over 60 days (tons/day)
Manila City, Philippines	14 million	47	280	16,880

Source: ADB (2020)

promulgated in the same year. The Joint AO is pursuant to the pertinent laws, rules, and regulations that govern HCWs.

Despite the aforementioned laws and policies, HCW management is still a pressing concern in the country (Cruz et al., 2014). Sañez (2008) reported that 86.79% of the hospitals in the country were unregistered. Public concern has also been growing over the disposal of wastes produced from health care facilities (Health Care Without Harm Asia, 2007). Cruz et al. (2014) cited several reports of large, but inconsistent, figures of the amount of infectious wastes hospitals in Metro Manila alone produce daily. However, little information is available on what is done with these wastes, especially after the banning of incineration in the country. These concerns are amplified by incidents of wastes ending up in open dumpsites and waterways or water bodies.

Table 4 shows the data regarding the healthcare waste in the Philippines before and during COVID-19 pandemic. Manila, with a population of 14 million, generated a total of 47 tons per day of medical wastes before the pandemic happened (Fernandez, 2020). Now, record shows that this data increased by almost six times. This is attributed to the number of hospitals found in Manila and its huge population. The perceived increase in healthcare waste in the Philippines is also attributed to the continued increase in the number of COVID-19 cases in the country. The Asian Development Bank (ADB, 2020) estimated that hospitals in Metro Manila—the region that accounts for at least 55% of the nation's cases—would generate 280 metric tons of medical waste in a single day. With the increase in COVID-19 cases, there is also an increase in the health waste being generated. It is expected that there will still increase in the use of medical gloves, surgical face masks, and many others. When blow up, the total medical waste to be generated could be as high as 16,880 tons in 60 days.

According to the DENR, data showed that the country has generated a total of 19,187.7 metric tons of infectious healthcare waste from April to July 20 of 2020. And from that number, only 29% has been treated and disposed of although the waste has already been hauled from hospitals. This means that management of HCW must be given priority. Hence, one of the current efforts of the DENR to address this concern is to streamline the permitting system for TSD of HCW (DOH-HFDB, undated).

The Healthcare Waste Management System has been established and presented in the most recent DOH Health Care Waste Management Manual (2020). This system consists of five key principles, namely: HCW Management Planning, HCW

Minimization, HCW Segregation, Collection, Storage, and Transport, HCW Treatment and Disposal, and Managing Wastewater Generated by Health Care Facilities.

6.1 HCW Management Planning

HCW management planning is a strategy employed to allocate the roles, responsibilities, and resources in waste management. It is a crucial part of the HCWM System because of legal and regulatory framework for HCWM. The HCWM Plan encompasses waste avoidance and minimization, proper segregation and containment, safe handling, storage, and transport until treatment and disposal.

As per the World Health Organization Basel Convention in 2005, the HCW Management Planning aims to:

- Develop the legal and regulatory framework for HCWM.
- Rationalize the waste management practices within HCFs.
- Develop specific financial investment and operational resources dedicated to waste management.
- Launch capacity building and training measures.
- Set up a monitoring plan.
- Reduce the pollution associated with waste management.

The HCWM Plan details the functions of each individual and describes the organizational structure of the HCW management. The success is naturally reliant on the commitment of the entire manpower involved in HCWM. The appointment or designation of a specific committee to handle HCWM in the HCF is critical on the part of the administrator or head of the facility.

6.2 HCW Minimization

The most preferable approach is to avoid producing waste as far as possible and thus minimize the quantity entering the waste stream. The manual states the following steps on the management of HCF to minimize waste:

- Establish an updated database for the waste generation rates, current hazardous waste management strategies, and current waste management costs;
- Institutionalize waste minimization and sustain the program in the long run;
- Have a written policy with established vision and mission to implement Waste Minimization Program (WMP);
- Be aware of their specific role in HCWM and be properly trained in waste minimization; and
- Adopt the Green Procurement Policy (GPP) pursuant to Executive Order No. 301, Series of 2009.

Some techniques that may be employed to adopt waste minimization are source reductions which may involve a product change, process change, or employ good practices. The more conventional Recovery, Reuse, and Recycling method is also suggested in the manual.

6.3   HCW Segregation, Collection, Storage, and Transport

The proper segregation of HCW at the point of generation, collection, storage, and transport for treatment prior to its final disposal is also important in the whole process of managing HCW. Segregation is the key to effective waste management and only implementation of proper waste management can ensure all HCW will be treated according to the hazards. The HCWM Manual dictates the following key principles to control waste flow from generation to disposal:

- segregation into different fractions, based on their potential hazard and disposal route, by the person who produces each waste item;
- storage, and transportation in a safe manner considering the risk and occupational safety and in accordance with existing laws, policies, and guidelines;
- separation of containers in each medical area;
- establishment of the appropriate labeling, signage, route, and segregation system;
- designation of storage areas based on the volume of waste generated by the HCF;
- HCF must register as waste generator with the DENR and secure a DENR waste generator identification number.

6.4   Health Care Waste Treatment and Disposal

The purpose of treatment is to reduce the potential hazard posed by HCW while endeavoring to protect the environment. Treatment should be viewed in the context of the waste management hierarchy.

The following conditions must be considered during the treatment and disposal of HCW:

- The quantities of waste produced daily at the PHC level;
- Availability of appropriate sites for waste treatment and disposal;
- Possibility of treatment in central facility or treatment facility within reasonable distance;
- Rainfall and level of groundwater (e.g., to take precautions against flooding of burial pits);
- Availability of reliable transportation;
- Compliance to the national policies and standards;
- The availability of equipment and manufacturers in the country or region;
- Social acceptance of treatment and disposal methods and sites;

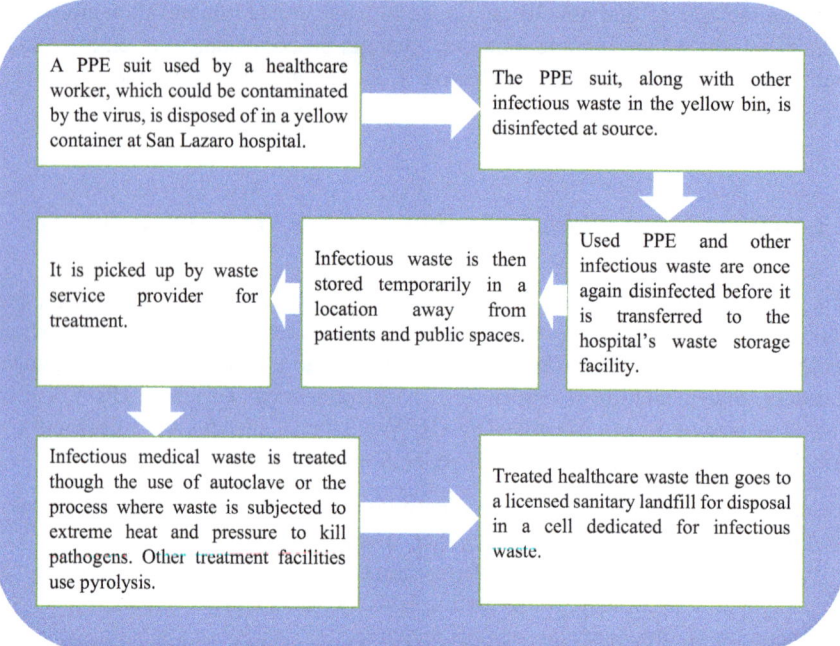

Fig. 2 Flow of Hospital Waste (Cabico, 2020a, b, c)

- Space available at the HCF;
- Availability of resources (human, financial, material);
- Estimate of capital and operating cost.

After treatment, the wastes should be stored in plastic liners/containers with the same color.

HCW generated at HCFs may pose harm and risks to the health care workers and to the environment if improperly managed. However, proper healthcare waste management has been gaining recognition slowly due to the substantial disease burdens associated with poor practices, including exposure to infectious agents and toxic substances. Ironically, the magnitude of the problem, practices, capacities, and policies in many countries in dealing with healthcare waste disposal, especially developing nations, is inadequate and requires intensification (Ananth et al., 2009). The Health Care Waste Management System is therefore imperative in addressing these growing concerns in the country and its proper and strict implementation, as well the compliance of HCFs to it, must be given utmost attention.

Below is an illustration of the infectious waste flow from the hospital (San Lazaro Hospital, Manila) to the final disposal facility.

Figure 2 shows the flow of hospital waste from the hospital to the sanitary landfill. The used PPEs and other infectious waste are disinfected at source and dispose in a yellow bin. It is once again disinfected before it is transferred to the hospital's waste

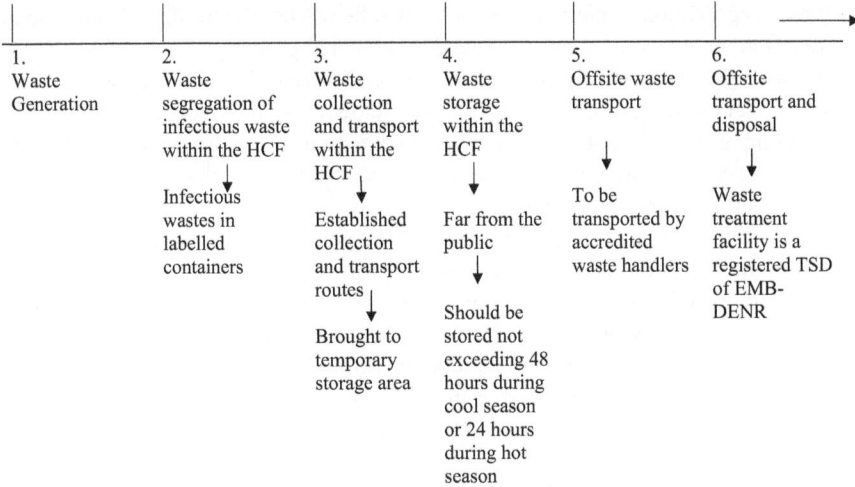

Fig. 3 Waste Management Plan of Healthcare Waste Generated from COVID-19 Vaccination

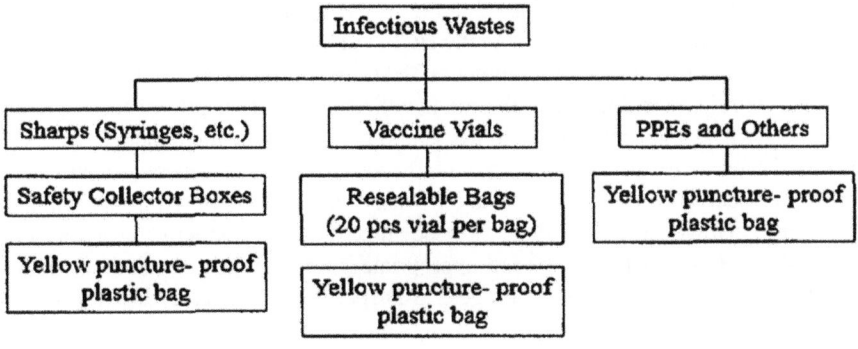

Fig. 4 Classification of infectious wastes (DOH Circular No. 2021–0031)

storage facility. It is temporarily stored away from patients and public spaces. It is picked up by waste service provider for treatment. Infectious medical waste is treated through the use of autoclave and pyrolysis. Treated healthcare waste then goes to a licensed sanitary landfill for disposal in a cell dedicated for infectious waste.

The waste flow in the hospital is related to the general guidelines on the management of HCW generated from COVID-19 vaccination (DOH Circular No. 2021-0031) as both generate infectious wastes. The guidelines required a waste management plan starting with generation; segregation in the HCF; waste storage within the HCF; offsite waste transport and offsite waste treatment and disposal (DOH Circular No. 2021-0031) (Fig. 3). Figure 4 shows the classification of infectious wastes as cited in the DOH Circular No. 2021-0031.

1. Waste segregation—refers to the waste handlers who ensure that all infectious wastes generated should be segregated and labeled properly.
2. Waste segregation within the healthcare facility (HCF)—waste handlers shall strictly follow the proper infectious waste segregation protocols for COVID-19. Labeling is very important as containers are required to segregate hazardous waste based on source, type of waste, or date and time of closure. Below is the classification of infectious wastes.
3. Waste collection and transport with the HCFs generally refers to the waste audit within the HCF. Waste containers with infectious wastes are to be handled with care in terms of protective gear of the handlers, containers tied tightly with three quarters full and a spared wheeled trolley should there be a container breakdown. The collected infectious wastes shall be brought to the temporary storage area of the HCF following designated routes of collection and transport to reduce contamination. Moreover, containers should be thoroughly cleaned and disinfected after every use.
4. Waste storage within the HCF should be properly secured and far from public spaces to minimize exposure. They should be properly labeled and more importantly they should be stored not exceeding 48 h during cool season and 24 h during dry season. Disinfection of storage areas is very necessary.
5. Off-site waste transport—refers to the HCF without waste service providers. The DOH Center for Health Development (CHD) or Provincial Health Office (PHO) shall coordinate and facilitate the waste collection and transport of the HCF for the COVID-19 vaccine wastes to the temporary's waste storage facility.
6. Off-site waste treatment and disposal—the last stage of the general guidelines is very critical as COVID-19 vaccination wastes are to be disposed in the sanitary landfill. However, they shall not be mixed with municipal wastes and non-hazardous wastes, following appropriate guidelines and regulations. More importantly, the Treatment, Storage and Disposal (TSD) facility is a registered TSD facility with DENR-EMB.

6.4.1 Healthcare Waste Generation Before Pandemic

Even before the onset of the COVID-19 pandemic, the Philippines already face unrelenting problems, issues, and concerns on health care waste production. In 2003, studies showed that majority (86.79%) of hazardous wastes generated in the country come from unregistered hospitals (Sañez, 2008). These toxic wastes from medical institutions account for about 12% of the total solid wastes generated in the country (SEPO, 2017). Specifically, healthcare facilities in Metro Manila were reported to produce 47 metric tons of medical waste daily. More than half (56%) of this medical waste was found to be potentially infectious (Cabico, 2020a, b, c).

Despite the growing concern over healthcare waste generation, the country lacks accurate, detailed, and consistent data on the overall amount of waste generated or produced by the health sector. In other words, there is little information available on the amount of infectious healthcare waste generated daily, the process these wastes undergo, and the facilities where the waste is disposed or discarded (Health Care

Without Harm Asia, 2007). Majority of the studies conducted were focused on the waste identification and characterization. There were few to none which studied the waste generation and disposal of the medical sector.

At present, the EMB's initial step toward acquiring reliable data on the country's healthcare waste generation is to require the medical facilities and hospitals to register as hazardous waste generators. Waste generator registration would help acquire useful data and information on the (1) number of healthcare waste generators; (2) the type and quantity of hazardous waste generated, produced, or transported outside; (3) amount of hazardous waste that has been certified by the waste treater as adequately treated, recycled, reprocessed, or disposed of; (4) number of facilities that have contingency plans for hazardous wastes accidents that conform with the guidelines issued by the EMB; (5) number of medical institutions who conduct personnel training on the implementation of the contingency plan, the hazards posed by the improper handling, storage, transport, and use of hazardous wastes and their containers (DENR AO 36 Series of 2004).

6.4.2   Healthcare Waste Disposal Before Pandemic

Regarding medical hazardous waste disposal, reports further showed that hospitals use sanitary landfill, safe burial on hospital premises, and septic/concrete vault in disposing their healthcare wastes (Molina, 2002; DOH, undated). Prior to the pandemic, the Department of Health has already set a standard procedure that would guide hospitals for safe healthcare waste disposal. For instance, hospitals that use sanitary landfills to dispose healthcare waste were required to ensure that the area is easily accessible for vehicles, there is personnel assigned for managing daily operations, and that the base and sides of the site are adequately sealed in order to avoid wastewater spill. Regarding safe burial of waste, the DOH has set strict standards for proper disposal (i.e. low permeability of burial site to prevent pollution of groundwater, chemical/pharmaceutical wastes not to exceed 1 kg, site not to be located in flood prone areas and is 50 m away from rivers and lakes, the height of the pit should be at least 1.50 m higher than ground water level, and must be used for only 1 to 2 years and contain 5 to 10 tons of waste). The septic and concrete vault, however, is used for disposing sharps and syringes. As advised by the DOH, the pit must follow the minimum size of 1 m $\times$ 1 m $\times$ 1.8 m depth; isolated 500 ft. away from groundwater supply; constructed with concrete walls and slaps with manhole extended a few centimeters above the soil surface to avoid water infiltration; and installed with security fence (DOH, undated).

Nonetheless, the study showed that while most hospitals practice waste segregation, only half conducts proper hazardous waste handling and segregation. In fact, only a few (40%) hospitals established a waste management committee and allot budget for waste management programs. Alarmingly, most medical facilities were reported to be in noncompliance to the Philippine Toxic Substances and Hazardous and Nuclear Waste Control Act. For instance, out of 304 registered healthcare facilities in Davao city, only 20 were found to use pyroclave technology in disposing their hazardous and infectious wastes. Furthermore, only 5.5 tons of 40 tons of infectious healthcare waste generated by Davao City every month were reported to be properly disposed (Carillo, 2013). Similar was the case of government hospitals in the Northern areas of the country. According to Cruz et al. (2014), while hospitals have relatively high levels of awareness regarding healthcare waste management requirements and display high levels of willingness to comply, there were various factors that impede their full compliance and adaptation of proper healthcare waste management practices. Among the influential factors that hamper the country's healthcare waste disposal management system include: (1) lack of logistics; (2) limited funding; (3) lack of training on healthcare waste management; (4) poor dissemination of rules and regulations; and (5) inadequate provision of policy and budget. The same was the findings of the WHO (2015) report. According to the report, some Philippine health-care workers have received training in health care waste management. However, many others still lack knowledge. Most importantly, financing health care waste management has consistently been a pressing problem in the country. Nonetheless, the country has been a recipient of Global Environment Facility (GEF) and the United Nations Development Programme (UNDP)'s support for health care waste management projects and training programs on the best environmental practices which include non-incineration and mercury-free alternatives technologies for the treatment of medical waste (HCWH, 2013). Specifically, the health care waste management program in CALABARZON Region was financed by Japan in 2012 (WHO, 2015).

Overall, "the Philippines has a comparatively comprehensive legal framework and a well-established healthcare waste management system but is lacking in implementation and enforcement" (WHO, 2015). Hence, prior to the pandemic the World Health Organization (2015) has already seen the need for the Philippines to (1) create a national HCWM strategy and action plan; (2) improve the monitoring and inspection system to enable the enforcement of the legal framework; (3) clarify responsibilities and development of a polluter-pays financing system for HCW; (4) strengthen the training system on healthcare waste; and (5) improve healthcare waste management most especially in smaller and more remotely located health-care facilities (WHO, 2015).

Nevertheless, the results of the World Health Organization's assessment on healthcare waste management in the Philippines were labeled good based in terms of management, training, regulation, and technology. Results further showed the country must improve its financing scheme and strategies for health care waste management.

On the bright side, the Philippines reported some best practices in medical waste management. Among the significant practices as reported by the Health Care Without Harm include: (1) "Materials recovery and recycling, banning of polystyrene foam packaging, composting, and encapsulation of autoclaved sharps waste" by the Hospital Waste Management Team at Maria Reyna-Xavier University Hospital in Cagayan de Oro City; (2) "Safe practices to prevent radiation exposure" by the Radiology Technology team in General Santos Doctors Hospital; (3) "Use of safer alternatives for surface cleaning and cleaning of non-critical items" by the housekeeping staff, autoclaving of waste, and maximizing natural lighting at St. Paul Hospital in Tuguegarao City; (4) "Installation of the first Philippine hospital biodigester designed by a biomedical waste worker to convert food and garden waste into methane gas for the kitchen and laundry, solar panels for water heating, green walls for cooling, and vermicomposting" at Perpetual Succor Hospital in Cebu City; (5) "Wastewater treatment and reuse of treated water for gardening" by the engineers of Philippine Heart Center and Our Lady of Peace Hospital in Paranaque City; and (6) "Mercury-free dental services" by the San Lazaro Hospital dentists" (HCWH, 2013).

6.4.3 Healthcare Waste Generation During Pandemic

Undeniably, the problem regarding healthcare waste generation, disposal, and management worsened amid the COVID-19 pandemic. For instance, a single government hospital reported to have generated an average of 10,000 kg of infectious medical waste per month. From March to June, the hospital declared that it generated an estimated 29,473 kg of infectious healthcare waste (Cabico, 2020a, b, c). According to research, "each person in a health care facility produces about 3.5 kilograms of health care waste due to COVID-19 per day" (Casilao, 2021). Cabico (2020a, b, c) further added that a single hospital patient consumes an average of 12 Personal Protective Equipment (PPE) sets every day. Thus, experts express their concern regarding the serious health impacts of improper disposal of accelerating amounts of used "coveralls, N95 mask, gloves, head cover, shoe cover, goggles, a surgical mask, and a surgical gown" (Cabico, 2020a, b, c; Subingsubing, 2020).

In relation to this, the Asian Development Bank predicted that hospitals in Metro Manila alone would generate 280 metric tons of medical infectious wastes which include "used personal protective equipment, dressings, swabs, blood bags, urine bags, sputum cups, syringes, test tubes, and histopathological waste." Overall, the DENR reported that the country generated more than 19,187.7 metric tons (MT) of health care waste during the pandemic (Cabico, 2020a, b, c). With this, the National Solid Waste Management Commission (NSWMC) stressed that "waste segregation at source has become more critical amid the pandemic" (Enano, 2020).

Alarmingly, only a few of the total infectious wastes have undergone proper treatment and disposal. Only 5600 MT have been subjected to proper treatment under the Department of Environment and Natural Resources–Environmental Management Bureau (DENR-EMB) accredited facilities. This denotes that only a few (29%) of the country's total infectious medical waste undergo the thermal process, chemical process, irradiation, biological process, encapsulation, and inertization prior to disposal to DENR-approved sanitary landfills (Pena, 2020; Cabico, 2020a, b, c).

6.4.4 Healthcare Waste Disposal During Pandemic

The waste treatment and disposal facilities were unprepared and overwhelmed with the unabated surge in the volume of infectious medical wastes produced during the pandemic. Added to this is the uneven distribution of the 26 registered treatment, storage, and disposal (TSD) facilities nationwide wherein most are in Luzon. Thus, hospitals express their difficulty in grappling with the medical wastes. The gravity of medical waste problem was further highlighted by the controversial issue caught by the CCTV footage where used rapid test kits were seen scattered along M. Dela Fuente St. in Sampaloc, Manila (Pena, 2020; Cabico, 2020a, b, c; Montemayor, 2020). Aside from the scattered rapid tests kits, another alarming news regarding medical waste is the 65 vials of blood specimen found along the shores of Subic Bay (Fermin, 2020). These contentious issues on medical wastes paved the way for the Department of Health to issue sanctions such as suspension from operation and revocation of license to "healthcare facilities and testing laboratories that fail to properly dispose infectious medical waste" (Cabico, 2020a, b, c).

Nonetheless, some hospitals have their own treatment system and employ color coding system to distinguish between infectious and non-infectious dry and wet types of wastes. San Lazaro Hospitals' infectious wastes, for instance, were sent to the Integrated Waste Management Inc.'s waste treatment facility. During the pandemic, among the popular waste disposal process includes autoclaves, pyrolysis, and incineration (Windfeld & Brooks, 2015). Typically, medical waste in the country undergoes steam sterilization treatment or autoclave, wherein wastes are put under intense heat and pressure to eliminate pathogens prior to disposal in the landfill. Autoclaves are advantageous in terms of its capacity to process 3000 kg of wastes per cycle. Aside from autoclaves, "pyrolysis, or the process of chemically decomposing organic materials at elevated temperatures without oxygen," is also used to process infectious medical wastes during the pandemic (Cabico, 2020a, b, c; Carillo, 2013; DOH, undated). The third most common waste processing is through biomedical waste incineration or the burning of wastes. However, environmental experts express their concern regarding its violation of the Clean Air Act of 1999 as the process emits poisonous and toxic fumes which put public health at risk especially amid the ongoing COVID-19 outbreak (DOH, undated).

Thus, the EMB-DENR, DOH, and other relevant agencies released various memorandum circulars that would guide for proper healthcare waste management as discussed in the early part of this report. In line with the healthcare waste management documents created, experts are now considering the use of reusable PPE to reduce the total waste volume. Specifically, the 110-year-old St. Paul Hospital in Iloilo promoted the use of plastic raincoats as an alternative for medical coveralls. The hospital eventually initiated the creation of their own reusable PPE made of nonpermeable cloth which saved the institution from ten million worth of expenses from acquiring protective suits as well as in disposing these disposable PPEs (Subingsubing, 2020).

Another important accomplishment of the medical sector is the operationalization of the newly completed PHP19-million medical waste treatment plant in Alabel town, otherwise known as the "Sarangani Medical Waste Treatment Facility" (Gubalani, 2021). The said technology will be using thermal disinfection to treat infectious, pathological, and pharmaceutical wastes from the six provincial government-run and supported hospitals as well as other local health facilities in Sarangani which produce a daily average of 120 kilos of medical wastes where 20 kilos are considered infectious (Gubalani, 2021).

6.5 Issues and Challenges on the Health Care Waste Management System

COVID-19 poses a very serious impact on the waste management capacity of the country. The increase in the amount of infectious wastes along with improper handling of this infectious waste triggers the already immense garbage problem in the country. The country not only lacks the capacity to handle this waste properly but there are also gaps in terms of financial resources as well as technological support and institutional mechanisms which can help address it concerns. Lastly, the low level of awareness of citizens on proper handling and segregating of health wastes brought about by the need for stronger information dissemination activities makes it more difficult to address the problem.

Below is the list of issues and challenges on the healthcare waste management in the Philippines due to COVID-19 pandemic.

6.5.1 Rapid Increase in the Amount of Healthcare Wastes

COVID-19 had led to increased amount of healthcare wastes including infectious wastes that are being generated in the country. On average, infectious wastes account to about 15,000 kg per month in the Philippines. Department of Environment and Natural resources has reported that in April to July 2020 alone, the country had already generated 19,187.7 of metric tons of infectious healthcare wastes. For an instance, the pandemic has led to an increase in the use of PPE in hospitals which include coveralls, n95 mask, gloves, head cover, shoe cover, goggles, surgical masks, and surgical gowns. The study of Sangkham (2020) showed that the country uses about 48,967,769 of facemasks every day which contribute largely to the increase in healthcare wastes in the country.

Additional wastes generation for healthcare wastes already dramatically increased in the Philippines. For example, one of the government hospitals at the forefront of managing COVID-19 has recorded additional waste generation of an average of 10,000 kg of infectious medical wastes every month. The Department of Health reported that for the months of March to June, the San Lazaro Hospital in

Manila City has already generated an estimated of 29,473 kg on infectious wastes which include personal protective equipment (PPE), dressings, swabs, blood bags, urine bags, sputum cups, syringes, test tubes, and histopathological waste.

Meanwhile, for the coming years, the Asian Development Bank in 2020 estimated an additional amount of healthcare waste in Metro Manila due to COVID-19 pandemic. The report estimated an increase of about 280 tons/day during the COVID-19 pandemic from the 47 tons/day before the COVID-19 pandemic. This figure shows a projected increase of about 496% increase in healthcare waste generated in Metro Manila. This projection is expected to pose more issues and concerns especially on the limited resources that the country already has.

6.5.2 Improper Handling and Management of Waste

Improper waste management of healthcare waste poses serious public health consequences in a developing country like the Philippines. This is one of the greatest issues brought about by the pandemic. This challenge could greatly affect the state of the environment in the country. The increase in the amount of infectious wastes (e.g., gloves, masks, gauzes) that are being mixed up make it harder to manage the already growing number of wastes in the country. This issue is also affected by the lack of equipped personnel who could safely and properly handle the healthcare wastes. The country also lacks the necessary equipment and facilities that will enable the proper handling of healthcare wastes. Department of Environment and Natural Resources (2020) has reported that in April to July 2020 alone, the country had already generated 19,187.7 of metric tons of infectious healthcare wastes. From this, only 29% are being treated and disposed properly.

The waste treatment and disposal facilities were unprepared and overwhelmed with the unabated surge in the volume of infectious medical wastes produced during the pandemic. Added to this is the uneven distribution of the 26 registered TSD facilities nationwide wherein most are found in Luzon only. In Eastern Visayas and Zamboanga Peninsula and Soccsksargen in Mindanao, LGUs do not have waste treatment centers. There are also no TSD facilities in the provinces of Cagayan Valley and Bicol region. Thus, hospitals do not express their difficulty in managing COVID-19 patients but are also grappling with the medical wastes.

Improper waste management is also coupled by the incapacity of some medical hospitals to practice proper waste segregation. Even households are having difficulty in segregating their infectious wastes from their other wastes. A case study from Ateneo de Manila University has mentioned specific instances where hospitals face difficulties in segregating infectious from non-infectious waste.

6.5.3 Lack of Institutional Capacity and Weak Coordination Among National Agencies

The Philippines and many other countries in the world face issues on their capacity to respond to the COVID-19 pandemic. Unfortunately, the waste problem brought about by the COVID-19 pandemic in particular is exacerbated by the fact that the country lacks the capacity to handle the huge increase in healthcare waste and more so it lacks the institutional mechanisms to address it.

Government is at the forefront of managing the issues and challenges brought about by the COVID-19 pandemic. However, the LGUs in the Philippines also lack the capacity and expertise to properly handle healthcare wastes that can pose substantial threat to the health of Filipinos. Before pandemic, some LGUs in the Philippines still have not come up with 10-year solid waste management plans. Hence, some LGUs still lack sanitary landfills that could somehow help manage the waste problem brought by the pandemic.

Conversely, the government also established the various responsibilities of each government office and agency to ensure the smooth implementation of its effort to ensure that wastes management of healthcare waste will be smooth. Multilevel coordination bodies were created to ensure that the government will be responsive to the issues and challenges. Interregional and event inter-municipal coordination are established to maximize joint efforts and responses.

By virtue of Executive Order No. 183, the Inter-Agency Task Force on Emerging Infectious Diseases (IATF-EID) led by the Department of Health (DOH) serves as the lead advisory board to the president on the management and implementation of necessary actions related to COVID-19. Under the IATF-EID, the DOH together with relevant government agencies comes up with the strategies needed to reduce transmission of COVID-19 in the Philippines (Philippine Humanitarian Country Team, 2020).

Meanwhile, Executive Order No. 2014–168 created the National task Force for COVID-19. The President of the Philippines is the National Command in Authority (NCA), supported by the Inter-Agency Task Force (IATF) led by the Secretary of the DOH. At the strategic level, the IATF is the policy-making entity and provides appropriate recommendations to the President related to COVID-19. At the operational level, the NDRRMC is organized as the National Task Force (NTF) for COVID-19 response with the DND Secretary as the Head (being the Chair of the NDRRMC), Secretary of DILG as co-chair, and OCD as Executive Director.

The three Task Groups that made up the government response include the Response Operations led by DOH, Resource Management and Logistics led by OCD, and Risk Communications by the (Presidential Communications Operations Office. Seven clusters were activated for COVID-19 response: Health (DOH), Governance (DILG), Law and Order (PNP), Economy, (NEDA), Food and NFI (DSWD), Logistics (OCD), Management of the Dead and Missing (DILG), and Crisis Communications (PIA). The NTF will be working at the NDRRMC

Emergency Operations Center (EOC) with a national incident commander (NIC) at the Secretary level, overseeing and managing its daily operations.

Due to the intricacy of the waste management issue brought about by the COVID-19 pandemic and the various levels of direction needed, the country also faces issues regarding coordination among number of government offices and agencies such as the DENR, DOH, and the LGUs. For example, the city of Baguio has already called for better coordination with DENR and DOH in managing healthcare waste. The LGU of Baguio mentioned that the Joint DENR-DOH Order No. 2 series of 2005 did not provide "cooperative linkages or coordination with local government units who significantly are easily seen to be blamed for any violation or non-observance of health care waste law, regulations and guidelines." This is despite the fact that the issuance was meant to harmonize the efforts of the DENR and the DOH on proper health care waste management (Refuerzo, 2020). This instance shows the complexity of the coordination among the various government agencies and offices in the country.

6.5.4   Lack of Awareness

Lastly, there is still lack of awareness regarding proper waste management especially healthcare wastes among Filipinos. Even before the pandemic, the country is already confronted with huge problems concerning the lack of awareness of Filipinos to proper waste management more so handling of healthcare wastes. The lack of awareness of some Filipinos regarding proper waste management affects the government's response. Since the information about COVID-19 is rapidly changing, there is also an urgent need for a continuous information dissemination that is spread all over the parts of the country. Winning the battle of waste pollution because of the COVID-19 pandemic requires a collective effort from all Filipinos. However, to do this, government has to step up its efforts to more proactively disseminate information and to also come up with strategies that would lead to change in behavior in Filipinos.

7   Post-Pandemic Countermeasures/Innovations in Managing Healthcare Waste: Lessons from Selected Cases

Healthcare management follows the principles of reduction, separation, and waste treatment flow in handling healthcare waste. As such, it needs collaboration of agencies and institutions in monitoring healthcare wastes in the Philippines. Institutions play an important role in curving the COVID-19 pandemic. Anchored on legal framework, these institutions collaborated to mitigate the spread of the disease.

Table 5 Some initiatives of the different institutions in response to the COVID-19 pandemic

DOH	DENR	DILG through LGUs	Healthcare without harm (NGO)
Development of the Healthcare management manual (DOH, 2020) It also serves as the lead for the IATF-EID	Policies and guidelines on effective and proper handling, collection, transport, treatment, storage and disposal of health care waste (denr. gov.ph)	Guidelines for LGU and DILG field support for the COVID-19 vaccine clinical trials (dilg. gov.ph)	Developed a webpage as a resource center for partners engaged at the intersection of environment, climate change, and health (noharm-global.org)

Table 5 shows some of the initiatives of the different institutions in curving COVID-19.

7.1 Department of Health: Mitigation, Preparedness, and Recovery

The Department of Health is the leading agency in informing, educating, and monitoring of COVID-19. The mitigation of infectious wastes and reduction of COVID-19 are anchored on Health Care Waste Management Manual in close collaboration with other government agencies, civil society, the academe and various professional groups. It provides practical information regarding safe, efficient, and environment-friendly waste management options. It also contains, in detail, safety procedures attendant to the collection, handling, storage, transport, treatment, and disposal of health care waste (DOH-HFDB, 2020). It is designed for use of different classifications of workers within the health care facilities, local government units (LGU), and private service providers who are involved in the generation, handling, storage, treatment, and disposal of health care waste. The major target groups of this manual are individuals responsible for overseeing the health care waste stream. The unit in-charge of health care waste management should be the first to become familiar with this manual so that they can oversee implementation of the program throughout the health care facility. The DOH also recommends calling 24/7 Telemedicine Hotlines for those who wish to consult health professionals from their homes.

The IATF, however, is instrumental in monitoring the spread of COVID-19 through Executive Order No. (E.O.) 168, s. 2014 created the Inter-Agency Task Force for the Management of Emerging Infectious Diseases (IATF) to facilitate inter-sectoral collaboration to establish preparedness and ensure efficient government response to assess, monitor, contain, control, and prevent the spread of any potential epidemic in the Philippines. Collaboration with DENR is through the implementation of RA 9003 or the Ecological Solid Waste Management Act,

Table 6 Classifications of the Community Quarantine during the COVID-19 pandemic

ECQ	MECQ	GCQ	MGCQ
100% stay at home	100% stay at home	Persons below 21 and 60 years old and above or those at risk for contracting the COVID-19 disease are required to stay home	Persons below 21 and 60 years old and above or those at risk for contracting the COVID-19 disease are required to stay home
Exception for workers in offices or industries permitted to operate	Restricted movement, only for assessing essential goods and services Exception for workers in offices or industries permitted to operate		
	Persons below 1 and 60 years old and above or those at risk for contracting the COVID 19 disease are required to stay home		

Source: Presidential Operations and Operations Office (pcco.gov.ph)

making sure that material recovery facilities are in place. These initiatives of the DENR put infectious wastes on its proper bin in collaboration with the LGUs. To realize the reduction of the pandemic in the country, the Department of Interior and Local Government enforces on the participation of LGUs up to the barangay level in the vaccine trial based on the guidelines adopted by the IATF-EID. Lastly, Health Care Without Harm (HCWH), an NGO is an active campaigner of health security developing a webpage that serves as a resource center for partners around the world, engaged at the intersection of environment, climate change, and health (noharm-global.org).

Inter-Agency Task Force using preventive activities initiated the quarantine protocol that serves as guidelines in isolating communities and mitigating the spread of COVID-19. The quarantine classification is an initiative to limit mobility of people and mitigate contamination. In 1 year's time, the country has undergone several classification of quarantine depending on the severity of the cases (Table 6).

7.2   Department of Environment and Natural Resources (DENR)-Mitigation, IEC

Proper waste segregation and disposal are important components in curving COVID-19. To further strengthen efficient waste management, the Environmental Management Bureau (EMB) of DENR assisted 598 barangays in establishing their respective materials recovery facility (MRF) and supported the operation of

24 functional MRFs (DENR website). The establishment of MRF entails sorting of wastes reducing wastes in the final disposal facility. Conversely, the DENR-led National Solid Waste Management Commission has approved a total of 106 solid waste management plans this year, bringing the total to 1063 since RA 9003 was enacted in 2001. It means that these plans would include the mitigation of infectious wastes. EMB also monitored the implementation of 670 closure and rehabilitation plans, as well as the actual closure and rehabilitation of 204 dumpsites. Moreover, the closure of these dumpsites would control the spread of infectious waste carelessly disposed. Webinars were also conducted by DENR to capacitate environment workers, particularly garbage collectors to enhance awareness on handling of infectious wastes. DENR Secretary Roy Cimatu, on the other hand, has temporarily halted the shipment of all forest products and wildlife throughout the nation in order to contain the spread of the corona virus disease (COVID-19). The DENR has taken this precautionary action in order to mitigate the spread of illegal wildlife trade and the contamination of zoonotic disease transmission of disease and animal trading. While the origin of the virus is unknown, it is thought to have begun from bats and spread to humans via an dentified intermediary host in a wet market in Wuhan, China.

7.3 LGU Initiative: The Case of Pasig City Using Community-focused Strategy

7.3.1 Monitoring, Preparedness, Recovery, Strengthening of RA 9003, and Elimination of Single Use Plastic

Pasig City is one of the cities that has been proactive in responding to COVD-19 as well as preparing constituents for the new normal. The city prioritizes economic recovery, health security, and environmental protection. The city has initiated a loan program to revitalize the economy of small and medium enterprises. The National Task Force COVID-19 has cited Pasig City as a model city for the entire country with its plans and programs to fight the COVID-19.

It implemented the Fast track Cities (FTCI), which aims to scale up the response to the growing HIV epidemic in the context of the COVID-19 pandemic. Pasig City proactively delivers quality services to everyone, including hard-to-reach populations by increasing investments to scale up HIV programs, establishing more treatment facilities and integrated social hygiene clinics, and training more healthcare providers to deliver HIV and COVID-19 services. It has established quarantine facilities providing food and medical care and other essential needs.

According to Mayor Sotto, around P300 million from the city's 2021 budget has been allotted for the procurement and purchase of COVID-19 vaccines, an addition to those that will be provided by the Department of Health to local government units. In the early months of the pandemic in 2020, the city government of Pasig subsidized the people with groceries and cash coupons to augment what they lost due to lockdown. The city has huge savings by 150 million for curving corruptions in public biddings due to transparency policy. In fact Mayor Sotto was lauded by the US government as one of the "anti-corruption champions" committed to transparency and accountability.

Facilitated by the project ARK, the Pasig local government runs rapid antibody tests on 10% of residents in the barangay that has the highest ratio of coronavirus cases in the city. The city was able to slow down COVID-19 and no significant surge during the Christmas holidays. Pasig city has COVID-19 cases website constantly monitoring the plight of COVID cases in the city.

The city is prepared for the COVID-19 vaccination. The government has clustered the barangays into 16 vaccination sites. The city government identified public schools as sites for the people. The first phase of the program focuses on master listing, registration, validation, screening, and profiling of their constituents while also preparing the logistics and venues for the immunization program.

7.4   Healthcare Without Harm-IEC

Healthcare without Harm is an NGO dedicated in eliminating healthcare wastes. They promote segregation at source. For them, there is no need to burn or incinerate COVID-19 related waste because medical waste autoclaves that use pressurized steam at 30 psi at 30 min saturation time are known to kill any heat resistant pathogens without need of any chemicals.

COVID-19 related waste can then be treated as regular municipal waste after being sterilize in an autoclave. They believe that all technologies whether autoclave or microwave should be validated and tested on regular basis. Apart from being extremely costly, incinerators and incinerators-in-disguise like pyroclaves, gasification, pyrolysis, plasma arc produce dangerous dioxins and furans that further harm people's health rather than protecting it. Waste can be sent for disposal or recycling after being disinfected in an autoclave.

7.5   Opportunities/Lessons Learned (Direct and Indirect Opportunities)

1. Resource sharing is important in curving COVID-19. No single institution can mitigate the spread of COVID-19, thus clustering of agencies should be based on strength and availability of resources supporting the lesser agencies.
2. The Philippines should strengthen the implementation of RA 9003 as infectious disease goes hand in hand with waste disposal. Disposal of infectious wastes is not an issue if proper waste management can be implemented by the LGUs. However, their lack of technical and financial capacity remains a challenge.

Further, the institutionalization of environmental office in the city or municipality that manages waste management system of the LGU is needed. The provision of a permanent position for the City or Municipal Environmental Officer (C/MENRO) and the regularization of the waste workers in the LGU are also recommended to ensure sustainability of services.

3. Continuous information campaign drives on hazardous waste to waste workers, community frontliners, and community members. Community households should be vigilant with the new variants coming out. Complacency of community members could be lead to community contamination.

4. Researches should be an important tool in curving COVID-19. The emergence of collaborative research conducted by University of the Philippines and University of Santo Tomas should set an example to other academic communities.

5. COVID-19 is a slow onset disaster. The government could have been vigilant with foreigners coming in and out with the outbreak of the disease from China. Massive information and education campaigns could have curved the disease earlier to avoid claiming lives especially from the frontliners. The use of face masks and face shield could have been earlier implemented to prevent the spread of disease. Preventive is also an important component of curving COVID-19. Preventive means that people should be disciplined to stay at home and less consumerism to avoid malling.

8   Conclusion and Recommendations

As shown in the above discussions, the policies and regulations as well as the waste management system on healthcare waste in the country are in place even before the pandemic. Interim guidelines are also issued by different relevant agencies to include the COVID-19 related waste. The Philippines have crafted many very good laws; however, the main challenge is on the strict enforcement and monitoring of these policies and regulations. Hence, the following recommendations among others are proposed:

(a) *Policy support for the LGUs both in terms of technical and financial*: LGUs as the main implementers of these laws are always on the receiving end of these policies and guidelines. However, they are too burdensome for them without the necessary support from the national and other relevant agencies. Management of healthcare waste including the COVID-19 related waste requires technical knowledge and financial resources to be effectively implemented. This is even more challenging for small cities and municipalities with limited resources, hence, the second recommendation.

(b) *Promotion of local collaboration between and among LGUs and other non-state actors*: The LGUs do not need to carry the burden of implementing these laws by themselves alone (Atienza, 2020). As discussed earlier, one of the main challenges for the effective management of healthcare waste is the limited number of treatment, storage, and disposal facilities. Establishment of TSD facilities requires huge amount of resources, hence, inter-local collaboration between and among LGUs and/or partnership with other non-state actors could be the most feasible option.

(c) *Strengthen the monitoring and evaluation of health care waste management*: One of the challenges encountered even while writing this report is the difficulty of finding available and reliable data particularly on the health care waste generation before and during the pandemic. Most of the LGUs have no or incomplete data on health care waste (HCW) probably due to the limited capacity in terms of manpower and financial resources. As mentioned in this report, each HCF should have one pollution control officer (PCO) but most LGUs cannot afford to employ PCO. Further, PCO is not a permanent position but only a designated position. Hence, despite its demanding responsibilities it is just additional duties with no additional benefits. But his role is very important particularly in ensuring the proper implementation and monitoring of health care waste. Availability of credible data is a significant input in developing evidence-based approaches and practical solutions in managing the HCW more efficiently and effectively in a sustainable manner.

(d) *Strengthen the information, education, and communication campaign* on the relevance of the effective implementation of the healthcare waste management policies and guidelines. As mentioned, the country has very good policies and waste management system in place but without the support and participation of the general public, the threats to the environment and human health posed by the improper management of healthcare waste will remain. Hence, providing the right information not only to the LGUs and healthcare workers but to every member of the community will encourage them to participate and comply with these laws. One of the possible and practical ways to widen the coverage of information dissemination is to make use of technologies and social media. The LGUs can make use of their official websites to provide updated and reliable data and awareness campaign programs to its constituents. Awareness can bring change to the behavior of the people which can lead to a more effective and sustainable healthcare waste management in the country.

Acknowledgement The authors would like to acknowledge the following key informants for generously sharing information and insights on managing healthcare waste in the Philippines: Engr. June Philip Ruiz (Development Management Officer, Research and Performance Management Division, HFDB-DOH); Engr. Rolando I. Santiago (Supervising Health Program Officer, DOH); Engr. Arnul Alcain, (Provincial Health Officer of Cavite); Engr. Jerry Mogol (Former Sanitation Program Manager, DOH); Mr. Gilberto L. Tumamac (Assistant Division Chief, Local Government Monitoring and Evaluation Division, Department of Interior and Local Government IV-A); Engr. Eli Ildefonso (Chief Environmental Specialist, Environmental Management Bureau, DENR); Ms. Juvinia P. Serafin (Senior Environmental Management Specialist, EMB-DENR); and Mr. Alvin Asis (Environment Unit Head, League of Cities of the Philippines).

References

ADB. (2020). *Managing infectious medical waste during the COVID-19 pandemic*. Retrieved from https://www.adb.org/publications/managing-medical-waste-covid19

Ananth, A. P., Prashanthini, V., & Visvanathan, C. (2009). Healthcare waste management in Asia. *Waste Management, 30*(1), 154–161. https://doi.org/10.1016/j.wasman.2009.07.018

Atienza, V. (2020). Promoting local collaboration on waste management: Lessons from selected cases in the Philippines. In M. Kojima (Ed.), *Regional waste management—Inter-municipal cooperation and public and private partnership. ERIA research project report FY2020 no. 12* (pp. 122–134). ERIA.

Cabico, G. K. (2020a, August 15). 'Earth not healing': Medical waste piles up as COVID-19 cases rise. *Philstar Global*. Retrieved from https://www.philstar.com/headlines/2020/08/15/2034986/earth-not-healing-medical-waste-piles-covid-19-cases-rise

Cabico, G. K. (2020b, September 2). *DOH: Facilities breaking infectious waste disposal protocols may be suspended, shutdown*. Retrieved from https://www.philstar.com/headlines/2020/09/02/2039593/doh-facilities-breaking-infectious-waste-disposal-protocols-may-be-suspended-shut-down

Cabico, G. K. (2020c). *In the Philippines, medical waste piles up as COVID19 cases rise*. Retrieved from https://earthjournalism.net/stories/in-the-philippines-medical-waste-piles-up-as-covid-19-cases-rise

Carillo, C. A. (2013, November 30). Davao waste handler notes hazardous trash still improperly disposed. *Business World Online*. https://www.bworldonline.com/content.php?section=Economy&title=Davao-waste-handler-notes-hazardous-trash-still-improperly-disposed&id=80101.

Casilao, J. L. (2021, January 27). ADB: Metro Manila daily medical waste up by 280 tons due to pandemic. *GMA news* Online. Retrieved from https://www.gmanetwork.com/news/news/metro/773553/adb-metro-manila-daily-medical-waste-up-by-280-tons-due-to-pandemic/story/

Cruz, C. P., Garcia, R. C., Colet, P. C., Cruz, J. P., & Alcantara, J. C. (2014). Healthcare waste management of the government hospitals in northern Philippines. *European Scientific Journal, 10*(26), 114–122. Retrieved from https://www.researchgate.net/publication/274255167

DENR. (2004). Procedural Manual Title III of DAO 92–29 "Hazardous Waste Management" DENR AO 36 Series of 2004. Retrieved from http://www.env.go.jp/en/recycle/asian_net/Country_Information/Law_N_Regulation/Philippines/HW_DAO%2004-36.pdf

DOH. (undated). *Health care waste management manual*. Retrieved from https://doh.gov.ph/sites/default/files/publications/Health_Care_Waste_Management_Manual.pdf

DOH-HFDB. (2020). *Health care waste management manual* (4th ed.).

DOH-HFDB. (undated). *Health care waste management for Covid-19 pandemic*.

Enano, J. O. (2020, May 20). *Waste segregation more crucial amid pandemic*. Retrieved from https://newsinfo.inquirer.net/1277808/waste-segregation-more-crucial-amid-pandemic

Fermin, M. (2020, March 9). Vials with blood specimen washed Subic Bay ashore. *Philippine Lifestyle News*. Retrieved from https://philippineslifestyle.com/vials-with-blood-specimen-washed-subic-bay-ashore/

Fernandez, H. A. (2020). ADB: Coronavirus could leave major Southeast Asian cities with 1,000 extra tonnes of medical waste per day. Eco Business. https://www.eco-business.com/news/adb-coronavirus-could-leave-majorsoutheast-asian-cities-with-1000-extra-tonnes-of-medical-waste-per-day/

Gubalani, R. (2021, January 21). Sarangani to operate medical waste treatment plant. *Philippine News Agency*. Retrieved from https://www.pna.gov.ph/articles/1128064

Health Care Without Harm Asia. (2007). *Best practices in health care waste management: Examples from four Philippine hospitals*. Retrieved from http://noharm.org/lib/downloads/waste/Best_Practices_Waste_Mgmt_Philippines.pdf

Health Care Without Harm Asia. (2013). *Philippines | hospitals show the way to health care waste management.* Retrieved from https://noharm-global.org/articles/news/global/philippines-hospi tals-show-way-health-care-waste-management

Jambec, J. (2015). Plastic waste inputs from land into the sea. *Science, 347*(6223), 768–771. https://doi.org/10.1126/science.1260352

Kenworthy, K. (2017). *Top 10 facts on healthcare in the Philippines.* Retrieved from https://borgenproject.org/healthcare-in-the-philippines/

Malekahmadi, F., Yunesian, M., Yaghmaeian, K., & Nadafi, K. (2014). Analysis of the healthcare waste management status in Tehran hospitals. *Journal of Environmental Health Science & Engineering, 12*(116), 1–5. Retrieved from http://www.ijehse.com/content/12/1/116

Molina, V. B. (2002). Waste management practices of hospitals in metro Manila. *University of the Philippines Manila Journal, 7*(3–4), 17–22.

Montemayor, M. T. (2020, September 3). DOH warns of sanctions vs. improper disposal of healthcare wastes. *Philippine News Agency.* Retrieved from https://www.pna.gov.ph/articles/1114212

NSWMC. (2020). National Solid Waste Management Commission (NSWMC) Resolution No. 1364, Series of 2020: Adopting the Interim Guidelines on the Management of Covid-19 Related Health Care Waste.

Pena, R. (2020, September 10). Pena: Management of health care waste. *Pampanga.* Retrieved from https://www.sunstar.com.ph/article/1869782/Pampanga/Opinion/Pena-Management-of-health-care-waste

Philippine Humanitarian Country Team. (2020). *COVID-19 response plan.* Retrieved from https://reliefweb.int/sites/reliefweb.int/files/resources/HCT%20Covid-19%20Response%20Plan%20%283%20April%20update%29.pdf

Refuerzo, A. (2020). *City seeks better coordination with DENR, DOH in health care waste management efforts.* Retrieved from https://www.baguio.gov.ph/content/city-seeks-better-coordination-denr-doh-health-care-waste-management-efforts

Sangkham, S. (2020). Face mask and medical waste disposal during the novel Covid-19 pandemic in Asia. *Case Studies in Chemical and Environmental Engineering, 2,* 100052. Elsevier Ltd. https://www.sciencedirect.com/science/article/pii/S2666016420300505

Sañez, G. G. R. (2008). *Healthcare wastes management in the Philippines: Policies, rules and status of implementation.* Retrieved from http://s3.amazonaws.com/zanran_storage/www.3rkh.net/ContentPages/53917451.pdf

SEPO. (2017). *Philippine solid wastes at a glance.* Retrieved from http://legacy.senate.gov.ph/publications/SEPO/AAG_Philippine%20Solid%20Wastes_Nov2017.pdf

Subingsubing, K. (2020, September 6). PPE waste disposal a growing challenge. *Philippine Daily Inquirer.* Retrieved from https://newsinfo.inquirer.net/1331800/ppe-waste-disposal-a-growing-challenge

WHO. (2014). *Safe management of wastes from healthcare activities* (2nd ed.). WHO. Retrieved from https://www.euro.who.int/__data/assets/pdf_file/0012/268779/Safe-management-of-wastes-from-health-care-activities-Eng.pdf

WHO. (2015). *Status of health-care waste management in selected countries of the Western Pacific Region.* Retrieved from https://apps.who.int/iris/rest/bitstreams/1246999/retrieve

Windfeld, E. S., & Brooks, M. S. (2015). Medical waste management—a review. *Journal of Environmental Management, 163*(2015), 98–108.

Part III
Management of Non-Medical Waste Infected by SARS-CoV-2

The Immediate Italian Response to the Management of Non-medical Waste Potentially Infected by SARS-CoV-2 During the Emergency Phase of the Pandemic

Francesco Di Maria, Eleonora Beccaloni, Lucia Bonadonna, Carla Cini, Elisabetta Confalonieri, Giuseppina La Rosa, Maria Rosaria Milana, Emanuela Testai, and Federica Scaini

1 Introduction

After being officially detected in late December 2019 (CDC, 2020) in the city of Wuhan (China) and officially declared a pandemic disease (WHO, 2020a), the severe acute respiratory syndrome (SARS) disease named coronavirus disease 2019 (COVID-19) has infected >75,000,000 peoples across the world causing >1,500,000 deaths (WHO, 2020b).

As reported by de Wit et al. (2016) other two highly pathogenic human coronaviruses were already detected in the last 20 years, the SARS-CoV (in 2002–2003) and the Middle East Respiratory Syndrome (MERS-CoV, in 2012). Other known and diffused human coronaviruses infecting of the upper respiratory tracts belong to the HCoV-229E, HCoV-OC43, HCoV-NL63 and HCoV-HKU1 strains.

When symptomatic, SARS-CoV-2 infection is generally associated with respiratory diseases (i.e. pneumonia, dyspnea, fever) but in some cases also with other symptoms such as headache, diarrhea, vomiting, and nausea (Huang et al., 2020). The most credited transmission pathway for this infection is by respiratory droplets

F. Di Maria (✉)
LAR Laboratory, Dipartimento di Ingegneria, University of Perugia, Perugia, Italy
e-mail: francesco.dimaria@unipg.it

E. Beccaloni · L. Bonadonna · G. La Rosa · M. R. Milana · E. Testai · F. Scaini
Dipartimento Ambiente e Salute, National Institute of Health (ISS), Rome, Italy

C. Cini
AMA SpA, Rome, Italy

E. Confalonieri
Regione Lombardia, Milan, Italy

© The Author(s), under exclusive license to Springer Nature Singapore Pte Ltd. 2022
S. K. Ghosh, P. Agamuthu (eds.), *Health Care Waste Management and COVID 19 Pandemic*, https://doi.org/10.1007/978-981-16-9336-6_5

105

generated by infected patients (Lewis, 2020; National Research Council, 2020; Yu et al., 2018; WHO, 2020c). However self-transmission by hands after being in touch with contaminated fomites was also considered another possible transmission route, suggesting interpersonal distance, hand hygiene, and protective mask use as effective preventive measures for limiting the spreading of the infection (Tobias, 2020). Airborne virus transmission (by droplet <5 μm in diameter) at distances greater than 1 m is still a matter of debate (Anderson et al., 2020; Morawska & Cao, 2020, WHO, 2020c). Oppositely, no traces of these studies were reported by Faridi et al. (2020) in air samples captured from 2 m up to 5 m from beds of patients infected by the virus. Nevertheless a systematic review of possible airborne transmission of distances from hospital beds hosting patients infected by SARS-CoV-2 in indoor environment was recently published, suggesting that SARS-CoV-2 may be transmitted through air in poorly ventilated indoor environments (Noorimotlagh et al., 2020). In addition, different experimental studies reported positive results confirming presence and persistence of the virus in indoor air (Santarpia et al., 2020; Liu et al., 2020; Kenarkoohi et al., 2020; Razzini et al., 2020; Chia et al., 2020).

Saawarn and Hait (2021) reviewed the presence of the virus in stools of infected patients suggesting the fecal–oral transmission as a further possible route of transmission (Heller et al., 2020). Twenty-five studies have demonstrated the occurrence of SARS-CoV-2 in human feces, two of which also found live virus isolated in the feces samples (Zhang et al., 2020; Ding and Liang, 2020; Wang et al., 2020). Moreover, SARS-CoV-2 RNA has been identified in raw sewage worldwide (Foladori et al., 2020). Concomitant with the presence of SARS-CoV-2 viral RNA in feces and raw sewage, studies also indicated its occurrence in treated wastewater, primary sludge, and river water (Mohapatra et al., 2020). Although currently there are no studies available to indicate the fecal–oral transmission of SARS-CoV-2 via sewage or wastewater systems (Hindson, 2020; Gu et al., 2020), such transmission cannot be excluded, especially in areas with poor sanitation where contaminated environments can represent a potential medium of virus transmission (Mehraeen et al., 2020).

An ongoing discussion is also associated with the contamination (and decontamination) of inanimate surfaces and the persistence of SARS-CoV-2 on different surfaces and materials represent an important topic (Chin et al., 2020; van Doremalen et al., 2020). Among the available information about the persistence of human and veterinary coronaviruses on inanimate surfaces as well as inactivation strategies with biocidal agents, Kampf et al. (2020) reported that coronavirus can persist up to 9 days and can effectively be inactivated by different biocidal products. It was also noted that other environmental and climatic conditions as temperature, humidity, UV rays concurred to the persistence of the virus on these fomites. In experimental laboratory studies, infectious virus could not be recovered from printing and tissue papers after a 3-h incubation and from treated wood and cloth on day 2. By contrast, SARS-CoV-2 was more stable on glass and banknotes (4 days) and on stainless steel and plastic (7 days). A detectable level of infectious virus was found on the outer layer of a surgical mask on day 7. More recently, other studies (reviewed in Marquès & Domingo, 2020) on the stability and infectivity of SARS-

CoV-2 on inert surfaces have been published (Biryukov et al., 2020; Carraturo et al., 2020; Colaneri et al., 2020; Morris et al., 2020), confirming that SARS-CoV-2 can last on different surfaces for times ranging from some hours to few days. The potential persistence of the SARS-CoV-2 on different surfaces and materials represents a matter of concern for both public health and environment protection also for other areas like schools, cinemas, restaurants, bars, and other places different from healthcare facilities for which specific regulation for infective waste is already implemented (DPR, 2003; WHO, 2020d). In this perspective also household waste represents another potential risk of infection spreading.

In particular, wastes generated in houses with subjects affected by COVID-19 are of particular concern compelling dedicated attention for their management (Nghiem et al., 2020) as well as household wastes from patients in quarantine in private home.

This aspect involves also another important aspect of social life, that is the diffused perception among the general population and the operators about the risk associated with the management of wastes (Gomes Mol & Caldas, 2020). These problems resulted even more emphasized for the workers of the informal sector, a very diffused economic activity in some areas of the world, often operating with no personal protection equipment (PPE) and characterised by the lack of minimal safety and health protection requirements (Cruvinel et al., 2019).

Even if the criteria delivered by the European Commission (EC, 2020) concerning waste management during the COVID-19 crisis recommended not to interrupt the separated collection, the ISS WG considered that it was not possible to neglect the intrinsic complexity of the waste management and the prominent role played by the socio-economic implications.

Indeed, the lack of scientific evidences together with the widespread presence of infected subjects, in some cases not confined in hospitals and medical structures, makes this a very challenging issue (Nghiem et al., 2020).

The Association of Cities and Regions for sustainable resource management (ACR+, 2020) reported that different protocols and practices were implemented by the EU Member States to adopt the waste collection to ensure higher levels of safety and protection to workers.

As an example, higher amount of waste were temporarily stored when its removal was delayed due COVID-19 crisis: incineration of household waste potentially infected by the SARS-CoV-2 and proper procedures for the disposal of PPE used by health care workers treating patients at their home, were introduced in the UK without any modification to plant permission (ACR+, 2020).

In Germany, the National Government indicated to put organic, packaging, and paper waste generated by infected patients treated at home together with residual waste. Other countries (e.g. Spain, Portugal, Estonia) imposed a minimum storage time of 72 h to protect against the risks from waste handling, to allow at least some virus inactivation. Differently, in the USA no specific regulations were introduced for managing household waste suspected to be potentially infected by the SARS-CoV-2, besides the adoption of PPE and other precautions already being imposed by safety regulation for workers. Similar indications were also detected by the Israelian

Ministry of Environmental Protection, by the Ramallah Municipality in Palestine and by the Peruvian Ministry of Environment.

All these different attitudes were also attributable to a lack of knowledge about the virus persistence on surfaces and the role that solid waste and fomites in general can have on the spreading of the COVID-19. The choice of the different approaches were mainly driven by socio-economic and cultural perspective, based on the perception of the potential risk of the transmission of the SARS-CoV-2 infection by the waste management, rather than by scientific well supported evidence.

Based on these considerations, since the 15th March 2020, during the peak of first wave of COVID-19 pandemic, a multidisciplinary working group (WG) was appointed at the Italian National Institute of Health (ISS) gathering medical, biological, environmental, hygiene, and engineering experts also involving different stakeholders such as local authorities and waste management companies. The aim of the WG has been to provide pragmatic response and guidelines for the management of the waste generated by patients that resulted positive to the SARS-CoV-2 tests and/or in compulsory quarantine treated at home or in areas other than hospitals and medical centers (ISS, 2020).

2   Waste Management

The COVID-19 rises the attention on several aspects concerning public and private service, imposing new and challenging actions for their adaptation to the specific emergency situation. Waste management represents, among the other, one of those services on which particular attention was focused by several governments for limiting the spreading of the virus and giving adequate protection levels to both operators and citizens.

Starting from these evidences, the WG was charged to develop proper, feasible, and viable guidelines oriented to protect from the potential infection both the workers of waste collection and treatment and the general population that could be directly and/or indirectly in contact with these potentially infected waste. The guidelines adopted and published in its first version a few days after the start of the hard lockdown, were developed on the basis of the available scientific evidence on the virus persistence on different materials surfaces and on the length of inactivation, also based on the use of disinfectants and/or proper temperatures and length of disposal treatments.

In fact, as known, SARS-CoV-2 virion consists of a spherical particle of ≈100 nm diameter, characterized by a positive single stranded RNA genome enclosed by a fragile lipid envelope. The damage of the lipid envelope may inactivate the viral particles making the virus non-infectious even if traces of RNA can be detected also for several hours and/or days.

Based on this evidence the first aspect analyzed by the WG was the identification of the potential infection routes and exposure scenarios through the whole waste management chain, from collection to final treatments and disposal.

The main routes of infections identified were:

1. direct contact with contaminated objects and/or surfaces (Kampf et al., 2020; van Doremalen et al., 2020);
2. contact with airborne droplets at a distance <2 m (Faridi et al., 2020; Liu et al., 2020; Santarpia et al., 2020).

For the waste management, several phases were depicted and analyzed based on the potential involved route of infection and/or exposure to the SARS-CoV-2 (Fig. 1):

1. waste packing and delivering by the users;
2. waste withdrawal by workers and/or trained volunteers;
3. waste transport;
4. waste treatment.

2.1 Waste Packing and Delivering by the Users

Concerning the first step of the whole waste management process, the main related potential routes of infection can be identified as follows: (1) become in contact with surfaces and objects infected by the SARS-CoV-2 during the waste handling; (2) risk of generation of aerosol during withdrawal due to, e.g. waste humidity and bags compression. Due to the potential risk of contamination by the virus of typical objects and different materials of household waste (i.e. plastics, paper, metals, glasses, food, masks, other medical devices) it was suggested as a first general recommendation to interrupt separated collection and to put all the waste commingled in the non-differentiated waste bag. This was proposed also to reduce the risk of virus scattering by the potentially infected materials through several waste streams and facilities with a further potential risk of enhancing the diffusion of the virus. A second important recommendation concerned the proper and adequate packaging of the waste for protecting both the waste handlers and the general population from possible injury and diseases arising from the contact with this waste. Therefore, in order to prevent both waste leakage and their mixing with non-infected waste streams, the use of clearly identified (i.e. by specific color and/or by specific labels) double flexible plastic bags for packing potentially contaminated waste was recommended, together with proper wrapping of sharp materials before its insertion into the double plastic bag. Before their delivery to collection workers or properly trained volunteers, these bags should be adequately sealed by stripes and/or adhesive tape. This last precaution was also indicated by the German government (ACR+, 2020).

In packing infected household waste, citizens should wear single use gloves to be disposed of in a new bag, taking the utmost care to peel downwards, away from the wrist, turning the glove inside out. For preventing the damage of the plastic liner,

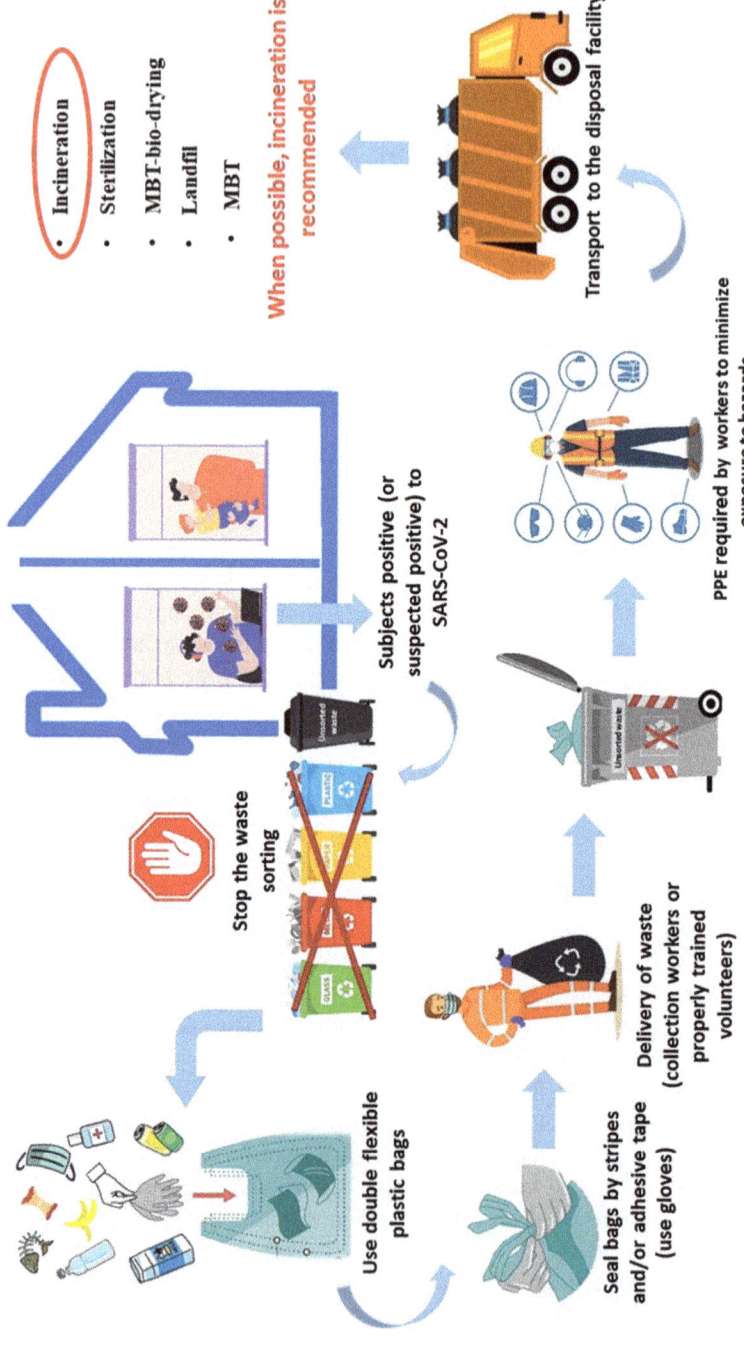

Fig. 1 Main management phases for household waste generated by subjects infected or in quarantine due to the COVID-19

both manual and mechanical compression (e.g. by collection vehicles) have to be avoided.

In addition, the WG also recommended the adoption of specific PPE and "behavior" to prevent contamination/infections of waste operators, as a specific group at risk. It was suggested (1) to wear facial filter masks FFP2 or FFP3 to protect the operator from splashes and sprays; (2) to use a pair of disposable gloves under normal non-disposable work gloves, in order to prevent skin contact with work clothes and masks; (3) a specific order in undressing at the end of the work shift; (4) the way of cleaning work clothes; (5) the replacement of non-disposable work gloves, in case a broken and/or open bag was handled; (6) sanitization of the driver's cab of the vehicles used for the waste collection, avoiding methods that produce splashes or aerosolize infectious material (ISS, 2020).

2.2 Waste Transport

After collection straightforward transport to the disposal facility, avoiding transference stations and other handling operation was also recommended. In this way a further possible contact with these wastes with other materials and with workers is minimized.

Avoiding the use of collection and transport vehicles with automatic compaction systems was also suggested. In fact, larger size vehicles used for the collection are equipped with hydraulic powered compaction systems able to reduce the volume of the loaded waste up to 1:4 (Di Maria & Micale, 2013). This can cause rips in the bags with consequent risk of leakage of waste material representing another potential route for the diffusion of the virus. This risk can also be emphasized if the same vehicle is successively used for other collection routes without preliminary adequate disinfection.

2.3 Waste Treatment

Together with use of disinfectants, temperature and time are identified as the two main parameters impacting the inactivation of the SARS-CoV-2. Chin et al. (2020) reported that temperature >70 °C maintained for not less than 5 min. is able to effectively destroy the infective potential of the virus. Kampf et al. (2020) reported that at lower temperature (up to 4 °C) the virus resulted degraded after >28 days. The same author reported that at temperature ranging from 30 to 40 °C the SARS-CoV survived for a period not longer than 96 h.

Based on this evidence, the most diffused treatment and disposal facilities adopted in the waste sector were ranked on their ability in achieving the inactivation of the virus based on the temperature and on the length of the treatment (Fig. 2).

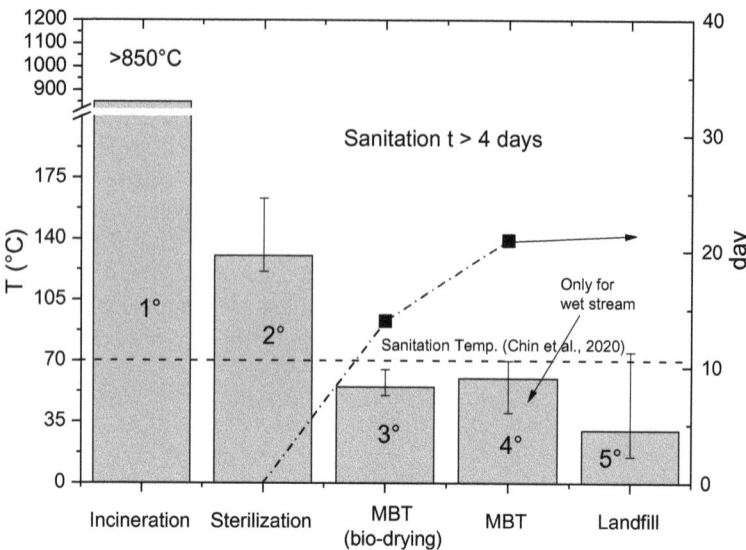

Fig. 2 Ranking of treatment/disposal facilities for household waste potentially infected by SARS-CoV-2 based on temperature and time of the treatment

Due to the high temperature levels (>850 °C) (Di Maria et al., 2018) the incineration was indicated to be the most suitable solution for managing such waste. Direct insertion in the combustion chamber, with priority respect to the other waste stored in the facility was also recommended. For this aim different colors of the bags can facilitate this job. The use of solid PVC bins were not recommended, in particular for lower size facilities (i.e. 100,000 tonnes/year) due to the risk of increase in HCl emission caused by their combustion and by the risk of hopper clogging.

For those areas lacking incineration capacity, sterilization resulted as another suitable treatment due to its ability of maintaining the materials at temperature from 121 °C up to 163 °C (Windfeld & Brooks, 2016) usually by saturated steam (DPR, 2003; WHO, 2020d). By the way, compared to incineration, sterilization shows two main limitations: a) narrowed treatment capacity generally limited to a few tonnes/day and b) necessity of further treatment of the sterilized waste as combustion since its recycling is not allowed by the current Italian legislation.

Another possible alternative can be also represented by the mechanical biological treatment (MBT). The most suitable MBT is the one based on those plants operating according to the bio-drying process, also known as mechanical and biological stability method (Di Maria, 2012; Di Maria et al., 2012). In this case, after shredding and eventually removal of bulky items, the whole waste undergoes to an aerobic treatment, lasting for about 2 weeks and achieving maximum average temperature of about 55 °C. The combination of these two parameters leads to an effective inactivation of the virus. By the way it is strictly recommended to avoid any contact of the waste with the operators until the end of the treatment.

Another largely diffused group of MBT facilities operates according to a different treatment scheme, based on a preliminary size sorting of the inlet waste by the use of trommels and/or disk screens (Di Maria, 2012).

After the sorting operation, the oversize materials (i.e. >100 m average size), consisting mainly of paper and cardboard, plastics, textile, and other high calorific value materials, are generally moved to other recovery and/or disposal treatments/operations. The undersize, consisting mainly of the organic fraction, is moved to the aerobic biostabilization process lasting on average for 2–3 weeks during which average temperatures >40 °C op to about 70 °C can be achieved. For the MBT of this group of waste, the sanitation requirements are usually verified only for the undersize stream subject to the biostabilization process. For the oversize it is strongly recommended to avoid any contact between the waste stream and the operators, before and after the treatment. Furthermore, storage of the oversize fraction for adequate periods (e.g. >96 h) before any further operation is suggested, since it can provide an adequate level of virus inactivation (van Doremalen et al., 2020; Kampf et al., 2020).

The last option can also be represented by landfilling if particular care is given to the waste handling, storage, and burial. In this case it is recommended the identify clearly the area in which these wastes are disposed, providing also adequate physical separation with other conventional waste streams and daily cover able to avoid any airborne dispersions. For landfilling the only parameter acting for the SARS-CoV-2 inactivation is the time even if temperature >70 °C was also detected in the landfill bodies (Di Maria et al., 2020). By the way it is worthy to highlight that according to the last EU Directive 2018/851/EC the landfilling of infective waste is not more allowed. This means that landfilling of these wastes can be permitted only in the case legal authorities declare the emergency status and just for limited periods.

3 Communication Strategy

Communication together with clarity and soundness is one of the strategic and relevant aspects for a successful implementation and enforcement of regulation and guidelines. The communication to the general population and workers was improved by adding to the guidelines some infographics able to summarize the main recommendations and suggestions by using pictures and bullet point lists written in layman language. Furthermore, the guidelines also respond to the multiple requests of safety procedures coming from stakeholders, including management companies, operators, and citizens during the COVID-19 crisis making the guidelines sound with respect to the current social and economic context.

This said the diffusion of the guidelines was enhanced by adopting different approaches considering also different target audiences (i.e. companies, operators, citizens). Briefly, the communication strategy was based on: (1) dissemination by media and press release; (2) delivering to local authorities; (3) activation of a dedicated section of the web site of the ISS "ISS per COVID-19" (https://www.iss.it/en/coronavirus) for free download; (4) delivery to specific associations.

This was also a keystone activity aimed to increase the awareness of the population on practical measures to contain spreading of the virus and on risk prevention.

4 Conclusion

The COVID-19 pandemic has transformed our way of life and also the production of our urban waste in terms of waste composition and amount. It is now clear that the management of hazardous waste is essential to minimize long-term risks to human and environmental health, but it was a challenge to develop specific and effective guidelines at the beginning of the hard lockdown in Italy for the multidisciplinary working group, involving public administration, waste management companies, universities, and health care competences activated by the Italian National Institute of Health (ISS).

The potential persistence of the SARS-CoV-2 on different surfaces and materials, including waste generated by (a) patients positive to the SARS-CoV-2 and treated at home, (b) subjects in preventive quarantine at home or in dedicated areas different from hospital or medical centers, was perceived as a risk and as a further possible route for virus spreading.

Beside citizens, particular concern was expressed by operators, trained volunteers, other subjects directly involved in waste disposal.

This situation gave rise the request of clear, sound, and adequate information about the proper procedures, practices, and precautions to adopt for avoiding this potential risk.

The group of experts at ISS provided specific guidelines to be adopted by all the subjects directly and indirectly involved in all the different phases of the waste management sector. The integrated approach gathering multidisciplinary expertise along with the involvement of main stakeholders resulted in a winning approach giving a pragmatic but scientifically sounded guidance document in the first emergency period.

The waste handling precaution and adaption varied in different countries, but a timely and dynamic measures are required to cope with this unprecedented situation. In Italy, asking for interrupting the separated waste collection for those patients confined at home or in quarantine was a difficult decision, but this was justified by the need to simplify the disposal procedure for the infected users and to avoid errors that may compromise the health of the waste operators: a risk benefit consideration that although "apparently" not environment-friendly can be considered as a measure to contain virus spreading and preventive for human health in an emergency situation.

References

ACR+. (2020). *Municipal waste management and COVID-19- Phase 1*. Retrieved June 6, 2020, from https://acrplus.org/en/municipal-waste-management-covid-19-phase1#international

Anderson, E. L., Turnham, P., Griffin, J. R., & Clarke, C. C. (2020). Consideration of the aerosol transmission for COVID-19 and public health. *Risk Analysis, 40*(5), 902–907. https://doi.org/10.1111/risa.13500

Biryukov, J., Boydston, J. A., Dunning, R. A., Yeager, J. J., Wood, S., Reese, A. L., Ferris, A., Miller, D., Weaver, W., Zeitouni, N. E., Phillips, A., Freeburger, D., Hooper, I., Ratnesar-Shumate, S., Yolitz, J., Krause, M., Williams, G., Dawson, D. G., Herzog, A., . . . Altamura, L. A. (2020). Increasing temperature and relative humidity accelerates inactivation of sars-cov-2 on surfaces. *mSphere, 5*. https://doi.org/10.1128/mSphere.00441-20

Carraturo, F., Del Giudice, C., Morelli, M., et al. (2020). Persistence of SARS-CoV-2 in the environment and COVID-19 transmission risk from environmental matrices and surfaces. *Environmental Pollution, 265*(Pt B), 115010. https://doi.org/10.1016/j.envpol.2020.115010

CDC (Centers of Disease Control and Prevention). (2020). *Coronavirus disease 2019 (COVID-2019)*. Retrieved May 3, 2020, from https://www.cdc.gov/coronavirus/2019-ncov/cases-updates/summary.html?CDC_AA_refVal=https%3A%2F%2Fwww.cdc.gov%2Fcoronavirus%2F2019-ncov%2Fsummary.html

Chia, P. Y., Coleman, K. K., Tan, Y. K., Ong, S. W. X., Gum, M., Lau, S. K., Lim, X. F., Lim, A. S., Sutjipto, S., Lee, P. H., Son, T. T., Young, B. E., Milton, D. K., Gray, G. C., Schuster, S., Barkham, T., De, P. P., Vasoo, S., Chan, M., . . . Marimuthu, K. (2020). Singapore 2019 Novel Coronavirus Outbreak Research Team. Detection of air and surface contamination by SARS-CoV-2 in hospital rooms of infected patients. *Nature Communications, 11*(1), 2800. https://doi.org/10.1038/s41467-020-16670-2

Chin, A. W. H., Chu, J. T. S., Perera, M. R. A., Hui, K. P. Y., Yen, H. L., Chan, M. C. W., Peiris, M., & Poon, L. L. M. (2020). Stability of SARS-CoV-2 in different environmental conditions. *The Lancet Microbe*. https://doi.org/10.1016/S2666-5247(20)30003-3

Colaneri, M., Seminari, E., Novati, S., et al. (2020). Severe acute respiratory syndrome coronavirus 2 RNA contamination of inanimate surfaces and virus viability in a health care emergency unit. *Clinical Microbiology and Infection, 26*(8), 1094.e1–1094.e5. https://doi.org/10.1016/j.cmi.2020.05.009

Cruvinel, V. R. N., Marques, C. P., Cardoso, V., Novaes, M. R. C. G., Araújo, W. N., Angulo-Tuesta, A., Escalda, P. M. F., Galato, D., Brito, P., & da Silva, E. N. (2019). Health conditions and occupational risks in a novel group: Waste pickers in the largest open garbage dump in Latin America. *BMC Public Health, 581*, 16–19.

de Wit, E., van Doremalen, N., Falzarano, D., & Munster, V. J. (2016). SARS and MERS: Recent insights into emerging coronaviruses. *Nature Reviews Microbiology, 14*, 523–534.

Di Maria, F. (2012). Upgrading of a mechanical biological treatment (MBT) plant with a solid anaerobic digestion batch: A real case study. *Waste Management & Research, 30*(10), 1089–1094.

Di Maria, F., Bidini, G., Lasagni, M., & Boncompagni, A. (2018). On time measurement of the efficiency of a waste-to-energy plant and evaluation of the associated uncertainty. *Applied Thermal Energy, 129*, 338–344.

Di Maria, F., & Micale, C. (2013). Impact of source segregation intensity of solid waste on fuel consumption and collection costs. *Waste Management, 33*, 2170–2176.

Di Maria, F., Sordi, A., & Micale, C. (2012). Energy production from mechanical biological treatment and composting plants exploiting solid anaerobic digestion batch: An Italian case study. *Energy Conversion and Management, 56*, 112–120.

Di Maria, F., Sisani, F., Contini, S., Ghosh, S. K., & Mersky, R. L. (2020). Is the policy of the European Union in waste management sustainable? An assessment of the Italian context. *Waste Management, 103*, 437–448.

Ding, S., & Liang, T. J. (2020). Is SARS-CoV-2 also an enteric pathogen with potential fecal-oral transmission: A COVID-19 virological and clinical review. *Gastroenterology, 159*(1), 53–61.

DPR. (2003). Decreto del Presidente della Repubblica 15 luglio 2003, n.254. Regolamento recante disciplina della gestione dei rifiuti sanitari a norma dell'articolo 24 della legge 31 luglio 2002, n. 179. (Regulation for the management of medical waste according to art. 24 of the legislation n. 179 of 31st July 2002). *Official Journal of Italian Republic* n. 211.

EC. (2020). *Waste management in the coronavirus crisis.* Retrieved May 3, 2020, from https://ec.europa.eu/info/sites/info/files/waste_management_guidance_dg-env.pdf

Faridi, S., Niazi, S., Sadeghi, K., Naddafi, K., Yavarian, J., Shamsipour, M., Jandaghi, N. Z. S., Sadeghiniiat, K., Nabizadeh, R., Yunesian, M., Momeniha, F., Mokamel, A., Hassanvand, M. S., & Mokhtari Azad, T. (2020). A field indoor air measurement of SARS-CoV-2 in the patient rooms of the largest hospital in Iran. *Science of the Total Environment, 725*, 138401.

Foladori, P., Cutrupi, F., Segata, N., Manara, S., Pinto, F., Malpei, F., Bruni, L., & La Rosa, G. (2020). SARS-CoV-2 from faeces to wastewater treatment: What do we know? A review. *Science of the Total Environment, 743*, 140444. https://doi.org/10.1016/j.scitotenv.2020.140444

Gomes Mol, M. P., & Caldas, S. (2020). Can the human coronavirus epidemic also spread through solid waste? *Waste Management and Research, 38*(5), 485–486.

ISS. (2020). *Gruppo di Lavoro ISS Ambiente e Rifiuti. Indicazioni ad interim per la gestione dei rifiuti urbani in relazione alla trasmissione dell'infezione da SARS-CoV-2.* Versione del 31Maggio 2020. Roma: Istituto Superiore di Sanità; 2020. (Rapporto ISS COVI-19, n.3/2020 Rev. 2). Available at: https://www.iss.it/documents/20126/0/Rapporto+ISS+COVID-19++3_2020+Rev2.pdf/4cbaa7b5-713f-da61-2cac-03e5d3d155b3?t=1591277298239

Gu, J., Han, B., & Wang, J. (2020). COVID-19: Gastrointestinal manifestations and potential fecal-oral transmission. *Gastroenterology, 158*, 1518–1519.

Heller, L., Mota, C. R., & Greco, D. B. (2020). COVID-19 faecal-oral transmission: Are we asking the right questions? *Science of the Total Environment, 729*, 138919.

Hindson, J. (2020). COVID-19: Faecal–oral transmission? *Nature Reviews Gastroenterology & Hepatology, 17*, 259.

Huang, C., Wang, Y., Li, X., Ren, L., Zhao, J., Hu, Y., Zhang, L., Fan, G., Xu, J., Gu, X., Cheng, Z., Yu, T., Xia, J., Wei, Y., Wu, W., Xie, X., Yin, W., Li, H., Liu, M., . . . Cao, B. (2020). Clinical features of patients infected with 2019 novel coronavirus in Wuhan, China. *Lancet, 395*, 497–506.

Kampf, G., Todt, D., Pfaender, S., & Steinmann, E. (2020). Persistence of coronaviruses on inanimate surfaces and their inactivation with biocidal agents. *Journal of Hospital Infection, 104*(3), 246–251.

Kenarkoohi, A., Noorimotlagh, Z., Falahi, S., Amarloei, A., Mirzaee, S. A., Pakzad, I., & Bastani, E. (2020). Hospital indoor air quality monitoring for the detection of SARS-CoV-2 (COVID-19) virus. *Science of the Total Environment, 748*, 141324. https://doi.org/10.1016/j.scitotenv.2020.141324

Lewis, D. (2020). Is the coronavirus airborne? Experts can't agree. *Nature, 580*, 175. https://doi.org/10.1038/d41586-020-00974-w

Liu, Y., Ning, Z., Chen, Y., et al. (2020). Aerodynamic analysis of SARS-CoV-2 in two Wuhan hospitals. *Nature, 582*, 557–560. https://doi.org/10.1038/s41586-020-2271-3

Marquès, M., & Domingo, J. L. (2020). Contamination of inert surfaces by SARS-CoV-2: Persistence, stability and infectivity. A review. *Environmental Research, 193*, 110559. https://doi.org/10.1016/j.envres.2020.110559

Mehraeen, E., Salehi, M. A., Behnezhad, F., Moghaddam, H. R., & SeyedAlinaghi, S. (2020). Transmission modes of COVID-19: A systematic review. *Infectious Disorders Drug Targets.* https://doi.org/10.2174/1871526520666201116095934

Mohapatra, S., Menon, N. G., Mohapatra, G., Pisharody, L., Pattnaik, A., Menon, N. G., Bhukya, P. L., Srivastava, M., Singh, M., Barman, M. K., Gin, K. Y., & Mukherji, S. (2020). The novel SARS-CoV-2 pandemic: Possible environmental transmission, detection, persistence and fate during wastewater and water treatment. *Science of the Total Environment, 6*, 142746. https://doi.org/10.1016/j.scitotenv.2020.142746

Morawska, L., & Cao, J. (2020). Airborne transmission of SARS-CoV-2: The world should face the reality. *Environment International, 2020*(139), 105730.

Morris, D. H., Yinda, K. C., Gamble, A., Rossine, F. W., Huang, Q., Bushmaker, T., Fischer, R. J., Matson, M. J., van Doremalen, N., Vikesland, P. J., Marr, L. C., Munster, V. J., & Lloyd-Smith, J. O. (2020). The effect of temperature and humidity on the stability of SARS-CoV-2 and other enveloped viruses. *bioRxiv*. https://doi.org/10.1101/2020.10.16.341883

National Research Council. (2020). *Rapid expert consultation on the possibility of bioaerosol spread of SARS-CoV-2 for the COVID-19 pandemic*. National Academies Press. https://doi.org/10.17226/25769

Nghiem, L. D., Morgan, B., Donner, E., & Short, M. D. (2020). The COVID-19 pandemic: Considerations for the waste and wastewater services sector. *Case studies in Chemical and Environmental Engineering, 1*, 100006.

Noorimotlagh, Z., Jaafarzadeh, N., Martínez, S. S., & Mirzaee, S. A. (2020). A systematic review of possible airborne transmission of the COVID-19 virus (SARS-CoV-2) in the indoor air environment. *Environmental Research, 9*, 110612. https://doi.org/10.1016/j.envres.2020.110612

Razzini, K., Castrica, M., Menchetti, L., Maggi, L., Negroni, L., Orfeo, N. V., Pizzoccheri, A., Stocco, M., Muttini, S., & Balzaretti, C. M. (2020). SARS-CoV-2 RNA detection in the air and on surfaces in the COVID-19 ward of a hospital in Milan, Italy. *Science of the Total Environment, 742*, 140540. https://doi.org/10.1016/j.scitotenv.2020.140540

Saawarn, B., & Hait, S. (2021). Occurrence, fate and removal of SARS-CoV-2 in wastewater: Current knowledge and future perspectives. *Journal of Environmental Chemical Engineering, 9*(1), 104870. https://doi.org/10.1016/j.jece.2020.104870

Santarpia, J. L., Rivera, D. N., Herrera, V. L., et al. (2020). Aerosol and surface contamination of SARS-CoV-2 observed in quarantine and isolation care. *Scientific Reports, 10*, 12732. https://doi.org/10.1038/s41598-020-69286-3

Tobias, A. (2020). Evaluation of the lockdown for the SARS-CoV-2 epidemic in Italy and Spain after one month follow up. *Science of the Total Environment, 725*, 138539.

van Doremalen, N., Bushmaker, T., Morris, D. H., Holbrook, M. G., Gamble, A., Williamson, B. N., Tamin, A., Harcourt, J. L., Thornburg, N. J., Gerber, S. I., Lloyd-Smith, M. O., de Wit, E., & Munster, V. J. (2020). Aerosol and surface stability of SARS-CoV-2 as compared with SARS-CoV-1. *The New England Journal of Medicine, 382*, 12.

Wang, W., Xu, Y., Gao, R., Lu, R., Han, K., Wu, G., & Tan, W. (2020). Detection of SARS-CoV-2 in different types of clinical specimens. *Journal of the American Medical Association, 323*(18), 1843–1844. https://doi.org/10.1001/jama.2020.1585

WHO. (2020a). *COVID-19 2020 Situation summary*. Updated 19 April 2020. Retrieved May 15, 2020, from https://www.cdc.gov/coronavirus/2019-ncov/cases-updates/summary.html#covid19-pandemic

WHO. (2020b). *COVID-19 weekly epidemiologic update*. Retrieved December 10, 2020, from https://www.who.int/publications/m/item/weekly-epidemiological-update-8-december-2020

WHO. (2020c). *Modes of transmission of virus causing COVID-19: Implications for IPC precaution recommendations*. Retrieved May 3, 2020, from https://www.who.int/publications-detail/modes-of-transmission-of-virus-causing-covid-19-implications-for-ipc-precaution-recommendations

WHO. (2020d). *Laboratory biosafety guidance related to coronavirus disease 2019 (COVID-19)*. Retrieved May 3, 2020, from https://apps.who.int/iris/bitstream/handle/10665/331138/WHO-WPE-GIH-2020.1-eng.pdf

Windfeld, E. S., & Brooks, M. S. L. (2016). Medical waste management – A review. *Journal of Environmental Management, 163*, 98–108.

Yu, H., Afshar-Mohajer, N., Theodore, A. D., Lednicky, J. A., Fan, Z. H., & Wu, C. Y. (2018). An efficient virus aerosol sampler enabled by adiabatic expansion. *Journal of Aerosol Science, 117*, 74–84. https://doi.org/10.1016/j.jaerosci.2018.01.001

Zhang, Y., Chen, C., Zhu, S., Shu, C., Wang, D., Song, J., Song, Y., Zhen, W., Feng, Z., Wu, G., Xu, J., & Xu, W. (2020). Isolation of 2019-nCoV from a stool specimen of a laboratory-confirmed case of the coronavirus disease 2019 (COVID-19). *China CDC Weekly, 2*(8), 123–124. https://doi.org/10.46234/ccdcw2020.033

Part IV
Health Care Waste Management During Pre- and Post-COVID-19 Scenario

Overview of Infectious Healthcare Waste Management in Thailand in Pre- and During COVID-19 Context

Pawan Kumar Srikanth, Wipatsaya Srimanoi, Prakriti Kashyap, and Chettiyappan Visvanathan

1 Introduction: Healthcare Facilities, Healthcare Coverage, and Medical Tourism in Thailand

1.1 Healthcare Facilities

Thailand, an emerging economy and an upper-middle-income country with a population of 68 million, ranks world sixth and Asia's first in the 2019 Global Health Security Index (GHS Index) of global health security capabilities in 195 countries. It is important to note that it is the only developing country in the list of world's top ten countries (GHS Index, 2019). Thailand started seriously investing in healthcare infrastructure five decades ago. This investment enabled a rapid expansion of the basic public health infrastructure nationwide in the last three decades (Thaiprayoon and Wibulpolprasert, 2017). In 2015, Thailand had 26 regional hospitals, 71 provincial hospitals, 734 district hospitals, 322 private hospitals, 9768 government health centers, 17,671 private clinics, and 11,154 pharmacies (Fig. 1) (Patcharanarumol et al., 2018). The healthcare system in Thailand is dominated by public health facilities. The major private hospitals are in urban areas.

The Ministry of Public Health (MoPH) is the major healthcare provider and owns about 60% of total healthcare facilities in urban areas and more than 95% in the rural areas. Private hospitals are located in Bangkok and other large urban areas and are regulated by the Medical Registration Division. The affluent urban population and foreign nationals are treated by the private sector with around 20% of overall health resources (Thaiprayoon and Wibulpolprasert, 2017). Other government units and

P. K. Srikanth · W. Srimanoi · C. Visvanathan (✉)
School of Environment, Resources and Development, Asian Institute of Technology, Pathum Thani, Thailand

P. Kashyap
Kathmandu, Nepal

© The Author(s), under exclusive license to Springer Nature Singapore Pte Ltd. 2022   121
S. K. Ghosh, P. Agamuthu (eds.), *Health Care Waste Management and COVID 19 Pandemic*, https://doi.org/10.1007/978-981-16-9336-6_6

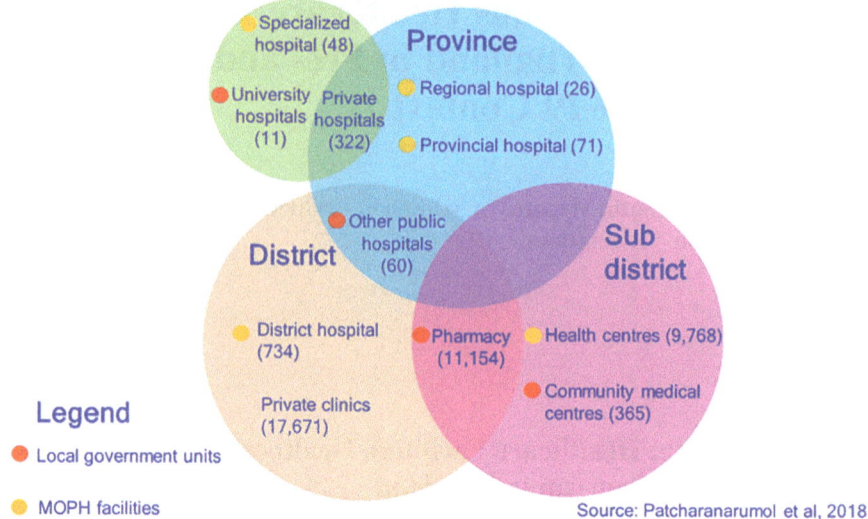

Fig. 1 Number of healthcare facilities in Thailand

public organizations also operate hospitals, including the military, universities, local governments, and the Red Cross.

1.2 Healthcare Coverage

Every Thai citizen is now entitled to essential preventive, curative, and palliative care for all age groups. Thailand is one of the few developing countries in the world that have successfully carried out the universal health coverage (UHC) to its citizens since 2002. There has been a good progress to UHC implementation despite political instability and underperforming economy (Sumriddetchkajorn et al., 2019). UHC has significantly contributed to the improvement of both the population's health and the overall health system of the country. However, it is also important to note that medical care in rural areas still lags far behind that in the cities.

Before 2002, Thailand's health coverage was a patchwork of arrangements for different population groups: the tax-financed civil servants' medical benefit scheme for public employees; the contributory social security scheme for private employees, the tax-financed medical welfare scheme for people in poverty, and the contributory voluntary health card scheme for households. The government's attempt to merge enabled the country to provide health coverage to the whole Thai population. The universal coverage scheme unified medical welfare and voluntary health card schemes. The healthcare system is empowered mainly by public health and complimented by private healthcare providers in urban and rural areas. In the decade after UHC was initiated (2001–2011), life expectancy at birth rose from 71.8 to

74.2 years with an increase of only 70.3–71.8 years during the decade before (1991–2001). UHC has significantly contributed to the improvement of both the population's health and the overall health system of the country.

1.3 Medical Tourism

According to the ASEAN Post Report, medical tourism generates at least one-third of private hospitals' revenue in most ASEAN countries (Thomas, 2019). The availability of the world-class healthcare facilities, trained doctors with Joint Commission International (JCI) accredited healthcare centers, which is considered the "Gold Standard" in healthcare and patient safety is the key to Southeast Asia's success in attracting medical tourists (Fig. 2). Cost-competitiveness and strong aviation network providing desired connectivity with other countries are other supporting factors.

Thailand tops the list of medical tourism destination with holding about 50% of the patient share, leaving Malaysia and Singapore on the second and third place, respectively (The ASEAN Post Team, 2017). Thailand is staying ahead of the game due to largest number of JCI accredited institutions and cost-completive advantage as compared to Malaysia and Singapore. Thailand is a preferred medical tourism destination for procedures like cosmetic or plastic surgeries, dental, cardiac, orthopedic, infertility treatment, bariatric surgery, ophthalmology, and eye surgeries.

In Thailand, medical tourism is largely concentrated in Bangkok, Phuket, and Chiang Mai. In 2019, Bangkok's Samitivej Hospital was fifth in the top-10 list of popular medical destination. Thailand's Bumrungrad International Hospital (BIH), one of the largest private hospitals in Southeast Asia, introduced a "Medication Tourism" program, serves over 1.1 million patients annually, including over 520,000 international patients (accounting for nearly a quarter of all medical tourists to the country) from over 190 countries. It was forecasted that in 2017, international patients will generate THB 48–49 billion income for private hospitals, representing a 3–4% growth from the previous year. Moreover, medical tourism is forecasted to support the growth of the tourism sector at around 16% per year during 2017–2020 (KPMG, 2018). This figure reflects the importance of the MT to the Thai economy (Kaewkitipong, 2018). From the study by Noree et al. (2016), out of 104,830 medical tourists, nearly 40% were from Eastern Mediterranean countries. Figure 3 shows the origins of medical tourists attending five hospitals in Thailand during 2010.

The prospering medical tourism in Thailand is a result of the Thai Institute of Small and Medium Enterprise Development (ISMED) initiation to develop a Medical Tourism cluster with financial support from the government in 2010. The key stakeholders were the Thai Hotels Association, the Private Hospital Association, the Association of Thailand Travel Agents and the Translators and Interpreters Association of Thailand. The main drivers are high healthcare costs and long waiting periods in the USA and several European countries made medical tourists seek

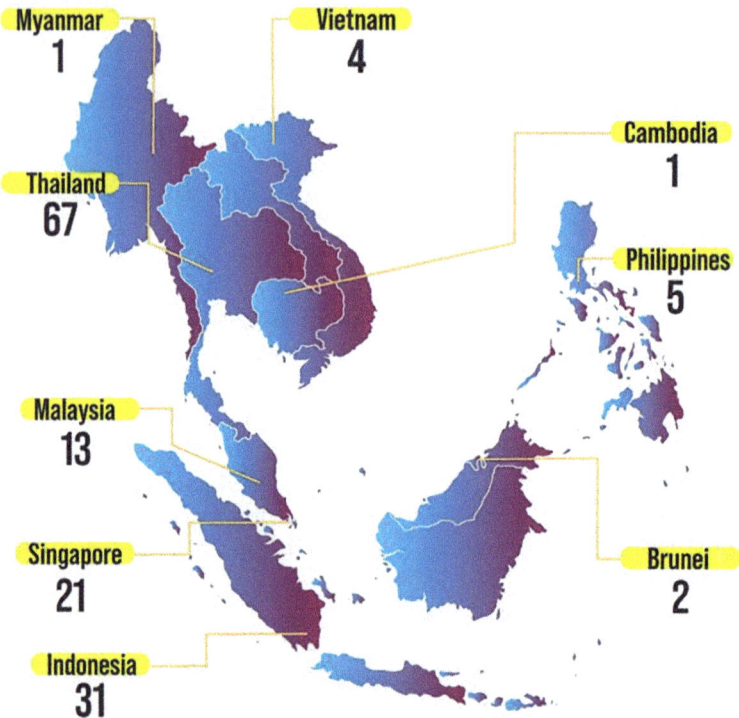

Fig. 2 JCI accredited healthcare facilities in ASEAN (Source: JCI, from the ASEAN Post)

care in Asia and Latin America. The credibility, quality, and higher standards of medical treatment are the critical factors which made Thailand as a favorable medical hub among the other Asian counterparts (Kaewkitipong, 2018).

2 Healthcare Waste (HCW): Definition, Institutional, and Policy Frameworks

2.1 Defining Healthcare Waste

HCW is the waste generated from the treatment, diagnosis, or immunization of humans and animals. This waste is comprised of two different categories, namely hazardous and non-hazardous. According to World Health Organization (WHO),

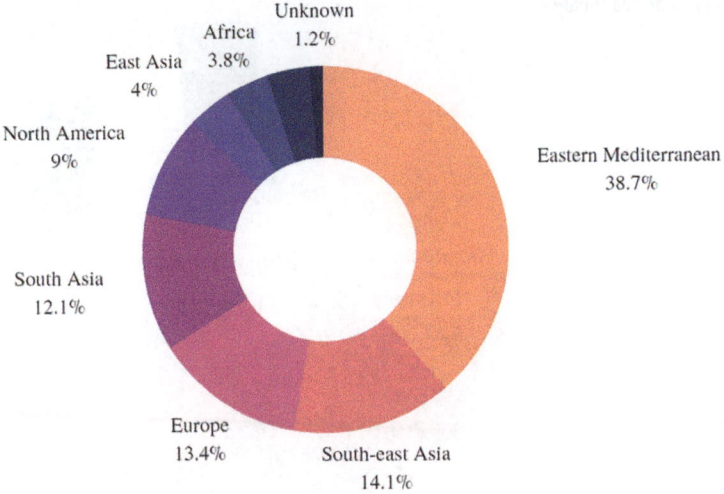

Fig. 3 Origins of medical tourists attending five hospitals in Thailand

75–90% of healthcare-related waste is non-hazardous, while the remaining percentage is considered as hazardous (Chartier et al., 2014). Hazardous waste is further categorized into infectious waste. Furthermore, infectious waste from healthcare institutions is categorized into sharps (scalpel blades and contaminated needles) and non-sharps (infectious waste contaminated with blood, body fluids, dressings, and intravenous (IV) lines, pathological waste, including microbiological cultures, blood samples, and anatomical body parts).

In Thailand, the Regulation of Ministry of Public Health B.E. 2545 (2002) defines infectious waste as "any waste that contains pathogens (e.g., bacteria, viruses, parasites, or fungi) in sufficient concentration and quantity to cause diseases in susceptible hosts." These are further described as (Fig. 4):

1. Body parts or carcasses of humans and animals generated from surgery, autopsies, and research.
2. Sharps such as needles, blades, syringes, vials, glassware, slides, and cover slides.
3. Discarded materials contaminated with blood, blood components, and body fluids from humans or animals, or discarded live and attenuated vaccines, such as cotton, other clothes, and syringes.
4. Discarded waste from wards.

The common sources of HCW in Thailand are healthcare facilities, veterinary clinics, health research centers, medical laboratories, clinics, polyclinics, government and private hospitals, educational institutions, Red Cross centers, detention centers, medical units, medical institutes, biotechnology units, home healthcare, medical manufacturing, and others produced during treatment, diagnosis, immunization of humans and/or animals. Various HCW are segregated using color-coded storage bins as shown in Table 1.

Fig. 4 Types of healthcare wastes

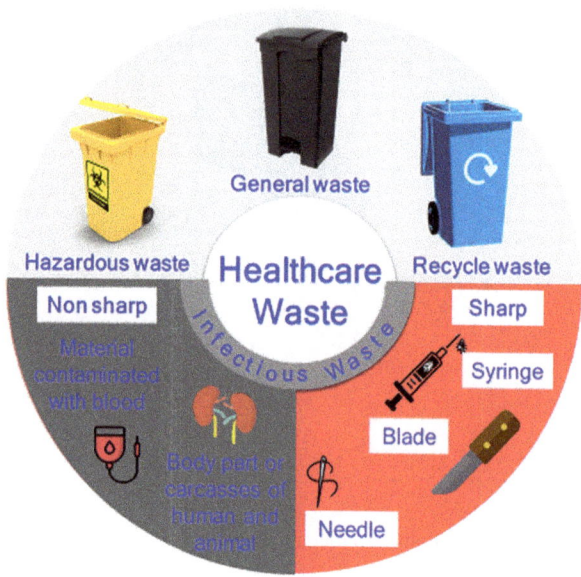

Table 1 Color code used for healthcare waste segregation

	Type of waste	Storage	Color Code
1	Infectious waste		
	Sharps	Bins	Red
	Non-sharps	Plastic bags	Red
2	Hazardous waste		Yellow
3	General wastes	Plastic bags	Black
4	Recyclable wastes	Bins	Blue

In Thailand, general waste accounts for 68% of the HCW generated, followed by 23% infectious waste (1% sharps and 22% non-sharps), 8% recyclables and hazardous waste representing only 1% of the HCW (3RKH, 2008).

2.2 Institutional Arrangements for HCW Management in Thailand

At the national level, MoPH is primarily responsible for framing legislations like the Public Health Act and the Public Health Ministerial Policy on Infectious Waste Management. The Department of Public Hygiene, of MoPH, oversees the disposal of hospital wastes from hospitals under the Ministry's jurisdiction and develops/procures required facilities. Technical matters on hospital waste management are dealt by the Pollution Control Department (PCD), DoH, and the Bangkok Metropolitan Authority (BMA). The PCD oversees the performance of all waste-related management facilities. The Ministry of Natural Resources and the Environment (MoNRE) and Ministry of Health develop the National Environmental Management Plan (NEMAP). The management of HCW is generally heavily regulated from the point of generation to disposal in Thailand (PCD, 2015).

As depicted in Fig. 5, every institution carries out its functions independently and have specific regulatory burden Institutions governing infectious waste management in Thailand adopt both compulsory and voluntary approach. The compulsory approach is administered by the MoPH, Public Health Committee and the Local authorities. Hospitals under the MoPH are to conduct their activities based on the ministry's policy, which mentions that healthcare facilities are to manage healthcare waste properly at source and are to comply with all regulations. Private healthcare facilities need to comply with the Healthcare Facility Act 1998, through a permit system in which solid waste management is one of the criteria set for the license. The Department of Medical Service under MoPH and the Healthcare Facility Committee takes care of the permit system. In the voluntary approach, each healthcare facility is to manage their waste based on the criteria set by specific systems, such as hospital accreditation system. The Institute of Hospital Quality Improvement and Accreditation sets up conditions for evaluation and assesses hospital accreditation. One of the criteria set for evaluation is that the management of healthcare waste shall comply with all regulations. To encourage a voluntary approach, the MoPH motivates hospitals to apply for accreditation and to manage all kinds of HCW through the Division of Public Cleansing Services (DPCS). The Local Government Agency (LGA) controls the collection, transfer, and disposal of waste. They must provide storage places for the transfer of waste and set appropriate disposal methods. They provide adequate personnel for the management of waste from collection to disposal. For every non-compliance, penalties apply to those who do not manage waste in line with the regulations.

2.3 Legal Frameworks for HCW Management in Thailand

The principal legislations governing infectious waste management in Thailand are the Public Health Act 1992, the Healthcare Facility Act 1998, and the Ministerial Rule on Disposal of Infectious Waste, 2002.

The Public Health Act 1992 and the Amendment 2007 specify that the collection, transportation, and disposal of waste within the area of any local government shall be the authority of such local government. In case of reasonable cause, the local government may assign to any person on its behalf under its control and supervision or may permit to carry out the activities of collecting, transporting, or disposing of healthcare waste.

The Ministerial Rule on Disposal of Infectious Waste, 2002, and the Public Health Act specify that healthcare facilities include all kinds of hospitals (for both human and animals), clinics, and laboratories. Every healthcare facility is responsible for the collection and storage of infectious waste within their facility in

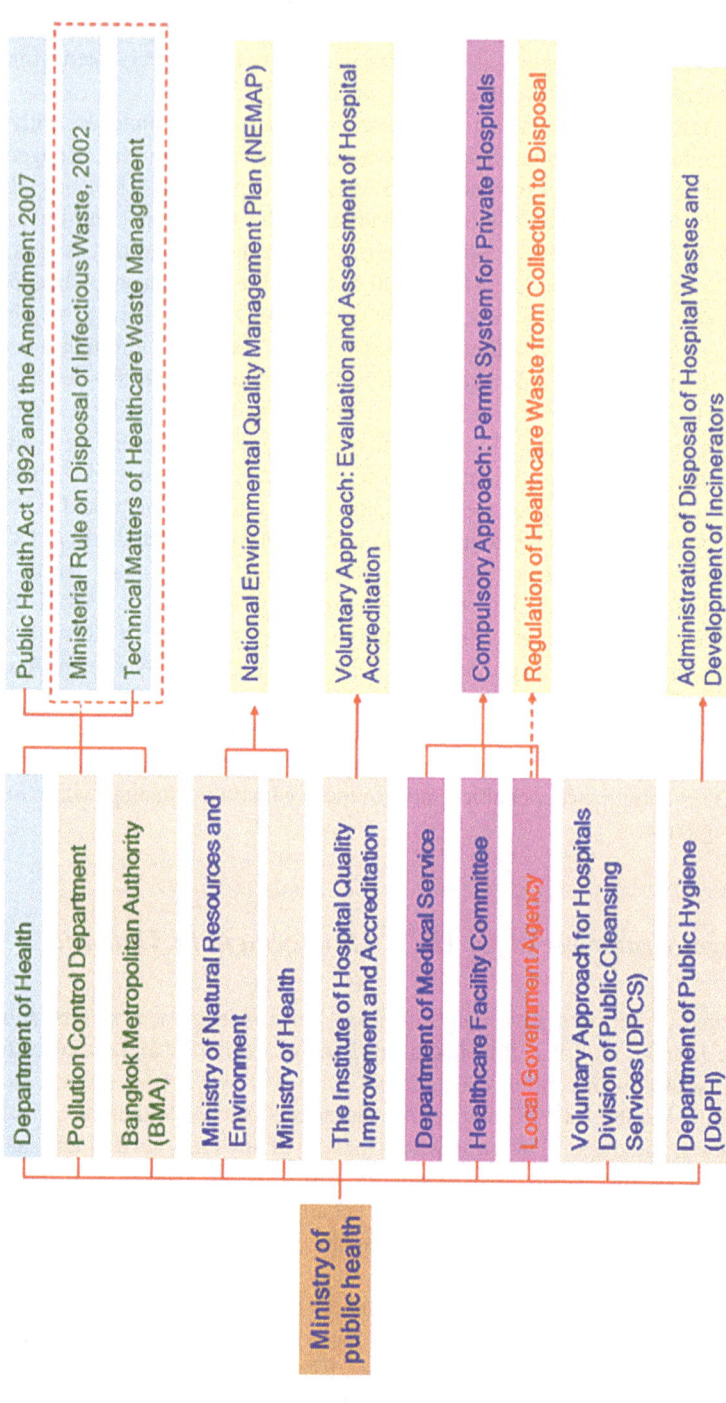

Fig. 5 Institutional arrangements and legal framework for HCW in Thailand

compliance with the Ministerial Rule. These include procedures for waste handling, the designation of the personnel responsible for waste management, provision of equipment and other facilities in collection, transportation, and disposal. The treated infectious waste is to be checked to ensure the destruction of germs before disposal. According to the Ministerial Rule, healthcare facilities are to be designated to the personnel for the collection, transportation, and disposal of infectious waste. These persons are to be trained in handling infectious waste at different stages of collection, transportation, storage, and disposal. Hospitals with their own treatment/disposal facility are to inform the local authority and are to operate their treatment facility. Private companies that conduct business on hospital infectious waste management are to apply for a license from the local authority.

The Healthcare Facilities Act 1998 specifies that the owners of the healthcare facilities have to apply for the license before the operation. Healthcare Facilities Committee assesses the facility based on criteria set by the Committee, among which the solid waste management system is one. Another plan relevant to the infectious waste management is the National Environmental Quality Management Plan (NEMAP). Solid waste management is one part of the NEMAP. The Plan encourages construction or utilization of central treatment plants.

3   Pre-COVID-19 Healthcare Waste Generation in Thailand

As the medical facilities grow in tandem with increase in population and demand, the amount of waste from operating the healthcare facilities also increases proportionally. It is therefore essential to establish an appropriate and efficient collection and treatment system to handle such magnitude of HCW.

The Thai institutions that deal with technical matters of medical waste generation include the PCD, DoH, and BMA. However, due to the differences in the definition of medical waste used by each institution, there are discrepancies in the basic data about hospital wastes, particularly concerning the types and generation rate of wastes. In 2000, PCD estimated the total quantity of infectious waste generated by both the private and public hospitals as 13,250 tons or 36.1 tons/day (Panyaping and Okwumabua, 2006). Out of 36.1 tons/day, Bangkok and the nearby areas generated about 24.1 tons. Predominantly, the bed capacity of the hospitals and the overall waste generation per year are the indicators to understand the magnitude of healthcare waste generated.

In 2012, the total number of public health facilities was more than 37,000, and the bed availability was around 140,000. The total number of public health facilities in Bangkok in 2012 was 2352, and bed availability was 28,143. Public health facilities are the major sources of infectious waste. Table 2 shows the infectious waste generation from 2012 till 2018 in Thailand.

Table 2 Infectious waste generation from 2012 to 2018

Year	Infectious waste generation (tons/year)	Remarks	Source
2012	42,000 28,000 tons/year (generated by governmental health facilities) 14,000 tons/year (generated by private healthcare facilities)	814 tons/month or 10,190 tons/year was generated from healthcare centers in Bangkok. The share of contribution of Bangkok healthcare facilities to the overall HCW generation has decreased from almost 50–25% due to improved healthcare facilities in other parts of the country while the Bangkok's net HCW generation increased by 10–15%	Pharino (2017)
2015	53,868	30,591 tons (57%) generated from public hospitals 9183 tons (17%) from private hospitals 10,349 tons (19%) from clinics 3431 tons (6%) from health-promoting centers and hospitals 311 tons (0.6%) from animal healthcare centers, and 3.1 tons (0.006%) from dangerous pathogen laboratories	Ministry of Natural Resources and Environment & Pollution Control Department (2019)
2016	55,646	Out of the total waste generated, nearly 49,056 tons (88%) of the wastes is disposed properly	
2017	57,954	Out of the total waste generated, nearly 51,300 tons (89%) of the wastes is disposed properly. Infectious waste management in large infirmaries has been improved but managing system from private medical clinics, veterinary clinics, and pet hospitals still needs to be implemented	
2018	55,497	The amount of wastes decreased from 57,954 tons in 2017 to 55,497.22 tons (4.2% decrease). 50% of the infectious waste was from public hospitals under the MoPH and 24% was from private hospitals and clinics. Nearly 49,897.86 tons (89.91%) of the wastes treated by infectious waste incinerators and autoclaving at the infirmaries	

4 Healthcare Waste Management Practices in Pre-COVID Times

4.1 Collection, Storage, and Transportation

In Thailand, there are four stages of infectious waste disposal in public hospitals. The MoPH determined the processes of HCW as (1) segregation and collection, (2) storage, (3) transportation, and (4) treatment and disposal. Both public and private hospitals in Thailand have established a clear guideline and implemented rigorous measures of HCW management on the collection, sorting, treatment, transport, up to the final disposal of waste materials as per the government's rules and regulations. All measures are to prevent infectious or toxic waste from contaminating the environment. The healthcare facilities have to be certain that the waste management company under its contract has been maintaining a high standard of practice if they do not have a treatment facility on their own. The storage rooms are ascertained to contain provisions for separate collection and containment of various types of wastes from healthcare facilities. The operation is overseen by personnel whose responsibility is to assure of the proper flow of hospital waste in and out of all rooms. A pollution prevention plan should be prepared, and hospital staff should be trained on the plan's implementation.

4.1.1 Segregation and Collection

HCW from public and private healthcare facilities is collected and segregated at the point of generation into different color-coded containers based on the type of waste. Collection and storage are the responsibility of the healthcare facility. Infectious waste must be segregated and collected in the place of generation well-specified containers for infectious waste storage, following the below listed procedures:

1. Containers must be visibly labeled as "Infectious Waste" and given a biohazard symbol. Red boxes and drums must be made of strong materials. They should be capable of resisting perforation or the erosion of chemical solutions to prevent fluid leakages. Red bags must be opaque as well as resistant to chemicals, laceration, leakages, and loading capacity.
2. All types of infectious waste excluding sharps must be packed in red bags but must not exceed two-thirds of the total volume.
3. Sharps must be packed in red boxes or drums. It must not exceed three-quarters of the total volume.

The handling of medical waste should be properly carried out within hospitals and sent offsite for disposal. Necessary infrastructure has to be set in place for the collection and transportation of the healthcare waste within and from each hospital to treatment facilities. The general practice in healthcare facilities includes segregation of wastes into hazardous, infectious, recyclable, and general wastes.

Fig. 6 Red bins for infectious waste

1. Hazardous Waste:

    The healthcare facilities have established a clear policy to minimize the use of hazardous wastes. Electric appliances containing hazardous substances such as light bulbs for use in the facilities, for example, are bought from a company that has agreed to take back for recycling in some private healthcare facilities. Another example, in some healthcare facilities, chemicals for photograph film processing were replaced with the Picture Archiving and Communication System (PACS).

    (a) General hazardous wastes such as used batteries and carbon papers are collected by special waste trucks operated by the Bangkok Metropolitan Authority or Provincial Authorities.
    (b) Toxic chemicals such as chemical solutions or expired chemicals are collected and treated by professional waste management companies that have been certified by the government.
    (c) Wastes with chemical contamination are put in separate bags and disposed of by outside waste management companies.
    (d) Wastes with radioactive contamination are segregated in a special container and kept in a lead-lined room. The radioactivity must be at a safe level for a human before the wastes can be transported out of the hospital facilities for disposal.
    (e) Radioactive wastes will be properly transported to the Office of Atoms for Peace for proper disposal in accordance with specified safety standards.
    (f) A manual for handling hazardous waste should be prepared to cope with this waste in hospitals. The small quantity of hazardous waste should be

accumulated until sufficient quantity is stored for transportation to the treatment and disposal facility that is located in the central part of Thailand, a long distance away, to avoid the high cost of disposal.

2. Infectious Waste:

There is a constant evolution of the definition of infectious waste in the private and public sectors to ascertain that it is sufficiently comprehensive to prevent any opportunities for infectious waste to leak to the outside environment. Special handling is required to dispose of infectious waste as sharp corners of infectious wastes can cut through disposal bags. All infectious wastes are moved to an outside professional waste management company for proper handling. Figure 6 shows the infectious waste bins used in a public health facility in Thailand.

3. General Waste:

General wastes are separated into dry and wet wastes. Attempts are made to separate recyclable materials from the waste as much as possible in many healthcare facilities. It is also important to note that a significant quantity of wastes such as PVC bags used for dialysis solution, which tend to be dumped in the infectious waste pile, can be processed into the material to material processing or energy recovery.

4.1.2 Storage and Training of Healthcare Waste Management Personnel

After segregating and collecting infectious waste, the next step is to transfer it to gathering or storage areas to wait for transfer for further disposal. The transfer of infectious waste is carried out as specified by law using infectious waste-containing trolleys and following specific routes to transfer infectious waste to gathering or storage areas. During the transfer of infectious waste, the waste-containing vehicles must not stop or pause anywhere. Infectious waste containers must not be thrown or dragged. Pliers or thick rubber gloves should be used to pick up dropped waste instead of picking up bare-handed. Carts or trolleys used for transferring infectious waste are made up of materials that are easy to clean and can be cleaned with water. They must have opaque floors and walls. When infectious waste containers are transferred to carts or trolleys, their lids must be tightly closed to prevent spillage. They should be labeled with "Only for the Transfer of Infectious Waste." Storage areas must have enough loading capacity, smooth floors, and walls; they must also be free of moisture, rails, or sewers connected to wastewater treatment systems. They must facilitate the transfer of infectious waste and be easy to clean. They must be visibly labeled with "Gathering or Storage Area for Infectious Waste Only." In the Bangkok Metropolitan area, the BMA provides services for the collection, transport, and disposal of healthcare wastes to both the public and private healthcare facilities. It is mandated that the workers who handle HCW have to undergo training programs involving detailed examinations. This is done to train them in the aspects of

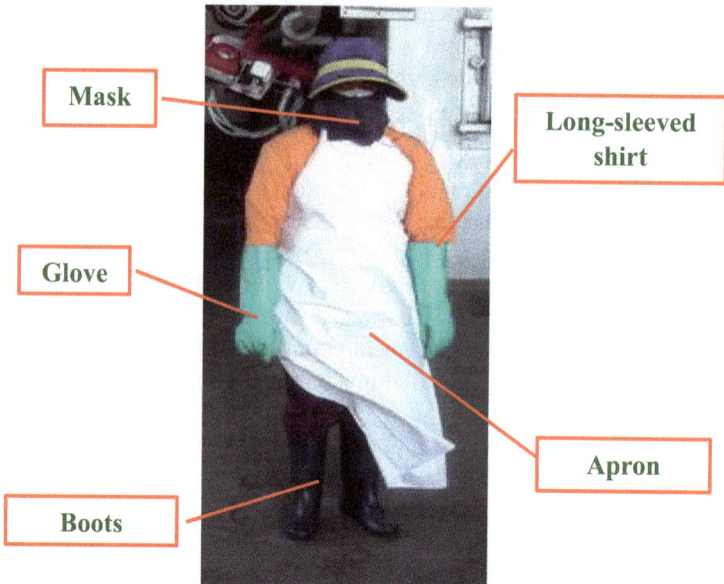

Fig. 7 Personal protective equipment for healthcare waste management personnel

prevention of harmful diseases and inhibition of outbreaks of communicable diseases caused by infectious wastes (Chartier et al., 2014).

Most hospitals are very conscious of the health of their workers. The frequency of training programs run by most hospitals ranged from one to four times per year. In some hospitals, it was as high as 12 times per year. Some hospitals provided incentives in the form of compensation and welfare to motivate their staff to improve infectious waste management. Most hospitals paid attention to each step of infectious waste management from the source of infectious waste generation to collection and transfer, storage, treatment, and disposal by setting up training programs. By doing so, it was observed that the overall efficiency of the healthcare waste management improved and frequency of damage or accidents from infectious sharps and wastes drastically reduced.

4.1.3   Personal Protective Equipment

Personal protective equipment (PPEs), gloves, particulate masks, boots for safety and soap and water, alcohol hand rub for hygiene (Fig. 7) are provided to personnel handling HCW. The most effective PPE in reducing the risk of injury are gloves to protect from exposure to blood and potentially infectious chemicals. The particulate masks (respirators) are to protect from respiratory infections, hazards, and particulates from burning waste, and boots for waste handlers to protect from sharp injuries to the foot. Availability and access to soap and water, and alcohol hand rub, for hand

Fig. 8 Four-ton truck for HCW transportation

hygiene, are also necessary to maintain cleanliness and inhibit the transfer of infection via dirty hands.

4.1.4 Transportation

Transportation refers to the transfer of infectious waste from gathering or storage areas to disposal facilities by infectious waste-containing vehicles which have a controlled temperature of not more than 10 °C. A 4-m^3 capacity 4-ton truck (Fig. 8) with a controlled temperature below 15 °C is used for HCW transportation. Drivers and workers must undergo also training programs and take exams in the prevention and inhibition of outbreaks of harmful diseases caused by infectious waste. On average, in 2005, BMA collected and transported 8.5 tons of healthcare wastes every day (0.11 kg/bed/day) from 582 hospitals, public health service centers, and healthcare institutions. The waste chutes are not recommended as they can increase the risk of airborne infections.

4.1.5 Transport Trolleys

Healthcare waste is generally bulky and heavy. They are transported using wheeled trolleys or carts that are not used for any other purpose. To prevent injuries and infection transmission, trolleys and carts are designed in such a way that it becomes easy to load and unload, easy to clean, labeled, and dedicated to a particular waste type. They have no sharp edges that could damage waste bags or containers during loading and unloading. In private and public health facilities in Thailand, trolleys are labeled and dedicated to a particular waste type and secured with a lock (for hazardous waste). The size of the container varies according to the volumes of waste generated at a healthcare facility. Hazardous waste should not be transported by hand due to the risk of accident or injury from infectious material or incorrectly disposed sharps that may protrude from a container. Spare trolleys are made available in case of breakdowns and maintenance. It is mandated that the vehicles should be cleaned and disinfected daily.

4.2 Treatment and Disposal of HCW

4.2.1 Healthcare Waste Management Methods and Treatment Technologies

Different management methods are followed by healthcare centers in managing healthcare waste generated. Table 3 describes these management methods.

Based on the physicochemical composition of HCW, suitable equipment specifications or operating parameters for treatment technologies are chosen. For example, some steam and microwave treatment systems rely on a minimum amount of moisture to be present in the waste. Some chemical systems are affected by the organic load and water content. Incineration is influenced by the percentage of incombustible (ash), heating (calorific) value, and moisture content of waste. After the disposal of infectious waste by these technologies, there needs to be an examination to see whether infectious waste successfully eliminated pathogens as specified in biological standards and regulations. Physical properties, such as bulk density (uncompacted mass per unit volume), are used to estimate storage, transport, and treatment chamber capacities, as well as specifications for compactors, shredders, and other size-reduction equipment. Common to any waste classification, the physicochemical characteristics of healthcare waste will vary from country to

Table 3 Different management methods for HCW

S. No.	Management method	Details
1.	Onsite hospitals	Some hospitals treat infectious waste using their own incinerators
2.	Local administrative organization	Some hospitals and public health facilities that have no incinerators transport infectious waste to incinerators of the local administrative organization
3.	Other hospitals	Some hospitals and public health facilities that have no incinerators transport their infectious waste to incinerators of other hospitals
4.	Private sector disposal service	Some hospitals and some public health facilities use the private sector for collection, transportation, treatment, and disposal of their infectious waste

Table 4 HCW treatment processes adopted by Thailand

S. No.	Type	Details
1.	Mechanical process	Changes the physical form of waste to facilitate waste handling. It consists of compaction and shredding. Compaction involves compressing the waste into containers to reduce its volume. Shredding is used to break waste into smaller pieces
2	Thermal process	Uses heat at low temperature (150 °C) and high temperature (600–5500 °C) to decontaminate infectious waste. Thermal processes include autoclaving and incineration. (a) Incineration uses high-temperature (800–1050 °C) combustion under controlled conditions to convert waste containing infectious and pathogenic materials to inert material residues and gases. It results in significant volume and weight reduction, and it sterilizes the waste. (b) Autoclaving is a steam sterilization technique that uses steam to directly disinfect waste. Steam under pressure is used to obtain a temperature of at least 121 °C
3	Chemical process	Uses chemicals (e.g., ozone [gas], chlorine, formaldehyde, ethylene oxide [gas], propylene oxide [gas], and peracetic acid) for disinfection. The effectiveness of each chemical agent depends on temperature, pH, and the presence of compounds that can interfere with disinfection
4	Irradiation process	Uses ultraviolet or ionizing radiation for irradiating and sterilizing infectious waste. This method includes microwave irradiation. Microwave irradiation is designed to use frequencies in the electromagnetic radiation spectrum between 300 and 300,000 MHz to inactivate microbial organisms

country. As depicted in Table 4, HCW in Thailand are managed using mechanical, thermal, chemical, and irradiation process.

Within 30 days of collection and transportation, infectious waste must be disposed scientifically. Monitoring and operating reports should be submitted monthly to the local government. Among all these different processes, thermal processes (incinerators and autoclaves) are the most commonly used technologies for HCW treatment and disposal in Thailand. Incineration and autoclave combinedly contribute about 80–90% of the overall treatment of HCW generated in Thailand. In 2018, the amount of infectious waste that was disposed of by waste incinerators was 35,014 tons or 65%, which included approximately 2493 tons disposed of by hospital waste incinerators, 14,926 tons were disposed of by ten disposal facilities of local authorities and 17,595 tons disposed of by seven private disposal sites (Ministry of Natural Resources and Environment & Pollution Control Department, 2019).

The Department of Public Hygiene, Ministry of Public Health, which oversees the disposal of wastes from hospitals under the Ministry's jurisdiction, has developed standard incinerator designs of 25, 50, and 100–150 kg/h. capacities. As of 2005, the MoPH had installed incinerators at all large hospitals, covering 87% of the total hospital beds under its jurisdiction (3R Knowledge Hub, 2008). According to

Table 5 Incinerators in different provinces that comply with the standard criteria required by DoH

	Hat Yai	Nonthaburi	Samut Sakorn	Phuket	Khon kean	Suphanburi	Akkhie Prakarn Public company limited
Size of the incinerator (tons/day)	5	5	5	400	1.2	100 (kg/h)	48–75
Daily storage capacity	1.5–1.6 tons	0.3 tons	100 kg	400 kg	0.7 tons	0.4 tons	N/A
Frequency of incineration	5–6 time/week	1–2 time/week	1 in 2 weeks	Everyday	N/A	1 time/week	N/A
Amount of waste being incinerated in a week (tons)	2	0.3–0.6	1.5	N/A	N/A	0.7	400–500 kg/h

Source: MONRE, PCD (2019)

DoH, 750 medical incinerators have been installed in hospitals under the jurisdiction of the MoPH, Thailand. About 30% of the hospitals under MoPH have onsite incinerators, 50% disposes through private companies, and the rest 20% manages by local authorities. Table 5 lists the incineration facilities in different provinces (Ministry of Natural Resources and Environment & Pollution Control Department, 2019). Small incinerators used can dispose 25, 50, and 100–150 kg/h. Temperature required is about 700 °C. However, this type of incinerator has not installed dust and air pollution prevention. Hence, only 41% operates well, 45% perform inadequately, and 12% could not be used. The cost of disposing infectious waste in general hospital is 9700 baht/month/hospital while the cost of disposing infectious waste in central hospital is approximately 17,400 baht/month/incinerator.

The BMA operates two incinerators of capacity 10 tons/day for the disposal of healthcare wastes in the Bangkok area. The basic treatment of the wastes is done before transporting them to municipalities and local sanitation administrations for those regional healthcare institutions that do not have waste disposal systems of their own.

5 Issues Related to Healthcare Waste Management

Even though Thailand is an outstanding example in multiple aspects of health security and infectious waste management strategies, the countries' HCW management issues are not fully addressed. Most of the incinerators for infectious waste were not equipped with an appropriate air quality assessment system. High-tech incinerators take care of the final treatment and disposal, constituting only a part of the HCW management chain. Large part of the HCW chain is relatively ignored. Holistic management of HCW through compositional analysis of the HCW fractions, separating the recyclables, and finally treating and disposing the HCW with the right treatment method are essential. Turning all potential recyclables to ashes may not be the best management strategy. Thailand's healthcare industry should adopt a lifecycle approach to healthcare plastic waste management, aiming for recovery, reuse, and recycling plastic products and packaging and promoting circularity in healthcare plastics, medical devices, and medical device packaging applications.

Another loophole in Thai HCW management is the improper segregation resulting a potential mix with municipal solid waste and improper disposals. Out of the total HCW generated in 2015, approximately 18,854 tons or 35% of infectious waste was missing from the system and some of this was disposed of by hospitals themselves. This is due to the small-scale healthcare service providers and their limitations with respect to their collection, transportation and disposal, onsite collection and disposal. Therefore, infectious waste from some of the small-scale healthcare service providers was transported and disposed of at a public hospital within the network, and some may have been unsuitably disposed of with municipal waste. The illegal medical practitioners and beauty clinics in Thailand contribute a minor quantity of medical wastes which get discarded in the environment. Figure 9 is one of many examples of improper disposal of infectious waste posted in Thai Facebook Page named Mor Lab Panda (หมอแล็บแพนด้า) @MTlikesara on February 16, 2020.

Pharino (2017) undertook a case study into the situation of infectious waste management in hospitals in Bangkok. Most hospitals use special and general rigid plastic containers which are strong and resistant to laceration and perforation by sharp objects for collection and disposal. These containers are intended to help prevent or at least reduce accidents to workers from infectious sharp accidents when collecting and transferring infectious sharps and waste for treatment and disposal. It was observed that most of the hospitals had very low or in some cases no infectious sharp accidents. It was inferred that strict following of rules in this aspect is the significant factor. The study also carried out a detailed survey across 65 different health facilities in Bangkok regarding the accidents to waste management workers from infectious diseases, and it was found no such cases were reported. In Thailand, only the storage aspect of HCW management is partially complied with WHO standards while the other aspects such as segregation, transport, treatment, and disposal are fully complied in accordance with WHO standards. In some hospitals, problems regarding general waste mixed with infectious waste are commonly observed. This increases the amount of infectious waste sent to disposal companies and causes an increase in costs for collection, transfer, treatment, and disposal. There is not sufficient information available regarding the involvement of informal sector in the management of healthcare wastes in Thailand.

The mismanaged HCW is majorly from clinics even though the amount of infectious waste in clinics is less significant than from other hospitals. In addition

Fig. 9 Improper disposal of HCW near a beach in Tambon Chong Samae San, Chon Buri, Sattahip district, Thailand (Thailand Today Blog, 2020)

to that, it is found that there is a need to control the improper medical waste disposal and offsite waste transportation to the final disposal destination (Akkajit et al., 2020).

6 Impacts of COVID-19 Pandemic on Thai Healthcare Management in Thailand

COVID-19 caused by SARS-CoV-2 (coronavirus) is probably one among the deadliest pandemics the world has faced in a century. Like the whole world, COVID-19 has impacted social-economic situation of the country and have added burden to country's healthcare system. Healthcare waste generation and its proper management is definitely one of the critical areas that COVID-19 has left the country to deal with.

On January 13, 2020, the Ministry of Public Health (MoPH), Thailand, reported the first imported case of lab-confirmed novel coronavirus (2019-nCoV) from Wuhan, Hubei Province, China. After declaring COVID-19 as a 14th dangerous communicable disease under the Disease Communicable Act B.E.2558 on February 26, 2020, any hospital that detected a case would have to admit the case and report the case admission to the Ministry of Public Health for its records. According to the MoPH record as of November 21, 2020, there are 3902 confirmed COVID-19 cases and 60 confirmed deaths. With better healthcare facilities and both mandatory and voluntary measures implemented to contain the outbreak, including the cancelation of public gatherings, remote working measures, the closure of entertainment and sport venues, and social campaigns to "stay home, stop the virus, save the nation," Thailand has been successful to contain the COVID-19 cases to one of the lowest. Recognizing that the demand for laboratory services would increase drastically, the MoPH began preparing surge capacity for COVID-19 diagnostic capabilities. As of July 8, 2020, 205 COVID-19 laboratories were established: 79 laboratories

(35 government and 44 private labs) in Bangkok and 126 laboratories (102 government and 24 private labs) across the country. There was also no lack of health facility and health professional shortage. During the COVID-19 outbreak, healthcare facilities have been repurposed to accommodate COVID-19 patients while maintaining the other essential services (Sirilak, 2020). WHO praised Thailand as one of the model countries that have had a good handle on the spread of COVID-19. Bloomberg ranked Thailand in fourth place among countries that have dealt with the COVID-19 outbreak very well. Thailand also ranks first in the Global COVID-19 Recovery Index (PR Thai Government, 2020). However, the COVID-19 cases surged to over 3000 per day in May–June 2021 ("Thailand Economic Monitor July 2021: The Road to Recovery," 2021). Stricter containments were imposed that reduced mobility and negatively affected consumption and business sentiment.

In fact, Thailand's track record in successfully handling epidemics in the past has been exemplary. Thailand through quality and robust healthcare facilities, promptness, and commitment showed success in stopping outbreaks in the past including severe acute respiratory syndrome (SARS) in 2003, 2009 H1N1 influenza Middle East Respiratory Syndrome (MERS) in 2012. These epidemics in the past have taught some lessons that it is completely necessary to maintain a high level of vigilance for early case detection, develop healthcare capacities and strengthen infection control systems, as well as practice the best possible healthcare waste management practices to further limit the disease transmission and to protect healthcare workers as well as general public.

The mode of transmission of coronavirus is the significant factor to determine COVID-19-related healthcare waste. COVID-19 impacts on HCW are to be seen from two perspectives: (1) *lowered quantity of HCW* due to postponement of non-emergency medical operations and other treatments, people avoiding to go to hospitals for non-emergency cases, and (2) *increased quantity of HCW* due to use of disposal PPEs, and as COVID-19 spread by contact, even general waste from medical areas may potentially have been classified as infectious healthcare waste.

Disposal plastic PPEs usage on the healthcare systems are probably the additional burden to HCW during COVID-19. Based on WHO modeling, an estimated 89 million medical masks are required for the COVID-19 response each month. For examination gloves, the figure goes up to 76 million, while international demand for goggles stands at 1.6 million per month. To meet such demand, WHO requested a 40% escalation of disposable PPE production (WHO, 2020a). Another estimate from WWF shows that even if just 1% of the masks were disposed of incorrectly and dispersed in nature, this would result in as many as ten million masks per month polluting the environment (WWF Italy, 2020).

Estimates show that there has been an increase in per capita HCW all over the world. During the COVID-19 outbreak in Hubei Province, People's Republic of China (PRC), infectious medical waste increased by 600% from 40 tons per day to 240 tons per day at the peak of the pandemic (February to March 15, 2020), nearly six times more than before the pandemic. Such surge in medical waste overwhelmed the existing medical transport and disposal infrastructure around hospitals (Jiangtao & Zheng, 2020). In order to manage the COVID-19 waste, the local authority

Table 6 Estimated additional amount of HCW in each city due to the COVID-19 pandemic

City	Population (world population review)	HCW generated before COVID-19 (tons/day)	Additional HCW during COVID-19 (tons/day)	Percentage of increase due to COVID-19
Manila	14 million	47	280	496
Jakarta	10.6 million	35	212	506
Kuala Lumpur	7.7 million	26	154	492
Bangkok	10.5 million	35	210	500
Ha Noi	8 million	27	160	493

Source: ADB (2020) and UNEP (2020)

decided to involve four companies specialized in solid waste management including Gient, to build a 30 tons/day capacity emergency treatment plant by February 22 to treat around 25% of total medical wastes generated in the city during the COVID-19 pandemic (Wei, 2020). Based on this PRC experience, the Asian Development Bank estimated probable volumes of infectious waste in major cities, including Bangkok (Table 6).

It is astonishing to note that there is an estimated 500% increase in the HCW generation per day during COVID-19 for Bangkok. From 35 tons/day to 210 tons/day increase, the burden of dealing with excess waste is insurmountable. Another estimate from UNEP 2020 shows that approximately 2.85 kg/bed/day of COVID-19 healthcare waste is being generated in Bangkok. Waste generated from healthcare facilities include (1) cohort ward hospital, (2) state quarantine, and (3) laboratory analysis with the quantity of 7.50 kg/person/day, 1.19 kg/person/day, and 0.08 kg/person/day, respectively.

However, the infectious waste in Thailand (Fig. 10) has almost remained at the same level during pre-COVID period at an average of 3850 tons/month (between October 2019 and January 2020). During COVID-19, infectious waste has decreased in Thailand to an average of 3500 tons/month (between February 2020 and May 2020). As COVID-19 became the priority, people avoided going to the hospitals and non-emergency medical operations, and other treatments were postponed during COVID lockdown period.

Not only the varying quantity of HCW waste due to COVID-19, safe collection, disinfection, and transportation were other challenging issues. The risk of virus transmission badly interrupted regular recycling activities of municipal waste as well as general and recyclable HCW.

To deal with HCW during COVID-19, WHO has released an interim guidance document entitled "Water, sanitation, hygiene, and waste management for the COVID-19 virus." It calls for following the best practices for safely managing healthcare waste, including assigning responsibility and sufficient human and material resources to segregate, recycle, and dispose of waste safely. The guidance further elaborates the collection and safe disposal of infectious waste produced during patient care, including those with confirmed COVID-19 infection (e.g., sharps,

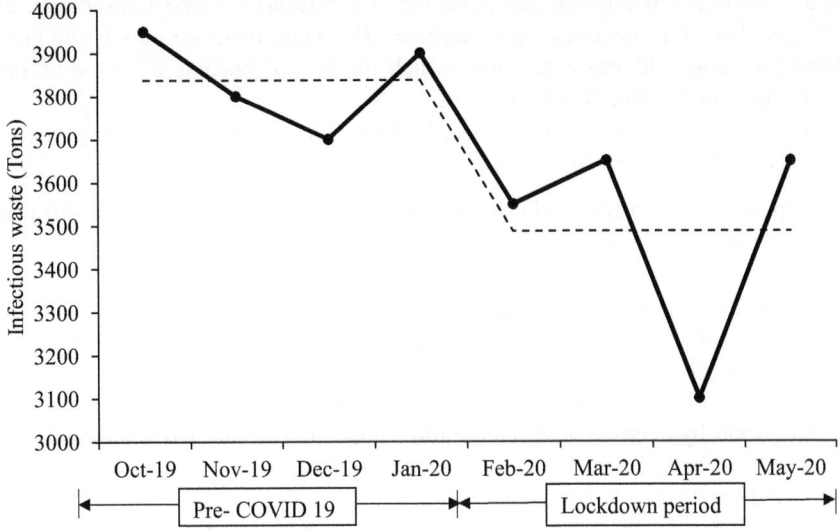

Source: Kongkratoke, S. (2020)

Fig. 10 Trend of infectious waste in Thailand pre- and during COVID-19. (Source: Kongkratoke, 2020)

bandages, pathological waste). These infectious wastes should be collected safely in clearly marked lined containers and sharp boxes, and this waste should be treated, preferably on-site, and then safely disposed. Preferred treatment options are high temperature, dual chamber incineration or autoclaving. If waste is moved off-site, it is critical to understand where and how it will be treated and disposed. Waste generated in waiting areas of health-care facilities can be classified as non-hazardous and should be disposed in strong black bags and closed completely before collection and disposal by municipal waste services. If such municipal waste services are not available, as interim measure, safely burying or controlled burning may be done until more sustainable and environmentally friendly measures can be put in place. All those who handle healthcare waste should wear appropriate PPE (long-sleeved gown, heavy-duty gloves, boots, mask, and goggles or a face shield) and perform hand hygiene after removing it. Additional waste treatment capacity, preferably through alternative treatment technologies, such as autoclaving or high temperature burn incinerators, may need to be procured and systems may need to be put in place to ensure their sustained operation. As an interim measure safely burying health care waste may be done until more sustainable measures can be put in place. Manual chemical disinfection of waste is not recommended, as it is not regarded as a reliable and efficient method (WHO, 2020a, 2020b).

Thailand has also prepared a three-phase contingency plan for infectious waste disposal—for normal, moderate, and severe conditions. Thailand has released three major guidelines for infectious waste disposal: (1) guidelines for healthcare facilities and hazardous laboratories, (2) guidelines for infectious waste transporters, and (3) guidelines for infectious waste agencies. However, there are no specific guidelines for household waste disposal, which might also become a source of virus transmission (Kaigate, 2020).

During the COVID-19 pandemic, Thailand practiced the following steps to manage HCW (UNEP, 2020):

- Separate into two types: (1) sharp items; (2) non-sharp items (COVID-19 waste under the non-sharp items).
- Disinfect and double bags.
- Designate a specific storage area.
- Send waste from community healthcare facilities to district healthcare facilities once a week.
- Temperature-controlled storage available at district level.
- Transport by licensed WMSPs (require temperature-controlled vehicle).
- Treat within 48 h after being transported.
- Disinfect vehicles and bins daily with NaClO.

In 2020, 47,962 tons of wastes caused by hospitals under the Ministry of Public Health, hospital under the Department of Academic Affairs under the Ministry of Public Health, Sub-district Health promotion hospital, hospitals affiliated with other ministries, private hospital, private clinic, animal hospital, and a dangerous infection laboratory. This is 10% less in comparison to the wastes produced in 2019. Out of 17.89 tons of used facemasks produced during 2020, about 10–15% of total quantity of masks is disposed by hired private company for safe disposal. Pollution Control Department has prepared *preliminary introduction to the management for handling used masks* to guide the epidemic outbreak via local authorities to the general public. In addition, a survey of the number of used masks was conducted by coordinating with the local authorities to report the number of used masks using QR code-based application which is reported monthly. About 8–9 tons of the face masks were collected by disinfected, burned, and sent to public health unit of Sub-District Administrative Organization for safe disposal (MoNRE, PCD, 2021).

For infectious COVID-19 waste management in other than healthcare facilities, MoPH also resealed a Guidance for Integrated Management of State Quarantine Facilities Ministry of Public Health on June 3, 2020 (MoPH, 2020). The Guidance for Operations and Management of State Quarantine Facilities lists an "Area for waste management purposes" as Components of the State-run quarantine facilities. The Quarantine site environment and sanitation management task group are also assigned.

1. Scope of responsibilities include waste management, monitoring, and counseling on sanitation management.
2. Operational procedure: Hotel should provide a completely enclosed, environmentally friendly temporary waste storage area; work with waste management unit of each district or BMA to request support with respect to waste collection

and disposal in the same manner as waste management for medical clinics in Bangkok area as per details below.

(a) Cleaning schedule for guest rooms and communal spaces is communicated to quarantined travelers.
(b) Clean the areas where there is frequent people traffic and frequently touched surfaces by mopping and scrubbing. Use of spray disinfectants is not recommended as spraying of disinfectants, if not done properly and without appropriate personal protective equipment (PPE), will put cleaning staff at risk for infection.
(c) In guest rooms, quarantined travelers will be asked to sort waste materials by themselves using opaque, chemical-resistant, durable, waterproof, and leak-proof garbage bags and no more than two-thirds of its capacity should be used. Garbage bags must be tied using rubber bands which will be made available to quarantined travelers before collection and disposal.

Waste materials should be sorted and disposed of as follows:

  (i) Potentially infectious waste (e.g., used tissue paper, sanitary napkins, spoon and fork, leftover food, and its container) should be contained in a red garbage bag.
 (ii) Other general, dry waste (e.g., snack bags, paper, coffee packets) should be discarded in a black garbage bag.
(iii) Collect and move waste material every day using a trash cart and using the path designated separately from other areas. Always clean the cart after use.
 (iv) As for black garbage bags, they should be dried in the sun and kept at the hotel's existing waste storage (of each hotel) and handled as per standard procedures.
  (v) Red garbage bags should be kept in the area designated specifically for infectious waste pending collection and disposal in the same manner as those from health facilities.

  Thailand has also shown digital innovation in tracking both COVID-19 patients and COVID-19 waste.

The DDC Care Application The Department of Disease Control has developed "the DDC Care" application to monitor and track patients under investigation. The "Thai Chana" web application was also developed by the Ministry of Digital Economy and Society to record population movement data for the benefit of contact tracing among risk groups and bringing them into the disease surveillance and investigation process (Sirilak, 2020).

D-Germ Application The D-Germ application was developed by the Economy for Waste free Thailand (CEWT) Research Center at Mae Fah Luang University. This mobile application development was inspired by the necessity to trace the COVID-19 wastes in Thailand. D-Germ is a mobile application-based innovative healthcare

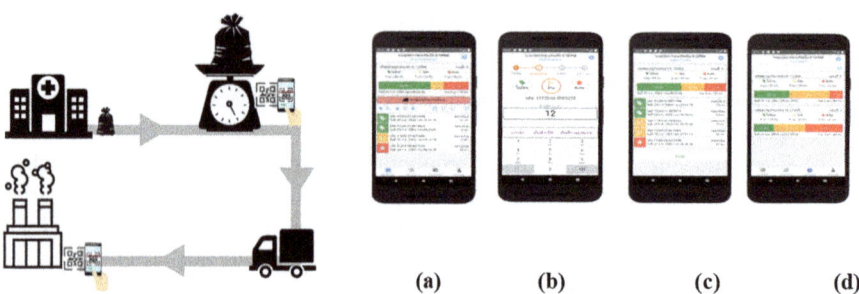

(a)   (b)   (c)   (d)

Fig. 11 Schematic diagram and the steps for using D-Germ application

waste management tool, which records and traces the amount of healthcare waste produced from the hospital to final treatment and disposal site.

The procedure for the D-Germ application (Fig. 11) is to (a) is the press to begin, (b) record the weight and type of waste, (c) confirm the infectious waste transportation, and (d) record the report on type and weight of infectious waste. This tool assists in monitoring and preventing the COVID-19 outbreak caused by the mismanagement of healthcare waste. Currently, this D-Germ application is on trial run. It involves tracking volume of healthcare waste throughout the process and location of medical waste to ensure that medical waste is safely collected and disposed. The mechanism is mainly the QR code which records the weight of the waste at the hospital, before transportation, and when it reaches the incineration facility (Kongkratoke, 2020; Buddhasri, 2020).

Thailand has shown a remarkable efficiency in both containing the COVID-19 confirmed cases to one of the lowest and management of HCW. However, learning from this pandemic, it is advised that hospitals should be prepared for potential post-COVID waste such as COVID-19 vaccine, viles, and injections. Also, waste management staffs should be included in the lists of essential workers and should be trained to handle waste management during a pandemic. It is also advisable to have an onsite treatment plants in order to avoid transmission of infectious waste to a centralized treatment plants away from healthcare facilities. MoPH in its contingency plans to address medical emergencies should address the use of healthcare waste engineering advice, realistic transportation and disposal arrangements, and the regular vaccination of waste workers (Chartier et al., 2014). These strategies and efforts could prove prudent in maintaining a sufficient level of public health protection through prompt waste removal and processing during future pandemics.

7 Conclusions

The current COVID-19 pandemic has raised challenges regarding waste generation, management, and recycling practices. It has one more time reminded the world of how these highly communicable diseases can increase the volume of healthcare waste and overwhelm the existing HCW treatment facilities. It also reminds that if not safely disposed, how these HCW create not only an immediate threat to environmental health but also conduce to the spread of infectious diseases. As a silver lining to any adversaries, this COVID-19 pandemic has shown us the opportunities to work toward better Healthcare Waste Management in case of pandemics in the future.

Past epidemics and current COVID-19 pandemics have proved that effective HCW management is a critical aspect of overall healthcare system. Ensuring the continuity of the existing services, healthcare institutions should be prepared to manage exponential increase in HCW in an epi/pandemic outbreaks and be prepared with safe and environmentally sound handling, treatment and final disposal of such infectious waste in a timely manner. Thailand should continue the best practices of effectively managing healthcare waste through appropriate identification, collection, separation, storage, transportation, treatment, and disposal. Other than centralized treatment facilities, increasing onsite and mobile treatment (incineration, chemical, and other technologies) facilities as contingency plans to handle HCW during pandemics need to be explored as well. Burning HCW into an incinerator may not always be the best waste management strategy. Placing and practicing rigorous waste minimization programs, and exploring opportunities for Circularity for Healthcare waste, especially healthcare plastics should be given due importance. While planning and constructing HCW treatment facilities, the plan of further expansion in the future depending upon the yearly increase in HCW due to medical tourism and unfortunate epidemics/pandemics must be considered in the design aspect itself as a strategic backup capacity for future. In order to minimize the risks of transmission of virus, expansion of storage area is critical. For example, it is said that coronavirus is likely to be active up to 2–5 days over various material surface (72 h on plastics); it might be helpful to have storage area to store segregated HCW recyclables for longer time prior to recycling activities. Similarly, increased storage area will safely store highly infectious waste while waiting for final treatment and disposals when the existing treatment facilities are suffering from huge volume of waste to be disposed of.

One of the learning from this COVID-19 experience is that in future we should be ready with a system that prepares safe recycling system even at the times of highly communicable diseases. This might mean establishing a HCW management system with proper disinfection of the HCW recyclables, training of the waste collection staffs, actors in recycling facilities, as well as breaking the false dogma that HCW is too dangerous to be recycled. In fact, medical plastics such as a blue wrap (made up of polypropylene and used to wrap sterilized materials), cardboard, glass, office paper, drink cans, newspapers, magazines, PETE and HDPE plastics, batteries, and electrical and electronic equipment if properly segregated and disinfected have a good recycling potential. Use of automated systems using the technology of Internet of Things for operating HCW treatment and recycling facilities with a minimum of workers involved, real-time tracking, and surveillance and controlling of the process without much human contact will also help in minimizing the risk of exposure.

Funding Information The authors received no specific funding for this work.

References

3R Knowledge Hub (3RKH), Asian Institute of Technology. (2008). *Healthcare waste in Asia: intuitions and insights.* Retrieved 11 November, 2020, from http://faculty.ait.ac.th/visu/wp-content/uploads/sites/7/2019/01/HCWA.pdf

Akkajit, P., Romin, H., Assawadithalerd, M., & Al-Khatib, I. (2020). Assessment of knowledge, attitude, and practice in respect of medical waste management among healthcare workers in clinics. *Journal of Environmental and Public Health, 2020*, 1–12. https://doi.org/10.1155/2020/8745472

Asian Development Bank (ADB). (2020). *Managing infectious medical waste during the COVID-19 pandemic.* Retrieved 11 November, 2020, from https://www.adb.org/sites/default/files/publication/578771/managing-medical-waste-covid19.pdf

Buddhasri, P. (2020). *MFU's CEWT researcher presents infectious waste management according to COVID-19 in the global arena.* Retrieved 16 November, 2020, from https://en.mfu.ac.th/en-news/en-news-detail/detail/News/7836.html

Chartier, Y., Emmanuel, J., Pieper, U., Prüss, A., Rushbrook, A., Stringer, A., Townend, A., Wilburn, S., & Zghondi, R. (2014). *Safe management of wastes from health-care activities* (2nd ed.). World Health Organization.

Jiangtao, S., & Zheng, W. (2020). Coronavirus: China struggling to deal with mountain of medical waste created by epidemic. *South China Morning Post.* Retrieved 12 November, 2020, from https://www.scmp.com/news/china/society/article/3065049/coronavirus-china-struggling-deal-mountain-medical-waste-created

Kaewkitipong, L. (2018). The Thai medical tourism supply chain: Its stakeholders, their collaboration and information exchange. *Thammasat Review, 21*(2), 60–90.

Kaigate, B. (2020). *Summary report online joint dialogue on waste management in the context of the COVID-19 pandemic.* Retrieved 16 November, 2020, from https://www.eria.org/uploads/media/!Summary-Report-on-Joint-Dialogue-on-Waste-Management-in-the-Context-of-COVID19-Pandemic.pdf

Kongkratoke, S. (2020). *Waste management according to COVID-19 in Thailand.* Retrieved 11 November, 2020, from https://jsmcwm.or.jp/international/files/2020/09/4_SitangKongtoke_Thailand.pdf

KPMG. (2018). *Medical tourism. industry focus.* Retrieved 11 November, 2020, from https://assets.kpmg/content/dam/kpmg/th/pdf/2018/03/th-medical-tourism-industry-focus-secured.pdf

Ministry of Natural Resources and Environment (MNRoE) and Pollution Control Department (PCD). (2015). *Thailand State of Pollution Report 2015.* Retrieved 14 November, 2020, from http://infofile.pcd.go.th/mgt/PollutionReport2015_en.pdf

Ministry of Natural Resources and Environment (MNRoE) and Pollution Control Department (PCD). (2021). *Thailand State of Pollution 2020 (B.E. 2563).* Retrieved 3 August, 2021, from https://www.pcd.go.th/wp-content/uploads/2021/03/pcdnew-2021-03-31_09-59-54_127801.pdf

Ministry of Natural Resources and Environment and Pollution Control Department. (2019). *Booklet on Thailand State of Pollution*. Retrieved 14 November, 2020, from http://www.oic.go.th/FILEWEB/CABINFOCENTER3/DRAWER056/GENERAL/DATA0001/00001462.PDF

Ministry of Public Health (MoPH). (2020). *Guidance for integrated management of state quarantine facilities*. Retrieved 23 November, 2020, from https://ddc.moph.go.th/viralpneumonia/eng/file/guidelines/g_guarantine_190620new.pdf

Noree, T., Hanefeld, J., & Smith, R. (2016). Medical tourism in Thailand: A cross-sectional study. *Bulletin of the World Health Organization, 94*(1), 30–36.

Nuclear Threat Initiative, John Hopkins University and Economist Intelligence Unit. (2019). *GHS Index Global Health Security Index*. Retrieved 14 November, 2020, from https://www.ghsindex.org/wp-content/uploads/2019/10/2019-Global-Health-Security-Index.pdf

Panyaping, K., & Okwumabua, B. (2006). Medical waste management practices in Thailand. *Life Science Journal, 3*(2), 88–93.

Patcharanarumol, W., Pongutta, S., Witthayapipopsakul, W., Viriyathorn, S., & Tangcharoensathien, V. (2018). Thailand. In H. Legido-Quigley & N. Asgari-Jirhandeh (Eds.), *Resilient and people-centred health systems: Progress, challenges and future directions in Asia* (pp. 346–373). World Health Organization, Regional Office for South-East Asia.

Pharino, C. (2017). Infectious waste management in Thailand. In C. Pharino (Ed.), *Challenges for sustainable solid waste management* (SpringerBriefs on Case Studies of Sustainable Development) (1st ed., pp. 77–91). Springer.

PR Thai Government. (2020). Retrieved 20 November, 2020, from https://www.facebook.com/thailandprd/posts/3437600819596641

Sirilak, S. (Ed.). (2020). *Thailand's experience in COVID-19 response*. Ministry of Public Health, Government of Thailand. Retrieved 10 November, 2020, from http://www.mbdsnet.org/wp-content/uploads/2020/08/Thailand-s-experience-in-the-COVID-19-response.pdf

Sumriddetchkajorn, K., Shimazaki, K., Ono, T., Kusaba, T., Sato, K., & Kobayashi, N. (2019). Universal health coverage and primary care, Thailand. *Bulletin of the World Health Organization, 97*(6), 415–422.

Thailand Today Blog. (2020, February 23). *Infectious waste must be properly disposed of*. Retrieved 11 November, 2020, from https://www.thailandtoday.co/23/02/2020/infectious-waste/

Thaiprayoon, S., & Wibulpolprasert, S. (2017). *Political and policy lessons from Thailand's UHC experience, ORF Issue Brief, Issue 174*. Observer Research Foundation. Retrieved 10 November, 2020, from https://www.orfonline.org/wp-content/uploads/2017/04/ORF_IssueBrief_174_ThailandUHC.pdf

The ASEAN Post Team. (2017, September 19). Medical tourism booming in Southeast Asia. *The Asean Post*. Retrieved 12 November, 2020, from https://theaseanpost.com/article/medical-tourism-booming-southeast-asia

Thomas, J. (2019, November 23). Medical tourism on the rise. *The Asean Post*. Retrieved 11 November, 2020, from https://theaseanpost.com/article/medical-tourism-rise

United Nations Environment Programme. (2020). *Waste management during the COVID-19 pandemic from response to recovery*. Retrieved 11 November, 2020, from https://reliefweb.int/sites/reliefweb.int/files/resources/WMC-19.pdf

Wei, G. (2020, June 3). Medical waste management experience and lessons in COVID-19 outbreak in Wuhan. *Waste, 360*. Retrieved 10 November, 2020, from https://www.waste360.com/medical-waste/medical-waste-management-experience-and-lessons-covid-19-outbreak-wuhan

World Bank Group. (2021). Thailand economic monitor July 2021: The road to recovery. Retrieved 3 August 2021, from https://documents1.worldbank.org/curated/en/260291626180534793/pdf/Thailand-Economic-Monitor-The-Road-to-Recovery.pdf

World Health Organization (WHO). (2020a, March 3). *Shortage of personal protective equipment endangering health workers worldwide.* Retrieved 11 November, 2020, from https://www.who.int/news/item/03-03-2020-shortage-of-personal-protective-equipment-endangering-health-workers-worldwide

World Health Organization (WHO). (2020b). *Water, sanitation, hygiene, and waste management for SARS-CoV-2, the virus that causes COVID-19.* Retrieved 11 November, 2020, from https://apps.who.int/iris/bitstream/handle/10665/333560/WHO-2019-nCoV-IPC_WASH-2020.4-eng.pdf?sequence=10&isAllowed=y

World Wide Fund for Nature Italy. (2020, April 29) *Responsibility is required when disposing of masks and gloves.* Retrieved 9 November, 2020, from https://www.wwf.it/scuole/?53500%2FNello-smaltimento-di-mascherine-e-guanti-serve-responsabilita

Pawan Kumar Srikanth is currently working as Research Associate at the Department of Energy, Environment, and Climate, School of Environment, Resources, and Development, Asian Institute of Technology, Thailand. His areas of interest include dumpsite mining, air quality, and noise impact assessment and solid waste management.

Wipatsaya Srimanoi is currently working as Research Associate at the Department of Energy, Environment, and Climate, School of Environment, Resources, and Development, Asian Institute of Technology, Thailand. Her areas of interest include climate change, solid waste management, and marine plastics abatement.

Prakriti Kashyap is an environmental consultant with excelling research, documentation, knowledge management, and project management ability. She has a decade-long experience in integrated waste management (including plastic waste and marine litter issues) in the ASEAN region.

Chettiyappan Visvanathan is Professor at the Department of Energy, Environment, and Climate, School of Environment, Resources, and Development, Asian Institute of Technology, Thailand. He lectures in Environmental Engineering and Management Program. His expertise includes solid–liquid separation technologies for water and wastewater treatment, clean technologies, circular economy for plastic waste management, and industrial pollution control.

Health Care Waste Management in Nepal: Pre- and Post-COVID-19 Scenario

Binaya Sapkota ⓘ

1 Background

1.1 Health Care Waste Management (HCWM)

The World Health Organization (WHO) defines healthcare waste (HCW) as all wastes generated as the byproduct of healthcare procedures in healthcare facilities (HCFs), research institutions, and laboratories (Chartier et al., 2014). HCW also includes waste produced during healthcare procedures even at home or community settings with dialysis, insulin injections, dressings, and others (Paudel & Pradhan, 2010; Sapkota et al., 2015). Composition-wise, usually 75–90% of HCWs are of general (i.e., non-hazardous) type and 10–25% hazardous waste (i.e., infectious, toxic, or radioactive) (Chartier et al., 2014; Karki et al., 2020a; Phengxay et al., 2005; Sapkota et al., 2015; Soares et al., 2013). Similarly, about 5% of hospital liquid wastes are non-infectious but hazardous, 10% infectious and 85% actually non-hazardous (Bakobie et al., 2018). These facts indicate that proper waste segregation alone may minimize the bulk of hazardous waste to make the whole waste mass easy to be managed (Sapkota et al., 2014).

The United Nations Conference on Environment and Development (UNCED) included environment-friendly management of hazardous, solid, radioactive wastes in its agenda as the "Agenda 21" in Chapters 20–22, respectively, in the Rio Conference in 1992 (UNCED, 1992). Healthcare waste management (HCWM) is a coordinated and holistic approach starting from the generation of wastes to their safe disposal to protect the health professionals, patients, caretakers, and the community people from their health and environment hazards (Joshi, 2013; Paudel &

B. Sapkota (✉)
Faculty of Health Sciences, Nobel College, Sinamangal, Kathmandu, Nepal

Kathmandu, Nepal
e-mail: binaya@nobelcollege.edu.np

Pradhan, 2010; Sapkota et al., 2015). Globally, HCWs are the most dangerous or hazardous, second to the radiation waste (Karki et al., 2020b; Manyele & Anicetus, 2006; Wafula et al., 2019). Global solid waste has been estimated to increase twofold by 2025, 53% of which will be solely generated from the lower- and lower–middle-income countries (LMICs). Global cost of such waste management has been computed to be approximately USD 375 billion by that period (Dangi et al., 2017). Hence, solid waste management (SWM) has turned to be problematic, especially for the urban areas of developing countries (Das et al., 2018).

All the personnel engaged in various points of waste, from generation to disposal, (e.g., doctors, nurses, medical laboratory staff, pharmacists, waste handlers, and others) are potentially at risk of health hazards. Therefore, HCFs should implement stringent rules, regulations, and mechanisms of training all staff for the proper HCWM (Singh et al., 2018).

1.2 Coronavirus Disease 2019 (COVID-19) and Nepal

The first case of coronavirus disease 2019 (COVID-19) was reported on December 31, 2019 and the WHO declared it as a Public Health Emergency of International Concern (PHEIC) on January 30, 2020 and global pandemic on March 11, 2020 (Dhakal & Karki, 2020; Piryani et al., 2020; Shrestha et al., 2020). In Nepal, first case of COVID-19 was suspected in a 32-year-old Wuhan, China returnee male on January 13, 2020, who was studying at Wuhan University of Technology and who returned to Nepal during winter vacation on January 9, 2020. He was declared COVID positive on January 24, 2020 after his report of throat swab sent to Hong Kong came to Nepal (Banerjee et al., 2020; Dhakal & Karki, 2020; Joshi et al., 2020; Pun et al., 2020). On May 17, 2020, COVID-19 took life of a 29-year-old female from Sindhupalchowk district for the first time in Nepal (Banerjee et al., 2020). There were 976,984 COVID cases, 11,936 deaths, 956,813 recovered cases, and 8235 active cases in Nepal till March 1, 2022, and this graph was previously on sharp inclining trend every day, luckily now on gradually declining pattern. Nepal holds the 57th position in the world corona cases to this date (JHU, 2020; MoHP, 2020; Worldometer, 2022).

1.3 Impact of Second and Third Waves of COVID-19 in Nepal

Nepal faced two waves of COVID-19 infection—the first wave started from mid-October to mid-November of 2020, and the second wave in mid-April of 2021. The second wave spread with 1260 times more severe and transmissible strain, named Delta variant and later Delta plus variant, and devastated many

South Asian countries including Nepal, after its first detection in India in the region, and led to many hospitalizations and deaths even in Nepal. Hence, it affected not only the elderly and immunocompromised but also young and otherwise healthy people. There came an increased trend of facemask wearing, physical distancing, handwashing, and frequent sanitizing after the rapid surge in COVID-19 infections in the second wave, resulting in a reduced vaccine hesitancy among people compared to the first wave. Still, Nepal had to further strengthen the capacity of HCFs to tackle the exponential COVID cases by running COVID-19 specific wards at primary health care centers (PHCCs) and higher-level HCFs with trained and skilled health professionals and with sufficient oxygen cylinders and manufacturing plants, and hospital beds within HCFs. During the first wave, local governments in many provinces helped the local HCFs manage the infected at the community-level isolation and quarantine centers, but decided to allow all mild cases to remain in self-isolation at their homes during the second wave, which made their family members stay at risk of getting infected. This was a great lesson for the policy makers to strictly follow the guidelines during home or institutional isolation to curb the potential third wave in the nation (Lama et al., 2021).

1.4 COVID Vaccination Management in Nepal

Vaccination against COVID-19 could reduce serious illness and death even against the newly emerged Delta variants to contain the spread of pandemic and to mitigate the trespass to the potential third wave in the country (Lama et al., 2021). Nepal already received and purchased 13,357,590 doses of COVID-19 vaccines: 4,422,740 AstraZeneca type, 7,400,000 Vero Cell, and 1,534,850 doses of Janssen vaccine (Poudel, 2021), and this number has sharply increased in the following times amidst the nation-wide vaccination campaigns.

1.5 Chapter Objectives

At the end of the chapter, the readers will be able to get ideas about the following:

- Health care waste categories and their management scenario in Nepal.
- Pre- and post-COVID scenarios of health care waste management (HCWM) in Nepal.
- Important legislative measures for HCWM in Nepal.

2 Methods

The following search strategies were employed to extract relevant literatures (Table 1).

2.1 HCW Categorization in Nepal

HCWs have been categorized in the following ways in Nepal (Table 2) (GoN MoHP, 2014).

2.2 Significance of Proper HCWM

The HCWs are of serious concern from the public health and environmental lens, if not managed properly (Adhikari & Supakankunit, 2014; Joshi et al., 2017; Karki et al., 2020b; Soares et al., 2013; Wafula et al., 2019). Improper HCWM may adversely affect health professionals, patients, waste handlers, community people, and the whole environmental network including land, water, air systems (Joshi, 2013; Sapkota et al., 2014). Proper HCWM encompasses all the processes such as minimization, segregation, storage, transportation, treatment, and disposal practices in proper dimensions (GoN MoHP, 2014). Improper HCWM may lead to transmission of cholera, gastroenteric, respiratory, ocular, genital, and skin infections,

Table 1 Search strategies employed

Pre-COVID-19	Post-COVID-19
"Healthcare waste" OR "health care waste" OR "medical waste" OR "biomedical waste" OR "clinical waste" OR "healthcare waste management" OR "health care waste management" OR "medical waste management" OR "biomedical waste management" OR "clinical waste management" AND "Nepal"	"COVID-19" OR "coronavirus" OR "SARS-CoV-2" AND (. . . #) AND "Nepal"
"Healthcare liquid waste" OR "health care liquid waste" OR "healthcare liquid waste management" OR "health care liquid waste management" AND "Nepal"	
"Pharmaceutical waste" OR "pharmaceutical waste management" AND "Nepal"	
"Hazardous waste" OR "hazardous waste management" AND "Nepal"	
"Sharp waste" "injection waste" OR "sharp waste management" OR "injection waste management" AND "Nepal"	

(. . . #): Same search strategies as mentioned in the column 'pre-COVID-19'

Table 2 Health care waste categories in Nepal (GoN MoHP, 2014)

Waste category	Color code for container	Examples of wastes
(A) Non-risk HCWs		
(i) Biodegradables	Green	Biodegradables: leftover stuff foods, fruits peels, etc.
(ii) Recyclables	Dark blue	Non-biodegradable, which can be recycled: plastic bottles, cans, metals, glass, plastics, papers, rubber
(iii) Other non-risk HCWs	Light blue	All other HCWs that do not belong to biodegradable and recyclable
(B) HCWs requiring special attention		
(i) Human anatomical wastes	Red	Human body parts, organs, human tissues, removed organs, amputated parts
(ii) Hazardous sharps	Red	Needles, glass syringes with fixed needles, scalpels, blades, glass, etc. which may cause puncture and cuts
(iii) Pharmaceuticals	Red	Unused and date expired drugs
(iv) Cytotoxic pharmaceutical wastes	Pink	Waste with antineoplastic effects such as: Alkylated substances, antimetabolites, antibiotics, plant alkaloids, hormones, etc.
(C) Infectious and highly infectious wastes		
(i) Infectious wastes	Red	Discarded items contaminated with blood and body fluids such as dressing materials, swabs, syringes without needle, infusion equipment without spike, soiled plaster, linen, bedding, other materials contaminated with blood, dialysis equipment, blood from patients infected with human immunodeficiency virus (HIV), viral hepatitis, brucellosis, respiratory tract secretion from patients infected with tuberculosis (TB), anthrax, rabies
(ii) Highly infectious wastes	Brown	Waste generated from the microbiological cultures, laboratory waste, such as sputum cultures of TB laboratories, highly concentrated microbiological cultures
(D) Other hazardous wastes	Yellow	Waste with high content of heavy metals, such as batteries, pressurized containers, organic and inorganic chemicals
(E) Radioactive wastes	Black	Solid, liquid, and gaseous waste contaminated with radionuclides such as cobalt, technetium, iodine, iridium; generated from in vitro analysis of body tissue and fluid, in vivo body organ imaging and tumor localization

anthrax, meningitis, HIV/AIDS, viral hepatitis A, B, C, D, and E, and many other (GoN MoHP, 2014). All of these diseases have potential to lead to death from different clinical spectra. Hence, proper HCWM is an urgent necessity of the present times to safeguard public health from adverse effects posed by HCWs and their improper disposal and management (Joshi, 2013).

The HCWs from the landfills or burial sites may contaminate the groundwater channel and may spread *Escherichia coli* and other pathogens and even their spores. Fumes generated from the wastes burnt may pollute the air that we breathe for a prolonged period (Sapkota et al., 2014). Safe and proper HCWM practices improve patient and occupational safety of health professionals engaged in different cycles of waste, from its generation to the endpoint. These health professionals may be physicians, nurses, medical laboratory personnel, pharmacists, and also the waste handlers. With improper HCWM, all of these personnel remain at risk of potential infections, toxic effects, and injuries (Sapkota et al., 2015). For example, proper sharp waste management along with adequate precautions (including personal protective equipment (PPE) use, vaccination against hepatitis B) helps ensure safety of both the providers and the receivers (Gyawali et al., 2014a).

2.3 Global Pre- and Post-COVID-19 Waste Management Practices

2.3.1 Pre-COVID-19 HCWM Scenario

Globally, it is estimated that 56 million people work in dangerous conditions, including waste collecting, sorting, and disposing, out of which about 15 million are prevalent in developing countries. Engaging in the waste recovery is a major source of livelihood, mainly for the urban poor (known as the informal waste workers (IWWs)) in the resource-constraint countries. These activities also keep them at increased risk of transmission of various infectious diseases. Then the vicious circle of "poor–sick–unemployed–poor" keeps revolving. Wide range of occupational risks faced by these IWWs, including chemical hazards, infection, musculoskeletal injury, and environmental contamination, has been reported in Brazil, the Philippines, Argentina, and India from their unhygienic activities (Black et al., 2019). Soares et al. (2013) trailed life cycle assessment (LCA) and cost analysis to determine the eco-efficiency of various HCWM techniques to achieve the best environmental performance (BEP) (Soares et al., 2013).

Toktobaev et al. (2015) implemented a novel low-cost HCWM system in all rural hospitals in Kyrgyzstan, which included safe transport and storage, mechanical needle removers, autoclaving, recycling of sterilized plastic and metal wastes, cement pits for anatomical waste, composting of biodegradable wastes, documentation, and training (Toktobaev et al., 2015). The Kyrgyz model was similar to that implemented by Sapkota et al. (2014) in a tertiary care hospital in Nepal (Sapkota et al., 2014).

2.3.2 Post-COVID-19 Waste Management Practices

Wuhan city, the COVID-19 starting point in China, saw a massive production of medical waste from 40 to 50 tons/day before the outbreak to 247 tons by 1st March

2020. The HCWs produced in various cities including Manila (capital of the Philippines), Kuala Lumpur (capital of Malaysia), Hanoi (capital of Vietnam), and Bangkok (capital of Thailand) also increased similarly, producing 154–280 tons more medical waste per day than prior to the pandemic (You et al., 2020).

2.4   Nepalese HCWM Scenario

Healthcare waste generated from public and private HCFs is in exponentially rising trend in Nepal with the population growth (Sharma et al., 2010a). A total of 274 hospitals produce 10,520 tons of non-hazardous and 3094 tons of hazardous biomedical waste each year in Nepal (Karki et al., 2020a, b). Liquid wastes are generated from different departments of hospitals such as operation theaters, medical laboratories, medical wards, gynecology-obstetrics procedures, and other sources (Sharma et al., 2010a). Nepal has banned purchase and use of mercury devices in all HCFs to minimize or avoid the mercury hazards on health and environment. Government of Nepal has recommended that HCWs be managed via following manner in the country: (Table 3) (GoN MoHP, 2014)

The GoN has recommended various waste management techniques for hospitals, PHCCs, health posts, sub-health posts, PHC out-reach clinics (PHCORCs)/out-reach immunization posts, private/public teaching hospitals and nursing homes/research institute, clinic/pathology laboratory clinics, and pharmacy/medical stores (GoN MoHP, 2014).

2.5   Hazards of Improper HCWM Practice

Major setback of improper HCWM is that risks posed by the same on health and environment are immeasurable in majority of the cases. Most populous countries in the world such as China, India, Brazil, Pakistan, Bangladesh, and Nigeria are facing $\geq 50\%$ hazards posed by improper HCWM in the global scenario (Joshi, 2013). Improper HCWM may contaminate the ground and surface water, air, and soil (Pokhrel & Viraraghavan, 2005). Toxic fumes containing life-threatening particulate matter (PM), liberated from the open burning and incineration of HCWs, and even the municipal solid waste (MSW), may cause immediate and persistent adverse health effects and global warming (Das et al., 2018; Karki et al. 2020a, b). Practices of open burning of HCWs, including plastic material containing polyvinyl chloride (e.g., blood bags, fluid bags) and low standard incineration, may generate persistent organic pollutants (POPs) such as carcinogenic dioxins, furans, and other PMs such as nitrogen oxides (NOx), sulfur dioxide (SO_2), ammonia (NH_3), and carbon monoxide (CO) (Das et al., 2018; GoN MoHP, 2014). These POPs persist in the environment and bio-accumulate in the food chain (GoN MoHP, 2014). Similarly, these PMs may cause acute and chronic respiratory and cardiovascular problems, cancer and birth defects among the newborn babies (Das et al., 2018). So, open

Table 3 HCWM techniques in Nepal (GoN MoHP, 2014)

Waste management techniques	Process description	Examples of waste to be processed
(1) Biological procedure	Enzyme mixture to decontaminate HCWs	
(i) Composting	Natural, biological decomposition of organic matter by fungi, bacteria, insects, worms, and other organisms	Biodegradable waste
(2) Autoclave	Steam sterilization under pressure	Treatment of highly infectious waste, such as microbial cultures or sharps
(3) Chemical disinfection	• With aldehydes, chlorine compounds, phenolic compounds to kill or inactivate pathogens • Reaction with 5% sodium hypochlorite • Acid hydrolysis followed by alkaline hydrolysis • Reduction using zinc powder, degradation using 30% hydrogen peroxide • Destruction using heated alkali	Treatment of liquid (and sometimes solid) infectious and highly infectious HCWs such as blood, urine, stools and sewage
(4) Encapsulation	Filling of the containers with waste, adding an immobilizing agent (e.g., plastic foam, bituminous sand, cement mortar or clay), and sealing the container	Suitable for sharps and pharmaceutical wastes
(5) Sanitary landfill	Engineered method, designed and constructed to keep the waste isolated from the environment	Disposing of certain types of HCWs (infectious wastes and small quantities of pharmaceutical wastes)
(6) Burial	Recommended for HCFs with minimal programs for HCWM	Hazardous wastes
(7) Septic/concrete vault		Used for the disposal of used sharps and syringes
(8) Incineration	To convert combustible materials into non-combustible residue or ash	
(9) Inertization	• HCW is mixed with cement and other substances in a composition of 65% waste, 15% lime, 15% cement, and 5% water. • Used to minimize the risk of contamination of toxic substances migrating to surface water or groundwater and prevent scavenging	Suitable for pharmaceuticals and incinerated ash with heavy metal content

burning practice of HCWs is being gradually discouraged by the WHO (Gyawali et al., 2014b). Improper HCWM, including unsafe injection practices, may cause environmental pollution and also transmit more than 30 pathogens including drug-resistant microorganisms such as typhoid, hepatitis B, C, and D, HIV, and other blood-borne viral diseases (BBVDs), *E. coli, Staphylococcus aureus, Psudomonas aeruginosa* (Bakobie et al., 2018; Gyawali et al., 2013a; Sapkota et al., 2014). Even if the HCWs and MSWs are dumped and never burned, their gradual degradation may yield potent greenhouse gas, i.e., methane (CH4) (Das et al., 2018).

2.6 Economic Implications of HCWM

2.6.1 Cost of Inaction on HCWs (Hazards of Improper HCWM)

Various bacteria such as *Shigella, E. coli, Klebsiella, Vibrio, Salmonella, S. aureus, Micrococcus* spp.*, Pseudomonas* spp.*, Proteus* spp.*, Enterobacter,* fungi, and viruses have been detected in sewage drain water. Hospital sewage is even the reservoir of various multi-drug resistant (MDR) bacteria, and pharmaceuticals (e.g., antimicrobial, disinfectants), heavy metals, radioisotopes, which are very much hazardous to the health, ecosystem, and environment, if not treated appropriately (Mahato et al., 2019; Sapkota et al., 2016; Sharma et al., 2010b). Similarly, *Legionella, P. aeruginosa* may be spread via direct inhalation of the aerosolized droplets. Hence, the whole integrated environment matrix including air, soil, water, and other must be regularly monitored to make the HCFs and the surrounding safe from the microbial burden. Maintaining hygiene, PPE use during health care procedures, and elimination of the reservoirs laden with microbes help prevent spread of *Pseudomonas* and many other infectious agents (Sapkota et al., 2016). If left untreated, these microorganisms may aggravate our personal health, whole health system, and environment, which may have irreparable consequences, which can not be expressible in monetary values.

2.6.2 Cost of HCWM

Cost–benefit analysis (CBA) helps to compare the costs and consequences of HCWM to evaluate the practice from the economic perspective, whereas inappropriate CBA design can mislead the resource allocation. Economic evaluation helps in strategic planning of proper HCWM to decide its economic feasibility and sustainability (Adhikari & Supakankunit, 2014). Overall costs of managing the wastes were found to be USD5,079,191, at a unit cost of USD2.36/kg in developed countries in 2014. Disposal costs were USD2,274,980, i.e., 44.8% of the overall costs (Vaccari et al., 2018). However, researches are yet to be conducted to explore the data on HCWM costs and circular economy or cost recovery in Nepal.

2.7 Health Care Scenario in Nepal

Government of Nepal (GoN) spends 1.8% of its gross domestic product (GDP) on health in the fiscal year 2018/19, and health sector budget has been allocated to be

Table 4 Overview of health care facilities and HCWs in Nepal in the fiscal year 2018/19

Particulars	Number	References
Public hospitals	125	GoN DoHS (2020)
Primary health care centers (PHCCs)	196	
Health posts (HPs)	3806	
Non-public facilities	2168	
Hospital beds	26,930	GoN MoHP (2020a)
ICU beds	1595	
Hospitals with ICU facility	194	

Table 5 Quantity and composition of HCWs generated in Nepal

Particulars	Composition (%) (Joshi et al., 2017)	Total (tons/year) (from 274 HCFs) (GoN MoHP, 2014)	HCWM techniques (Sapkota et al., 2014, 2015)
(a) Non-hazardous waste (kg/day/patient): (Total = 2.01)		10,520	
(i) Degradable wastes: 1.6	53.7		Composting
(ii) Recyclable wastes: 0.41	13.7		Recycling
(b) Hazardous waste (kg/day/patient): (Total = 0.97)		3094	
(i) Infectious wastes: 0.47	15.8		Autoclaving, incineration, chemical neutralization with 0.5% sodium hypochlorite
(ii) Pharmaceutical wastes: 0.2	6.7		Incineration, chemical neutralization with 0.5% sodium hypochlorite
(iii) Sharps: 0.18	6.1		Manual needle cutter and sharp pits
(iv) Chemical wastes: 0.1	3.3		Chemical neutralization with 0.5% sodium hypochlorite
(v) Radioactive wastes: 0.02	0.7		Chemical neutralization with 0.5% sodium hypochlorite

NPR 78.4bn (approximately 650 million USD) in the year 2019/20 (MoHP and DFID/NHSSP, 2019). Ministry of Health and Population (MoHP) is the supreme regulatory authority responsible for delivery of health services in Nepal. The Environmental Health and Healthcare Waste Management Section under the Department of Health Services (DoHS) regulates the HCWM systems in Nepal (GoN DoHS, 2020). The GoN provides health services through PHCCs, outreach clinics (ORC), sub-health posts (SHP), health posts (HPs), and hospitals (district, provincial, and central) (Gyawali et al., 2014a). Tables 4 and 5 show the HCFs and HCWs in Nepal (Tables 4 and 5).

2.8 Pre- and Post-COVID-19 Waste Management Practices in Nepal

2.8.1 Pre-COVID-19

Large quantities of wastes were being disposed in rivers such as Bagmati and Bishnumati, and other rivers across the country despite the mandate of the Solid Waste Management Act 2011 in Nepal (Black et al., 2019; Das et al., 2018). Sapkota et al. (2014) conducted an interventional study on the HCWM practices at a tertiary care governmental hospital at Kathmandu, Nepal. The HCWM committee developed HCWM policy and standard operating procedure (SOP) for the first time in the hospital and provided training to all health care providers and waste handlers in phase-wise manner. Color coded buckets were installed at each ward with patient information flex chart and leaflet for the proper segregation of the waste. They developed a system of disposing of sharp wastes daily on sharp pit, autoclaving all wastes in yellow containers, and incinerating the waste in red buckets twice weekly via double-chambered incinerator with temperature from 900 to 1200 °C and chimney height of 20 m. The incinerated end product, i.e., fly ash was then land filled (Sapkota et al., 2014).

Generally, public hospitals generate more HCWs than the private hospitals due to the high patient flow in the former (Chaudhary et al., 2015). Therefore, taking average of waste generated, it was found to be 2.01 and 0.97 kg/day/patient in case of non-hazardous wastes and hazardous wastes, respectively, with average of 1.35 kg/day/patient (Joshi et al., 2017). Table 6 shows the HCWM scenario during pre-COVID period in Nepal.

2.9 Waste Management Legislations in Nepal

In Nepal, stringent implementation of rules and regulation for the proper HCWM are still lacking, though some rules have been formulated and partially implemented (Singh et al., 2018).

2.10 Environment Protection Act, 2019

The Act was promulgated by the Parliament to protect the Constitution-endorsed right of each citizen to live in a healthy environment, to compensate the victims by the polluters for any damage incurred by the environmental pollution and to mitigate adverse environmental impacts. The following section mainly focuses on HCWM: (EPA, 2019)

- Section 15(2): No person shall create pollution causing significant adverse impacts on public health and environment.

Table 6 HCWM scenario in Nepal in the pre-COVID-19 period

Date/ duration	Location	HCW category	Key findings related to HCWM	References
March to October 2018	10 sewages samples from different hospitals of Biratnagar	HCLW	Almost all positive samples contained both *E. coli* and *Klebsiella* spp. *E. coli* isolates were resistant to ampicillin, amoxicillin, cefoxitin, cefuroxime, and cefpodoxime.	Mahato et al. (2019)
January and September 2015	10 wards/units and 77 locations of Civil Service Hospital, Kathmandu	Air samples	*S. aureus, Micrococcus, CONS, Bacillus, Pseudomonas,* and *Acinetobacter* were commonest organisms on inanimate objects in various wards of the hospital.	Sapkota et al. (2016)
Sept 15 to Oct 28, 2012	54 community pharmacies in Pokhara	Injection	Needles and sharps were collected in a mineral water bottle or a hard plastic container with a wide opening.	Gyawali et al. (2014b)
February 12 to October 15, 2013	Gynecology, obstetrics, pediatrics, Medicine, and orthopedics wards at Civil Service Hospital, Kathmandu	All HCWs	• During pre-intervention period, hospital did not have HCWM Committee, policy, SOP for waste segregation, collection, transportation, and storage. • HCWM policy and SOP were developed in consistent with the national and international laws and regulations. • Health professionals, such as doctors, nurses and waste handlers were trained on HCWM practices.	Sapkota et al. (2014)
May 18 to June 16, 2012	VDCs of Baglung district	Injection	The used syringes (with recapped needles) and other sharp wastes were collected in open boxes or even in pits, and were burnt in the pits.	Gyawali et al. (2013b)
May to October 2006	Narayani Sub-Regional Hospital, Birgunj	All HCWs	• Total HCW generated was 128.4 kg/day, with 0.8 kg/patient/day, 96.8 kg (75.4%) general waste, 24.1 kg (8.8%) hazardous waste, and 7.5 kg (5.8%) sharps/day. • Hospital adopted two types of disposal systems,	Paudel and Pradhan (2010)

(continued)

Table 6 (continued)

Date/ duration	Location	HCW category	Key findings related to HCWM	References
			burial and dumping, and had one incinerator only for the treatment of placenta. Later, it was stopped on special request of nearby public and then these wastes produced during operation were burned openly at the disposal site of the hospital. • Needle and syringes were burned openly before disposal.	
May to December 2008	10 central hospitals of Nepal	HCLW	• Mean of total viable bacterial counts ranged from 12.3×10^6 to 56×10^6 cfu/ml • MDR bacteria were resistant to the groups of penicillins, cephalosporins, quinolones, aminoglycosides.	Sharma et al. (2010b)
May to December 2008	10 central hospitals of Kathmandu	HCLW	• None of the hospitals had HCLWM guidelines and committees. • Only two hospitals had treatment plants for HCLWM. • Five hospitals disposed their liquid wastes into sewerage system and the remaining disposed the same into nearby rivers.	Sharma et al. (2010a)

cfu colony forming units, CONS Coagulase negative *Staphylococcus, HCLW* Health care liquid waste, *HCLWM* health care liquid waste management, *HCW* health care waste, *HCMW* healthcare waste management, *MDR* multidrug resistant, *SOP* standard operating procedure, *VDC* village development committee

2.11 Constitution of Nepal, 2015

Article 30(1) of the Constitution asserts that every citizen shall have the right to live in a healthy environment (Constitution of Nepal, 2015).

2.12 Health Care Waste Management Guideline, 2014

Health Care Waste Management Guideline, 2014 defined and categorized HCWs into five major categories: non-risk HCWs, HCWs requiring special attention, infectious and highly infectious wastes, other hazardous wastes, and radioactive wastes (Table 2). The guideline directed provisions for waste minimization, segregation and collection, storage and transportation/handling, treatment, and disposal for the HCWM to be implemented at different levels of HCFs (GoN MoHP, 2014).

2.13 Solid Waste Management Rules, 2013

- Rule 3: Chemical or harmful solid waste to be managed by the concerned waste generator.
- Rule 5: Solid waste not to be discharged by mixing harmful, chemical, organic or inorganic wastes with other waste types.
- Rule 6: The HCFs should process and manage health institutions-related wastes generated and discharged by segregating at their respective sources (SWMR, 2013).

2.14 Solid Waste Management Act, 2011

This Act was promulgated by the Constituent Assembly to manage solid waste systematically and effectively by reducing at its source, reusing, processing, and to maintain a healthy environment via reduction of adverse effects to the public health and environment. Various provisions were arranged under the Act, some of which are: (SWMA, 2011)

- Section 5: Generation of solid waste to be reduced by any individual, organization, or body.
- Section 6: Segregation of solid waste into organic and inorganic at its source.
- Section 7: Harmful waste or chemical waste not to be discharged at solid waste collection center.
- Section 8: Solid waste collection center to be prescribed.
- Section 9: Prescribed transportation vehicle to be used for transportation of solid waste.

2.15 Environment Protection Rules, 1997 (EPR, 1997)

Chapter 3 of the Rules provided various provisions under Rules 15–20 for the prevention and control of pollutions, some of which are:

- Rule 15: Prohibition to emit waste against the standards.
- Rule 17: Complaints in case anyone is found to be causing pollution or emitting waste.

2.16 Solid Waste (Management and Resource Mobilization) Act, 1987

The Act was promulgated to manage solid waste, mobilize resources, and to ensure the health convenience by controlling the adverse impact from solid waste. The following provisions were important from the Act: (SW(MRM)A, 1987)

- Section 4.3: The Solid Waste Management and Resource Mobilization Centre (i.e., Center in the Act) may prohibit keeping, throwing, burning, burying, or storing, disposing of or destroying harmful solid wastes in public places.
- Section 4.9: The Center may make necessary arrangements for the prevention of pollution resulting from solid wastes.

Table 7 summarizes the legislative measures in Nepal made for the HCWM.

Table 7 Summary table of HCWM legislations in Nepal

Year promulgated	Legislations	Key features	References
2019	Environment Protection Act	No person shall create pollution causing adverse impacts on public health and environment.	EPA (2019)
2014	Health Care Waste Management Guideline	Defined and categorized HCWs into five major categories and directed provisions for waste minimization, segregation and collection, storage and transportation/handling, treatment and disposal for HCWM at different levels of HCFs.	GoN MoHP (2014)
2013	Solid Waste Management Rules	Chemical or harmful solid waste to be managed by the concerned generator	SWMR (2013)
2011	Solid Waste Management Act	Segregation of solid waste into organic and inorganic at its source	SWMA (2011)
1997	Environment Protection Rules	Prohibition to emit waste against the standards	
1987	Solid Waste (Management and Resource Mobilization) Act	The Center may prohibit keeping, throwing, burning, burying, or storing, disposing of or destroying harmful solid wastes in public places.	

2.17   Nepal as the Signatory to the International Legislations Related to HCWM

Nepal is signatory to following important international legislations related to HCWM: (GoN MoHP, 2014; Joshi, 2013)

- Basel Convention on the Control of Trans-boundary Movements of Hazardous Waste, 1989 (Basel convention, 1989).
- Stockholm Convention on Persistent Organic Pollutants (POPs), 2001 (Stockholm Convention 2001).
- Strategic Approach to International Chemicals Management (SAICM).

2.18   Strengths and Weaknesses of Health Care Waste Management Practice in Nepal

(a) **Strengths**

- Solid HCWs have been tackled by the HCWM guideline of the country. Sharma et al. (2010a, b) had surveyed the healthcare liquid waste management (HCLWM) practices at seven government hospitals and three teaching hospitals from May to December 2008, and found that none of the hospitals had HCLWM guidelines and committees for the management of such wastes. Besides, four out of five municipal wastewater treatment plants (WWTPs) of Kathmandu valley were non-functioning (Sharma et al., 2010a).
- Bagmati Area Sewerage Project (BASP) in Kathmandu was a modern WWTP in Nepal (Lamichhane, 2007).

(b) **Weakness**

- Still, liquid wastes, pharmaceutical and cytotoxic wastes, and other hazardous wastes are to be managed properly in the country. Previously, Gyawali et al. (2014b) also reported that Nepal had no official guidelines to involve community pharmacies and pharmacists for the pharmaceutical waste management (Gyawali et al., 2014b).
- Infectious wastes were still discharged directly without any prior treatment, which was causing public health hazards such as seasonal outbreaks of water-borne communicable diseases in the country (Sharma et al., 2010a).

2.19   HCWM During Post-COVID Period

2.19.1   Guiding Principles for COVID-19 Related HCWM in Nepal

The GoN has introduced the following guiding principles for COVID-19 related HCWM in the country: (GoN MoHP, 2020b)

- All COVID-19 related wastes including the PPE are infectious, and should be collected, transported, treated, and disposed according to the standard procedures.
- All the COVID-19 related HCWs should be managed by following the risk reduction principles.
- Best practices for HCWM should be followed, including assigning trained human and material resources, including techniques and technologies, to segregate at source, treat, and safely dispose of the wastes.
- All the steps of HCWM, including minimization, segregation, collection, transportation, storage, treatment, and disposal, remain the same as of other biomedical wastes.

The Health Emergency and Disaster Management Unit of the MoHP issued some interim guidance documents in Nepal for the management of HCWs in the COVID periods: (HEOC, 2020; UNEP, 2020)

- Health Care Waste Management in the context of COVID-19 Emergency.
- Health Sector Emergency Response Plan, COVID-19 Pandemic.
- Standard Operating Procedure (SoP) of Cleaning and decontamination of the ambulance used in COVID-19.
- Interim Clinical Guidance for Care of Patients with COVID-19 in Health Care Settings.
- Guidelines for use of PPE during COVID-19.

However, HCWs were mostly burned (openly or in the small-scale incinerators) or dumped on land, and sent to the municipal landfill sites during post-COVID periods in Nepal (UNEP, 2020).

2.20 Challenges in Waste Management in the Post-COVID-19 Periods

After the COVID-19 outbreak in Wuhan, China on December 2019, 175 Nepalese medical students were airlifted to Kathmandu, Nepal from there on 15th February 2019. Since the time they were in the airplane and in the quarantine at Kharipati, Bhaktapur, the waste management was considered problematic in Nepal. Brown paper bag was used to collect their wastes during flight time and disposal machines were used in the quarantine (Rajbhandari et al., 2020). Infectious medical waste in the hospitals in China was found to be increased by 600% during COVID-19 treatment periods (GoN MoHP, 2020b). Facemasks and gloves were found to be scattered in the streets, especially in city areas in Nepal (Giri et al., 2021). However, researches on quantification and management of HCWs during the COVID periods are yet to come.

3   Future Prospects

- Waste management has become a global concern, including in Nepal, these days (Black et al., 2019; Joshi, 2013; Sapkota et al., 2014).
- Sewage treatment plants (STPs) or WWTPs must be established in hospitals for the proper management of effluent and sludge generated from various units (Mahato et al., 2019). Joshi et al. (2017) also reported that there were no WWTPs in 74.12% of HCFs in Nepal (Joshi et al., 2017). Development of guidelines for HCLWM along with installation of effluent or WWTPs, development of SOP for environmental indicators are must (Sharma et al., 2010a).
- It is high time that effluent quality standards be set to get stringent environmental quality standards (Lamichhane, 2007).
- Circular economy or 5Rs principles (i.e., refuse, reduce, reuse, repurpose, and recycle) should be implemented by local communities in collaboration with the public (Pathak et al., 2020).
- Discouraging needle recapping and proper disposal of sharps and needles help prevent needle stick injuries (NSIs) (Gyawali et al., 2013b).
- Excess dental amalgam should be stored in a closed container to minimize its hazard and then recycled (Singh et al., 2018).
- Medicine take-back concept can also be feasible to be implemented by the nearby community pharmacies to collect unwanted, unused and expired (UUE) medicines (Sapkota et al., 2021).

4   Conclusions

Health care wastes are of serious concern from the public health and environmental lens, if not managed properly, because improper management may lead to transmission of cholera, gastroenteric, respiratory, ocular, genital, and skin infections, anthrax, meningitis, HIV/AIDS, viral hepatitis A, B, C, D, and E, and many other infections. Health Care Waste Management Guideline, 2014 defined and categorized health care wastes (HCWs) into five major categories in Nepal: non-risk HCWs, HCWs requiring special attention, infectious and highly infectious wastes, other hazardous wastes, and radioactive wastes. A total of 274 hospitals produce 10,520 tons of non-hazardous and 3094 tons of hazardous biomedical wastes each year in Nepal, and it is on increasing trend each year with population surge. Facemasks and gloves were found to be scattered in the streets, especially in city areas in Nepal. However, researches on quantification and management of HCWs during the COVID periods are yet to come.

References

Adhikari, S. R., & Supakankunit, S. (2014). Benefits and costs of alternative healthcare waste management: An example of the largest hospital of Nepal. *WHO South-East Asia Journal of Public Health, 3*(2), 171–178.

Bakobie, N., Sulemana, A., & Duwiejuah, A. B. (2018). Assessment of hospital liquid waste management in public and private hospitals in Tamale Metropolis, Ghana. *Journal of Waste Management and Disposal, 1*(1), 1–7.

Banerjee, I., Robinson, J., Kashyap, A., Mohabeer, P., Shukla, A., & Leclézio, A. (2020). The changing pattern of COVID-19 in Nepal: A global concern—A narrative review. *Nepal Journal of Epidemiology, 10*(2), 845–855. https://doi.org/10.3126/nje.v10i2.29769

Black, M., Karki, J., Lee, A. C. K., Makai, P., Baral, Y. R., Kritsotakis, E. I., Bernier, A., & Heckmann, A. F. (2019). The health risks of informal waste workers in the Kathmandu Valley: A cross-sectional survey. *Public Health, 166*, 10–18. https://doi.org/10.1016/j.puhe.2018.09.026

Chartier, Y., Emmanuel, J., Pieper, U., Prüss, A., Rushbrook, P., Stringer, R., Townend, W., Wilburn, S., & Zghondi, R. (2014). *Safe management of wastes from health-care activities* (2nd ed., pp. 1–308). World Health Organization.

Chaudhary, N., Mahato, S. K., Chaudhary, S., & Bhatia, B. D. (2015). Biomedical waste management in Nepal: A review. *Journal of Universal College of Medical Sciences, 2*(8), 1. https://doi.org/10.3126/jucms.v2i4.12070

Constitution of Nepal. (2015). (pp. 1–226). Retrieved October 3, 2020, from http://www.lawcommission.gov.np/en/wp-content/uploads/2018/09/constitution-of-nepal-2-2.pdf

Dangi, M. B., Schoenberger, E., & Boland, J. J. (2017). Assessment of environmental policy implementation in solid waste management in Kathmandu, Nepal. *Waste Management & Research, 35*(6), 618–626. https://doi.org/10.1177/0734242X17699683

Das, B., Bhave, P. V., Sapkota, A., & Byanju, R. M. (2018). Estimating emissions from open burning of municipal solid waste in municipalities of Nepal. *Waste Management, 79*, 481–490. https://doi.org/10.1016/j.wasman.2018.08.013

Dhakal, S., & Karki, S. (2020). Early epidemiological features of COVID-19 in Nepal and public health response. *Frontiers in Medicine, 7*, 524. https://doi.org/10.3389/fmed.2020.00524

EPA. (2019). *The Environment Protection Act 1–24*. Retrieved October 3, 2020, from http://www.lawcommission.gov.np/en/wp-content/uploads/2020/06/The-Environment-Protection-Act-2019-2076.pdf

EPR. (1997). *Environment Protection Rules 1–55*. Retrieved October 3, 2020, from http://www.lawcommission.gov.np/en/wp-content/uploads/2018/09/environment-protection-rules-2054-1997.pdf

Giri, A., Sapkota, B., Shrestha, R., Khatiwada, A. P., Tiwari, R., Aryal, M., Timilsina, M., Bhujel, B., Adhikari, M., Sah, R., Bhandari, D., Ozaki, A., Martellucci, C. A., Kotera, Y., Mousavi, S. H., & Shrestha, S. (2021). A narrative review of personal protective equipment uses in coronavirus disease 2019 and its disposable practices. *Japan Medical Association Journal, 4*(2), 86–90. https://doi.org/10.31662/jmaj.2020-0120

GoN DoHS. (2020). *Annual report 2018/19*. Government of Nepal Ministry of Health and Population Department of Health Services 1–411. Retrieved October 6, 2020, from https://drive.google.com/file/d/1LxC8CGUcNP57oM0EA3wAt_m0zSrM6aVb/view?fbclid=IwAR2Em1-sMfk_c5YFeTdLEHmo2kKejCT9aH6325e1yUqEekAY4kTSTFqmTeU

GoN MoHP. (2014). *Health care waste management guideline 2014* (pp. 1–62). Government of Nepal Ministry of Health and Population Department of Health Services.

GoN MoHP. (2020a). *Health sector emergency response plan: COVID-19 pandemic* (pp. 1–29). Government of Nepal Ministry of Health and Population.

GoN MoHP. (2020b). *Health care waste management in the context of COVID-19 emergency (interim guidance)* (pp. 1–11). Government of Nepal Ministry of Health and Population.

Gyawali, S., Rathore, D. S., Shankar, P. R., & Kc, V. K. (2013a). Strategies and challenges for safe injection practice in developing countries. *Journal of Pharmacology and Pharmacotherapeutics, 4*(1), 8–12. https://doi.org/10.4103/0976-500X.107634

Gyawali, S., Rathore, D. S., Kc, B., & Shankar, P. R. (2013b). Study of status of safe injection practice and knowledge regarding injection safety among primary health care workers in Baglung district, Western Nepal. *BMC International Health and Human Rights, 13*, 3. https://doi.org/10.1186/1472-698X-13-3

Gyawali, S., Rathore, D. S., Shankar, P. R., Maskey, M., & Kc, V. K. (2014a). Injection practices in Nepal: Health policymakers' perceptions. *BMC International Health and Human Rights, 14*, 21. https://doi.org/10.1186/1472-698X-14-21

Gyawali, S., Rathore, D. S., Adhikari, K., Shankar, P. R., Kc, V. K., & Basnet, S. (2014b). Pharmacy practice and injection use in community pharmacies in Pokhara city, Western Nepal. *BMC Health Services Research, 14*, 190. https://doi.org/10.1186/1472-6963-14-190

HEOC. (2020). *Coronavirus disease (COVID-19) outbreak updates & resource materials.* Health Emergency Operation Center October 14, 2020. Retrieved October 17, 2020, from https://heoc.mohp.gov.np/update-on-novel-corona-virus-covid-19/

JHU. (2020). Johns Hopkins University & Medicine: Coronavirus Resource Center. Retrieved October 10, 2021, from https://coronavirus.jhu.edu/region/Nepal

Joshi, H. D. (2013). Health care waste management practice in Nepal. *Journal of Nepal Health Research Council, 11*(23), 102–108.

Joshi, H. D., Acharya, T., Dhakal, P., Ayer, R., Karki, K. B., & Dhimal, M. (2017). Health care waste management practice in health care institutions of Nepal. *Journal of Nepal Health Research Council, 15*(35), 7–11.

Joshi, J., Mishra, P., Kamar, S. B., Sharma, N. D., Parajuli, J., Sharma, S., & Pandey, H. R. (2020). Clinical profile of cases of COVID-19 in far Western Province of Nepal. *Journal of Nepal Health Research Council, 18*(46), 135–137. https://doi.org/10.33314/jnhrc.v18i1.2602

Karki, S., Niraula, S. R., & Karki, S. (2020a). Perceived risk and associated factors of healthcare waste in selected hospitals of Kathmandu, Nepal. *PLoS One, 15*(7), e0235982. https://doi.org/10.1371/journal.pone.0235982

Karki, S., Niraula, S. R., Yadav, D. K., Chakravartty, A., & Karki, S. (2020b). Risk perception towards healthcare waste among community people in Kathmandu, Nepal. *PLoS One, 15*(3), e0230960. https://doi.org/10.1371/journal.pone.0230960

Lama, T. P., Manohar, S., Adhikari, B., Acharya, B., Fitzgerald, M., Kc, A., Neupane, D., & Dhakal, S. (2021). How Nepal must prepare for 3rd Covid wave. *Nepali Times* [updated on August 14, 2021]. Retrieved October 10, 2021, from https://www.nepalitimes.com/banner/how-nepal-must-prepare-for-3rd-covid-wave/

Lamichhane, K. M. (2007). On-site sanitation: A viable alternative to modern wastewater treatment plants. *Water Science & Technology, 55*(1–2), 433–440. https://doi.org/10.2166/wst.2007.044

Mahato, S., Mahato, A., Pokharel, E., & Tamrakar, A. (2019). Detection of extended-spectrum beta-lactamase-producing E. coli and Klebsiella spp. in effluents of different hospitals sewage in Biratnagar, Nepal. *BMC Research Notes, 12*(641). https://doi.org/10.1186/s13104-019-4689-y

Manyele, S. V., & Anicetus, H. (2006). Management of medical waste in Tanzanian hospitals. *Tanzania Health Research Bulletin, 8*(3), 177–182.

MoHP. (2020). *Nepal—Recent updates.* Retrieved October 10, 2021, from https://covid19.mohp.gov.np/

MoHP and DFID/NHSSP. (2019). *Budget analysis health sector* (pp. 8–43). Ministry of Health and Population and Department for International Development/Nepal Health Sector Support Programme.

Pathak, D. R., Mainali, B., Naga, H. A. E., Angove, M., & Kong, I. (2020). Quantification and characterization of the municipal solid waste for sustainable waste management in newly formed municipalities of Nepal. *Waste Management & Research, 38*(9), 1007–1018. https://doi.org/10.1177/0734242X20922588

Paudel, R., & Pradhan, B. (2010). Health care waste management practice in a hospital. *Journal of Nepal Health Research Council, 8*(17), 86–90.

Phengxay, S., Okumura, J., Miyoshi, M., Sakisaka, K., Kuroiwa, C., & Phengxay, M. (2005). Health-care waste management in Lao PDR: A case study. *Waste Management and Research, 23*, 571–581. https://doi.org/10.1177/0734242X05059802

Piryani, R. M., Piryani, S., & Shah, J. N. (2020). Nepal's response to contain COVID-19 infection. *Journal of Nepal Health Research Council, 18*(46), 128–134.

Pokhrel, D., & Viraraghavan, T. (2005). Municipal solid waste management in Nepal: Practices and challenges. *Waste Management, 25*, 555–562. https://doi.org/10.1016/j.wasman.2005.01.020

Poudel, A. (2021). Nepal all set to receive 100,000 doses of Pfizer's Covid-19 vaccine from Covax. *The Kathmandu Post* [Updated on September 13, 2021]. Retrieved October 10, 2021, from https://kathmandupost.com/health/2021/09/13/nepal-all-set-to-receive-100-000-doses-of-pfizer-s-covid-19-vaccine-from-covax

Pun, S. B., Mandal, S., Bhandari, L., Jha, S., Rajbhandari, S., Mishra, A. K., Chalise, B. S., & Shah, R. (2020). Understanding COVID-19 in Nepal. *Journal of Nepal Health Research Council, 18*(46), 126–127. https://doi.org/10.33314/jnhrc.v18i1.2629

Rajbhandari, B., Phuyal, N., Shrestha, B., & Thapa, M. (2020). Air medical evacuation of Nepalese citizen during epidemic of COVID-19 from Wuhan to Nepal. *Journal of Nepal Medical Association, 58*(222), 125–133. https://doi.org/10.31729/jnma.4857

Sapkota, B., Gupta, G. K., & Mainali, D. (2014). Impact of intervention on healthcare waste management practices in a tertiary care governmental hospital of Nepal. *BMC Public Health, 14*, 1005. https://doi.org/10.1186/1471-2458-14-1005

Sapkota, B., Gupta, G. K., Mainali, D., & Shrestha, N. (2015). Development and implementation of healthcare waste management policy at Civil Service Hospital, Nepal. *Journal of Pharmacy Practice and Research, 45*, 57–63. https://doi.org/10.1002/jppr.1054

Sapkota, B., Gupta, G. K., Shrestha, S. K., Pradhan, A., Karki, P., & Thapa, A. (2016). Microbiological burden in air culture at various units of a tertiary care government hospital in Nepal. *Australasian Medical Journal, 9*(1), 1–7. https://doi.org/10.4066/AMJ.2015.2558

Sapkota B, Giri A, Bhatta B, Awasthi K, Bhurtyal K, Joshi B, Joshi KR. Implementation of medicine take-back concept at community level in Nepal: a pilot study. J Public Health (Oxf) 2021:1–11. https://doi.org/10.1093/pubmed/fdab134

Sharma, D. R., Pradhan, B., Pathak, R. P., & Shrestha, S. C. (2010a). Healthcare liquid waste management. *Journal of Nepal Health Research Council, 8*(16), 23–26.

Sharma, D. R., Pradhan, B., & Mishra, S. K. (2010b). Multiple drug resistance in bacterial isolates from liquid wastes generated in central hospitals of Nepal. *Kathmandu University Medical Journal, 8*(1), 40–44.

Shrestha, A., Rajbhandari, P., & Bajracharya, S. (2020). Hospital preparedness for outbreak at Patan Hospital: Lesson learnt from COVID-19. *Journal of Nepal Health Research Council, 18*(46), 142–143. https://doi.org/10.33314/jnhrc.v18i1.2547

Singh, T., Ghimire, T. R., & Agrawal, S. K. (2018). Awareness of biomedical waste management in dental students in different dental colleges in Nepal. *BioMed Research International, 2018*, 1742326. https://doi.org/10.1155/2018/1742326

Soares, S. R., Finotti, A. R., da Silva, V. P., & Alvarenga, R. A. F. (2013). Applications of life cycle assessment and cost analysis in health care waste management. *Waste Management, 33*, 175–183. https://doi.org/10.1016/j.wasman.2012.09.021

SWMA. (2011). *Solid Waste Management Act 1-34*. Retrieved October 3, 2020, from http://www.lawcommission.gov.np/en/wp-content/uploads/2018/10/solid-waste-management-act%2020 68-2011.pdf

SWMR. (2013). *Solid waste management rules, 2013:1-24*. Retrieved October 17, 2020, from http://extwprlegs1.fao.org/docs/pdf/nep187145.pdf

SW(MRM)A. (1987). *Solid Waste (Management and Resource Mobilization) Act, 1987*. Retrieved October 17, 2020, from http://www.lawcommission.gov.np/en/wp-content/uploads/2018/10/solid-waste-management-and-resource-mobilization-act-2044-1987.pdf

Toktobaev, N., Emmanuel, J., Djumalieva, G., Kravtsov, A., & Schüth, T. (2015). An innovative national health care waste management system in Kyrgyzstan. *Waste Management & Research, 33*(2), 130–138. https://doi.org/10.1177/0734242X14565209

UNCED. (1992). *Agenda 21. United Nations Conference on Environment & Development, Rio de Janerio, Brazil 1992 June 3 to 14:1-351.* Retrieved October 5, 2020, from https://sustainabledevelopment.un.org/content/documents/Agenda21.pdf

UNEP. (2020). *Waste management during the COVID-19 pandemic: From response to recovery* (pp. 9–57). United Nations Environment Programme.

Vaccari, M., Tudor, T., & Perteghella, A. (2018). Costs associated with the management of waste from healthcare facilities: An analysis at national and site level. *Waste Management & Research, 36*(1), 39–47. https://doi.org/10.1177/0734242X17739968

Wafula, S. T., Musiime, J., & Oporia, F. (2019). Health care waste management among health workers and associated factors in primary health care facilities in Kampala City, Uganda: A cross-sectional study. *BMC Public Health, 19*, 203. https://doi.org/10.1186/s12889-019-6528-4

Worldometer. (2022). *Reported cases and deaths by country, territory, or conveyance.* Retrieved March 1, 2022, from https://www.worldometers.info/coronavirus/?utm_campaign=GenericU Ablogposts?

You, S., Sonne, C., & Ok, Y. S. (2020). COVID-19's unsustainable waste management. *Science, 368*(6498), 1438. https://doi.org/10.1126/science.abc7778

COVID-19 and Healthcare Waste Management (HCWM) in Myanmar: Perspectives from the Triple R (Response, Recovery, and Redesign)

Premakumara Jagath Dickella Gamaralalage, Matthew Hengesbaugh, and Ohnmar May Tin Hlaing

1 Introduction

Since the onset of the COVID-19 pandemic, the role of safe and reliable healthcare waste management (HCWM) has been receiving increasing global attention. Recognising its importance in protecting public health and the environment (Tsukiji et al., 2020; WHO and UNICEF, 2020; Scheinberg et al., 2020; Kapoor et al., 2020; ADB, 2020), sound HCWM also contributes to achieving several UN Sustainable Development Goals (SDGs), including good health and well-being (Goal 3), clean water and sanitation (Goal 6), decent work and economic growth (Goal 8), responsible consumption and production (Goal 12), and climate actions (Goal 13) (WHO, 2019; The Global Fund, 2020; You et al., 2020; Nor Faiza et al., 2019; Singh et al., 2020).

The World Health Organisation (WHO) defines healthcare waste (HCW) as all waste generated from healthcare facilities (HCFs), including hospitals, medical research centres, and any other laboratories related to medical procedures. This definition also includes HCW that originates from other, more widely dispersed sources, including healthcare activities conducted at the household level such as

P. J. D. Gamaralalage (✉)
IGES Centre Collaborating with UNEP on Environmental Technologies (CCET), Institute for Global Environmental Strategies (IGES), Hayama, Kanagawa, Japan
e-mail: premakumara@iges.or.jp

M. Hengesbaugh
Integrated Sustainability Centre (ISC), Institute for Global Environmental Strategies (IGES), Hayama, Kanagawa, Japan
e-mail: hengesbaugh@iges.or.jp

O. M. T. Hlaing
Environmental Quality Management Co., Ltd (EQM), Yangon, Myanmar

home dialysis, self-administration of insulin, and recuperative care (Chartier et al., 2014). A large percentage (approximately 85%) of HCW generated by HCFs is typically classified as non-hazardous, comparable in composition to domestic waste and therefore managed alongside general municipal solid waste (MSW). The remaining 15% comprised of hazardous waste includes infectious waste, representing 10% as well as chemical and radioactive waste, making up roughly 5%, all of which necessitate additional safety measures to prevent environmental contamination and associated health risks (Chartier et al., 2014; WHO and UNICEF, 2019).

While there is no scientific evidence that the handling of HCW has resulted in the transmission of the COVID-19 virus (SARS-CoV2 virus) (WHO and UNICEF, 2020; Scheinberg et al., 2020; Sinha et al., 2020), Interim Guidance provided by WHO recommended following basic principles for managing HCW safely, such as separating general waste from infectious waste at source (WHO and UNICEF, 2020). This includes the collection, treatment, and disposal of infectious waste, including those positively associated with confirmed COVID-19 cases, by making use of safely marked lined containers and sharp-boxes. In addition, the WHO suggests treating collected waste immediately, preferably on-site, using appropriate waste management options, such as high temperature, dual-chamber incineration or autoclaving (WHO and UNICEF, 2020). Where such treatment options are not available, safely burying the waste or treating it via controlled burning can serve as interim measures until more environmentally preferable methods are available. If HCW is transported and treated outside of designated HCFs, maintaining detailed records on waste generation and processing is crucial for monitoring progress and measuring results. Moreover, workers responsible for handling HCW are advised to wear appropriate personal protective equipment (PPE), such as long-sleeved gowns, heavy-duty gloves, boots, masks, and goggles or face shields as well as practice safe hand hygiene (WHO and UNICEF, 2020).

In contrast to these guidelines, a review of global data (WHO and UNICEF, 2019) indicates that as many as one-third of HCFs failed to practice primary HCWM, including the segregation of HCW into a minimum of three categories (general, infectious items, and used sharps), as well as the safe treatment and disposal of infectious and used sharps even before the emergence of COVID-19. These trends are even starker in developing countries, with only 27% found to possess basic HCWM systems (WHO and UNICEF, 2019). Inadequate HCWM including inappropriate treatment and disposal risks wider contamination (Jang et al., 2006) and poses significant health and environmental safety hazards (WHO, 2017; Chartier et al., 2014; Windfeld & Brooks, 2015), especially for formal and informal waste collectors lacking proper PPE.

This situation is not uncommon in Myanmar, where coupled with limited awareness among healthcare professionals, HCWM practices are currently deficient (Thien et al., 2020; Premakumara et al., 2019; Premakumara et al., 2016; Aung et al., 2019; Htut, 2019; Win et al., 2019; Karstensen et al., 2017). The COVID-19 pandemic has created additional challenges in managing HCW across the country due to uncontrolled fluctuations in generation quantities, changing composition of

HCW streams, gaps in technical capacity, and resource limitations. Enhancing the safe handling and disposal of HCW in Myanmar therefore remains critical for guiding the country's effective emergency response to COVID-19. However, Myanmar's management of HCW not only represents an urgent and essential public service for minimising possible secondary impacts on health and the environment; it also has implications for redesigning the country's HCWM system to be more resilient over the long term.

This paper discusses HCWM practices in Myanmar outlining key challenges faced by the HCWM sector both prior to and during the COVID-19 pandemic. Data was collected from various secondary sources, including the internet, journals, reports, and official documents covering national policies, institutional arrangements, and country-level HCWM practices. In addition to suggesting some innovative solutions for tackling existing HCWM challenges, the paper also presents a list of potential measures that policymakers can consider in building a localised HCWM system for addressing future pandemics based on the Triple-R framework developed by the Institute for Global Environmental Strategies (IGES) (Zusmn et al., 2020; Mori et al., 2020). The Framework is made up of three integrated components, which connect targeted "response" interventions with broader "recovery" policies and related stimulus spending together with key measures for "redesigning" systems aimed at achieving sustainable HCWM in Myanmar.

2   HCWM in Myanmar: Before and During the COVID-19 Pandemic

2.1   HCW Generation

According to the official data, Myanmar had approximately 1134 government hospitals, 296 traditional medicine clinics and medical centres as well as 2237 healthcare facilities with approximately 60,000 available beds, with an average of 67% of bed occupancy in 2018 (CSO, 2019). In addition, the country hosts roughly 206 private hospitals, 4778 private clinics and 855 private dental clinics (Htut, 2019). Although there is no systematic database or official data available on HCW generation from these HCF (WHO, 2017; MOHS, 2019; MOHS, 2016; Premakumara et al., 2019; Premakumara et al., 2016), SINTEF's study has estimated that upwards of 67 tonnes of HCW were generated daily from HCFs before the emergence of COVID-19 in Myanmar (Saha et al., 2020). This estimation also identified that approximately 15% of total HCW (10 tonnes/day) were associated with hazardous infectious and sharps wastes, in line with the following categories in Table 1 (MOHS, 2014; MOHS, 2019).

Since the onset of the COVID-19 pandemic, there has a rapid increase in HCW generation at the global level and especially in Asia and the Pacific (Agamuthu & Barasarathi, 2020). As shown in Table 2, Wuhan City in China experienced a

Table 1 Type of hazardous and infectious waste

Sharp waste	Includes needles, scalpels, knives, nails, bladders, broken glass and other sharp objects used in healthcare activities
Infectious waste	Contains pathogen like bacteria, viruses, parasites or fungi in sufficient concentration or quantity to cause disease in susceptible hosts. This also includes waste contaminated with blood or other bodily fluids, culture and stocks of infectious agents from laboratory work waste from infected patients in isolation wards.
Pathological waste	Consists of tissues, organs, body parts, blood, body fluids and other waste from surgery and autopsies on patients with infectious diseases
Pharmaceutical waste	Includes expired, unused, spilt and contaminated pharmaceutical products, prescribed drugs, vaccines that are no longer required, vials containing pharmaceutical residues, gloves, masks and connecting tubing
Chemicals	Consists of discarded solid, liquid and gaseous chemicals with toxic, corrosive, flammable, reactive, and genotoxic properties. Chemicals most commonly used in HCFs include formaldehyde, photographic chemicals, heavy metals such as mercury from broken clinical equipment, solvents, organic and inorganic chemicals, and expired, used or spilt pharmaceuticals. Hazards from chemical and pharmaceutical waste include intoxication as a result of acute or chronic exposure from dermal contact, inhalation or ingestion and contact burns from corrosive or reactive chemicals

Source: MOHS (2014) and MOHS (2019)

Table 2 Increase of HCW generation due to the COVID-19 pandemic

City	The volume of HCW generation before COVID-19 pandemic (tonnes/day)	The volume of HCW generation during the COVID-19 pandemic (tonnes/day)
Manila (Philippines)	47	280
Jakarta (Indonesia)	35	212
Bangkok (Thailand)	35	210
Hanoi (Vietnam)	27	160
Kuala Lumpur (Malaysia)	26	154
Wuhan (China)	40–50	247

Source: ADB (2020), Tsukiji et al. (2020), and Lie et al. (2021)

massive rise in HCW, growing from 40 to 50 tonnes/day before the outbreak to about 247 tonnes/ day in March 2020 (Lie et al., 2021). Many other cities in Southeast Asia, such as Manila (Philippines), Kuala Lumpur (Malaysia), Hanoi (Vietnam), and Bangkok (Thailand), have also experienced similar surge in HCW generation, with data indicating a five-fold increase from prior to the pandemic (ADB, 2020; Tsukiji et al., 2020).

Since the identification of two initial cases on 23 March 2020, COVID-19 has spread rapidly across Myanmar. Similar to other countries, Myanmar also experienced a second and third waves of COVID-19 infections within a short period of time (Win, 2020). As of October 2020, daily caseload numbers across the country reached more than 2000 affected persons, with cities such as Yangon becoming major epicentres for the outbreak (Nikkei Asia, 2021). According to the Coronavirus Disease 2019 (COVID-10) Surveillance Dashboard of Myanmar, as of 9 March 2021 an estimated 142,045 cases (MOHS, 2021) were reported in the country. The data also indicates that more than 77% of confirmed cases were identified in two cities, namely Yangon (67%) and Mandalay (10%). The second wave of the pandemic led to significant overcapacity in Myanmar's hospitals, and coupled with the re-imposition of lockdown measures, contributed to a larger economic downturn. According to a World Bank survey conducted in late 2020, nearly half of businesses in Yangon remained temporarily closed, with reopening complicated by greater labour and supply chain interruptions throughout the country (World Bank, 2020). These impacts have had expected and wide-ranging consequences on livelihoods and poverty reduction initiatives.

From May 2021, a third wave of the virus has ravaged Myanmar, brought on by general shortage of medical personnel and slow roll out of vaccinations (Nikkei Asia, 2021). Ongoing clashes between the military, which seized power in early February, and the Civil Disobedience Movement (CDM), many of them health workers, have caused further disruptions to relief efforts (Theresa, 2021). Coronavirus infections have since risen steadily, with the more infectious Alpha, Delta, and Kappa variants all detected (Theresa, 2021). As of July 2021, 296 of 330 townships nationwide had reported incidences of the virus, reaching 7000 new cases, comprising a total of 184,375 persons and 3685 fatalities since the onset of the pandemic (Frontier, 2021). This third wave has been worsened by increasing prices for pharmaceuticals as well chronic lack of testing, medicines, and supplies, including face masks, oxygen, and most critically, vaccines (Frontier, 2021). Border closures and outbreaks abroad have created bottlenecks for the import of vaccines and equipment. While the previous government is credited for initiating one of the region's earliest vaccination drives, as of May 2021 only 3.3% of the country's population had received their first vaccine dose (Nikkei Asia, 2021). The Ministry of Health set a target for 50% of Myanmar's population to be immunised in 2021 (Regan, 2021), and military medical units have been established to coordinate vaccination campaigns. According to state media, as many as six million doses of an unnamed Chinese vaccine (Sinopharm COVID-19 and Sinovac vaccines 15) are to be delivered within 2021, with the initial roll out starting in Yangon (Nikkei Asia, 2021). According to the Ministry of Health, a total of 341,300 are laboratory confirmed among 3,298,004 total specimens tested. Currently, 256,671 persons have recovered from the virus, with the death toll among lab-confirmed Covid-19 cases standing at 12,452 persons as of 11 August 2021 (MOHS, 2021).

The climbing number of positive COVID-19 cases has also been correlated with increasing volumes of HCW in Myanmar. The Yangon City Development Committee (YCDC) has estimated that approximately three tons of hazardous HCW were

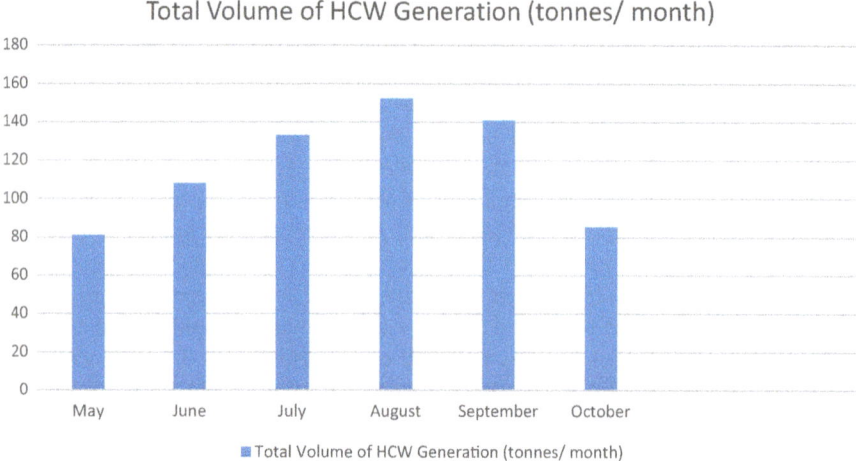

Fig. 1 Total volume of HCW generation during May–October 2020. Source: compiled by authors based on the data provided by the YCDC

generated daily from 72 government and private hospitals and approximately 4000 polyclinics in the Yangon region prior to the onset of COVID-19. However, HCW volumes also increased by 30% over the period of May to October 2020, averaging 3.9 tonnes/day (see Fig. 1), corresponding with growing rate of hospitalisation and additional use of temporary HCFs in Yangon. For example, auxiliary hospitals with extra beds, including Waibargi Special Hospital (90 beds), South Okkalapa Special Hospital (80 beds), and Phaung Gyi COVID-19 Treatment Centre (1200 beds), have established in the Yangon region. Due to overcrowding, four additional temporary facilities with a total capacity of approximately 4050 beds were set up in the Yangon Region aimed at accommodating patients with less severe COVID-19 symptoms. Similarly, the Myanmar Armed Forces (MAF) began treating 150 civilians for COVID-19 at its Military Medical Corps Centre in Hmawbi Township, Yangon Region, on 25 September (OCHA, 2020). In addition, patients under investigation (PUI) have also been quarantined in various facilities such as schools, monasteries, and stadiums, or have been instructed to remain confined at home in order to prevent contagion.

The situation is largely the same in other cities across Myanmar. According to the Mandalay City Development Committee (MCDC), municipal waste authorities are presently collecting as much as 10 tonnes of HCW (including both general and hazardous waste) from HCFs, including both COVID-19 treatment and quarantine centres (Khaing, 2020), compared to a previous baseline of 1.2 tonnes/day.

2.2   Practice of HCWM

2.2.1   3R (Reduce, Reuse, and Recycling)

Sound HCWM requires considering all stages of the waste hierarchy in line with 3R (reduce, reuse and recycle) principles (Chartier et al., 2014; WHO, 2019; Tsukiji et al., 2020). Waste prevention, which includes reducing waste generated at source and the adoption of circular solutions, represents a crucial first step for advancing sustainable HCWM practices. Such measures are central to achieving waste reduction targets, improving the efficiency of collection, as well as minimising the costs associated with waste treatment and disposal. While some HCFs have introduced concrete measures aimed at preventing waste generation by adapting purchasing and stock control strategies, such as adjustment of purchasing procedures, control of inventory, and production of less toxic materials, etc.), the application of 3R principles has yet to be fully mainstreamed across Myanmar.

Likewise, at the global level, there has been a significant and observable increase in the use of single-use plastic, especially in terms of packaging (44.8%) and other medical applications (13.2%) aimed at managing the spread of COVID-19 (Sharma et al., 2020). Large volumes of PPE, such as infected masks, gloves, and other protective types of equipment, as well as a higher percentage of similar non-infected items, have been found to be generated by HCFs. For example, some data suggests that the global market for PPE increased by 6.5% between 2016 and 2020, from approximately USD 40 billion to USD 58 billion (Singh et al., 2020). WHO (2020) estimates that the production of PPE must be scaled up by 40% in order to meet monthly demand and effectively address the COVID-19 pandemic.

Newly available research indicates that PPE disinfection and reuse is possible by applying an infusion of hydrogen peroxide vapour, ultraviolet or gamma-irradiation, ethylene oxide gasification, application of spray-on disinfectants, and infusion of base materials with antimicrobial nanoparticles (Singh et al., 2020; Price et al., 2020). Given that such research findings are still preliminary, current practices dictate that PPE materials are safely treated and disposed of in an environmentally sustainable manner rather than being reused and recycled. However, this brings additional challenges for the already-strained HCWM system in Myanmar. This is perhaps best captured by a recent image, widely circulated on the country's social media, showing a man removing a used medical hazmat suit (hazardous materials suite) from a roadside trash bin, potentially for reuse or recycle. Scenes such as this strongly underscore the need for more systematic disposal of used PPE in Myanmar (Gerin, 2020).

At the same time, recycling operations in Myanmar also face obstacles, given that the majority of recycling in the country occurs through the un-organised or informal sector. Protecting informal waste workers is especially challenging given that occupational safety and health is typically not seen as a priority. These issues are also accompanied by logistical complications: sharps and infectious waste need to be separately collected and transported directly to the disposal sites yet often enter informal recycling markets, increasing concerns over potential exposure to COVID-19. In some areas of Myanmar, recycling facilities have discontinued operation due to health risks and the falling market price for recyclable materials. These developments disproportionately affect waste pickers who are largely dependent on income generated from working at such facilities.

2.2.2 Waste Segregation, Storage, and Collection

Proper segregation, as well as the safe storage and collection of HCW are all critical for reducing potential risks associated with the handling of hazardous and non-hazardous waste components. Myanmar's Hospital Management Manual (HMM) of 2011 and related Standard Operating Procedures (SOP) from 2019 provide relevant guidance towards this end, including instructions on the segregation of HCW into colour-coded bins, such as general waste (black), infectious waste (yellow), and highly-infectious waste (red), dating prior to the COVID-19 pandemic. In fact, however, the segregation of HCW does follow a formal or standard procedure, with the exception of some hospitals that carry out such practices after receiving external support and training (MOHS, 2016). In Myanmar, the segregation of HCW, labelling and use of colour-coded bins for waste storage conducted in private (Aung et al., 2019) and non-hospital primary healthcare centres (PHC) are largely inadequate in comparison to government hospitals (Win et al., 2019). An assessment of HCWM practices in Mon State revealed that only 7% of non-hospital type PHCs carried out proper labelling, waste separation or followed a colour coding system for HCWM; in contrast, 36.4% in government hospitals implemented these practices (Win et al., 2019). Such findings were also corroborated by another study examining community-level clinics in Shan State, which found that only 6% of healthcare facilities practiced colour coding for HCW segregation (Taung, 2010).

In view of the associated health and safety risks associated with HCWM, government oversight on proper source separation, labelling, and collection has grown more stringent during the COVID-19 pandemic. For instance, the MOHS issued new guidelines on HCWM in view of the health crisis (updated as of July 2020) making reference to previous policy documents, including HMM, SOP, as well as the prior Healthcare Waste Management Guidelines from 2019, and Hospital Infection Control Guidelines from 2016. The 2020 guidelines outlined recommendations for carrying out immediate source segregation of all HCW. In addition, the guidelines suggest making use of three differentiated coloured bags, in line with previous waste instructions, with the additional precaution of doubling yellow bags for storing of infectious waste. Lastly, the guidelines recommend labelling waste properly with relevant information including the name and date of generation.

Assigning requisite HCF staff tasked with handling HCW is another important requirement for ensuring sound HCWM (WHO, 2014). Yet, even prior to COVID-19, the designation and allocation of such staff in Myanmar has largely been conducted in an ad-hoc manner, often without following necessary administrative or operating procedures. The situation continues to be further complicated because of the COVID-19 pandemic. As health workers in Myanmar continue to be overwhelmed by growing numbers of admitted COVID-19 patients, large numbers of volunteers have been recruited without adequate training and awareness on HCWM.

Moreover, many hospitals in Myanmar lack formally designated containment areas for storing HCW. As a result, non-hazardous waste and hazardous waste are often commingled at storage sites. The use of temporary HCW storage sites, especially those unequipped with safety or security features, may result in potential spillage and contribute to further spread of disease during the pandemic. In sum, existing knowledge gaps on HCWM, together with unsuitable management practices carried by waste generators and operators, the absence of appropriate equipment for waste separation, inadequate funding, and poor enforcement all present barriers to the effective implementation of source segregation, storage, and collection in Myanmar (MOHS, 2016; Aung et al., 2019).

2.2.3 Waste Transportation and Disposal

At present, Myanmar does not utilise private contractors for HCW removal; instead, municipal authorities are responsible for the collection and treatment of waste, including but not limited to HCW. Both Yangon and Mandalay City Development Committees (CDCs) collect HCW from central hospitals and special clinics on a daily basis. Collection services are also available for other HCFs either once a week or on an on-call basis. Infectious waste is incinerated and burned at cemeteries, while sharp wastes are buried underground in landfills (Thien et al., 2020). Non-hazardous waste is disposed of at landfills together with other domestic waste. Incineration facilities currently in operation in Yangon and Mandalay for treating HCW have been found to be inefficient, running at low temperatures and without proper emission control systems, which has contributed to severe environmental pollution in surrounding areas (Karstensen et al., 2017).

Currently, Myanmar hosts one privately managed hazardous waste management facility, called Golden Dowa Ecosystems Private Limited, which has set up a treatment and disposal facility in the Thilawa Special Economic Zone (SEZ) area of Yangon. The facility maintains a sorting/stabilisation shed, processing plant for managing oil and gas wastes, wastewater and leachate treatment plant as well as a resident laboratory. Golden Dowa currently accepts wastes from industries located in Thilawa SEZ, including oil and gas companies, as well as chemical industries from across Yangon, such as a nearby gas turbine factory (managing used glycol and used chemicals), and a local beverages manufacturer (treating ETP sludge). However, the facility does not comprehensively manage all hazardous waste streams in Yangon, including HCW. It is important to note that Yangon has also established its first waste-to-energy (WtE) power generation plant near Mingalardon temporary waste dumping site with a capacity of 60 tonnes per day aimed at managing a small fraction of its domestic waste.

In other areas, respective township development committees (TDC) are tasked with collecting HCW from local HCFs, which has resulted in many regional health centres and sub-centres taking up the responsibility of independently managing such waste. While some HCFs have on-site disposal facilities for HCW, the majority of these facilities are substandard. In Mon State, approximately 88.2% of Primary Health Centres (PHCs) were found to be lacking effective HCWM treatment and disposal facilities, with data suggesting that burning HCW in open pits was the predominant method (78.5%) of final disposal (Win et al., 2019). The study also confirmed that in addition to a majority of the pits in Mon State lacking proper fencing sharp waste was often mixed with general waste, despite waste segregation previously conducted at source. Observable safety issues such as these pose additional infection transmission risks due to poor waste decomposition conditions.

Since the onset of the COVID-19 pandemic, Myanmar cities have faced added challenges with providing effective HCWM services. According to the MCDC, Mandalay City has been collecting as much as 10 tonnes of additional HCW daily from COVID-19 treatment and quarantine centres, making use of specially designated waste vehicles. The city allocates four such vehicles for daily waste collection from COVID-19 centres, each assigned with one driver and two waste workers, who operate between the periods of 7 am and 8 pm. Collected HCW is thereafter directed to the municipal incinerator facility located at the Taung Inn-Myauk Inn area, which operates 24 h a day. MCDC reports that due to a range of technical issues, including inadequate operation and maintenance, the machine is often out of commission, resulting in much of Mandalay's HCW diverted to the local final disposal site where it is buried underground.

Yangon City faces a similar set of circumstances, experiencing a shortage of workers and vehicles and lacking sufficient facilities to treat HCW generated in the wake of the COVID-19 pandemic. The city reports collecting approximately 20 tonnes of HCW per day from designated quarantine centres, making use of 18 waste vehicles, which in turn direct such waste to the local incinerator facility, located in Hlaing Tharyar Township. Furthermore, the city utilises a waste-to-energy plant (originally used for domestic waste) as well as the Htein Bin final disposal site for managing HCW. Lastly, with support from the Yangon Regional Development Committee, YCDC has also established a furnace (crematorium) in Phaunggyi Township to treat HCW from a temporary facility located at the Central Institute of Civil Service.

2.3   Institutional and Policy Framework

2.3.1   Policy Setup

The Constitution of the Republic of the Union of Myanmar, formulated in 2008, outlines the country's legal framework and enshrines the fundamental rights and duties of its citizens, including justice, liberty, equality, and continuation of peace and prosperity (The Republic of the Union of Myanmar, 2008). Myanmar's Constitution also stipulates that every citizen is afforded a right to basic healthcare in line with the country's health policy (Section 367), regardless of race, birth, religion, official position, status, culture, sex or wealth. Moreover, the Constitution also requires all citizens to support the Union to preserve and safeguard environmental conservation (Section 390). Myanmar is also a signatory to various multilateral environmental treaties and agreements, including the Convention on the Control of Transboundary Movements of Hazardous Wastes and their Disposal (Basel Convention) (1989), the 2030 Agenda for Sustainable Development (SDGs) (2015), and the Paris Agreement on Climate Change (2016). The country is also a party to the Stockholm Convention on POPs, the Vienna Convention and the Montreal Protocol (Thien et al., 2020; Premakumara et al., 2019).

Myanmar has also enacted several national policies and strategies relevant to environmental protection and waste management. Relevant environmental-related legislation includes the country's National Environmental Policy (2018), the Environmental Conservation Law (ECL) (2012), Environmental Conservation Rules (ECR) (2014), the Environmental Impact Assessment Procedures (EIAP) (2015), and the National Environmental Quality (Emission) Guidelines (NEQG) (2015). According to Myanmar's ECL formulated in 2019, the Ministry of Natural Resources and Environmental Conservation (MONREC) is responsible for implementing environmental conservation policies, including those pertaining to waste management (MOHS, 2019).

In this context, Myanmar's National Waste Management Strategy and Master Plan (NSWMSMP), (2018–2030), sets out an overarching vision and framework for achieving sustainable waste management. The NSWMSMP aims to address all waste media (solid waste, liquid waste/wastewater, and gaseous emissions) by outlining strategic programmes and actions to improve solid waste collection, promote 3R principles (reduce, reuse, and recycling), and strengthen intermediate treatment and disposal of waste. The NSWMSMP also sets a number of targets aimed at ensuring the sustainability of waste management services over the long term including through supportive financial mechanisms, sound policies, institutional governance and monitoring frameworks (Thien et al., 2020). Similarly, Myanmar's draft Master Plan for Hazardous Waste Management (2019) identifies actions to maximise proper collection and disposal of industrial, medical, and other hazardous wastes (Premakumara et al., 2019; MOHS, 2019).

At present, Myanmar does not have a specific national policy or legislative measure in place for addressing HCWM. Several health policies and regulations, such as the Public Health Law (1972) and National Health Policy (1993), refer to the importance of protecting public health and the environment. Myanmar's National Health Policy framework is also aligned with the country's National Comprehensive Development Plan (Health Sector) (2011 to 2030–2031). The plan outlines seven priority programmes, including strengthening the country's health system through supportive policies and legislation, ensuring universal health coverage, and enhancing health information systems. The HMM (2011) also provides technical guidance to HCFs operating in Myanmar, aimed at directing the safe management of HCW.

The HMM recommends all HCFs maintain the appropriate infrastructure and necessary equipment for HCWM, which includes advising hospitals to establish relevant HCWM committees chaired by a medical superintendent, supported by a microbiologist or pathologist appointed as secretary together with other positions such as a ward-in-charges, pharmacist, and hospital engineer.

However, current legislation in Myanmar does not specify how HCFs are to adequately fund, enforce, coordinate, and supervise the implementation of HCWM at multiple levels. In response to the growing emergence of communicable diseases including COVID-19, Myanmar's Occupational and Environmental Health Division, Department of Public Health, MOHS developed a series of Healthcare Waste Management Guidelines and SOPs for healthcare providers, health workers, and waste handlers as listed below (MOHS, 2019):

- Case management of COVID-19 disease at District/Township hospitals without an ICU.
- Standard Operating Procedures for Managing COVID-19 Cases for Hospitals.
- Decontamination and Disinfection Procedures for COVID-19.
- Management Protocol for COVID-19 Acute Respiratory Disease.
- Clinical Management Guidelines for COVID-19 Acute Respiratory Disease.
- Covid-19 Specific Healthcare Waste Management Procedures.
- Management of COVID-19 Quarantines.
- Paediatric Clinical Management Guidelines for COVID-19 Acute Respiratory Disease.
- Novel Coronavirus COVID-19 Decontamination and Disinfection.
- COVID-19 Laboratory Sample Collection Guidelines.
- Hospital Infection Control Guidelines (2016).
- Biosafety and Biosecurity Protective Procedures for Mekong Basin Cross-border.
- COVID-19 Acute Respiratory Diseases SOP for Hospitals.

2.3.2   Institutional Setup

The MOHS is responsible for developing and implementing the country's legal framework on HCWM at the national level, with regulations specific to both public and privately managed HCFs. Headed by a Union Minster, MOHS is divided into six functional departments, including (1) Department of Public Health and Medical Services, (2) Department of Human Resources for Health, (3) Department of Medical Research, (4) Department of Traditional Medicine, (5) Department of Food and Drug Administration, and (6) Department of Sports and Physical Education. While the provision of healthcare is largely decentralised, with many responsibilities devolved to villages, at the upper level, Myanmar's state or regional health department is responsible for planning, coordination, training, supervision, monitoring, and evaluation of health services. For instance, radiology and the operation of clinical laboratories is conducted at the upper level, whereas community health services are implemented mainly at the peripheral level or the township level. Public sector-operated Rural Health Centres (RHCs), sub-RHCs, Maternal and Child Health Centres (MCHCs), and Urban Health Centres (UHCs) provide ambulatory care and are the first line of access to formal healthcare services. Ultimately, however, the implementation of HCWM in Myanmar remains inadequate at all levels due to financial, technical, and human resource constraints. Continued efforts are needed to expand access, improve communication,and strengthen understanding among primary level health staff especially with regard to relevant HCWM guidelines. Over the long term, addressing these challenges will require dedicated support for capacity building, and resource mobilisation in order to improve implementation and enhance the quality of services.

3 Measures to Mitigate Impacts of COVID-19 in HCWM: Perspectives from the Triple R (Response, Recovery, and Redesign)

3.1 HCWM Sector Became more Vulnerable to the COVID-19 Pandemic

As discussed in the previous sections, HCWM in Myanmar faces a number of challenges many of which preceded the COVID-19 health crisis. At the policy level, the country lacks any specific policy or legal/ regulatory mechanism to support HCWM, with the exception of recent guidelines issued by the MOHS. The absence of an effective legal, regulatory, and policy framework presents obstacles to ensuring the efficient allocation of national budget for HCWM and the introduction of necessary enforcement mechanisms. Moreover, current legislation does not adequately address coordination issues such as the need for supervisory oversight over HCWM functions shared between national ministries and at different administrative levels. Although some technical and operational guidance has been issued by MOHS to all HCFs with regard to developing the supportive infrastructure, procuring the appropriate equipment, and allocating requisite budgets for HCWM, many HCFs in Myanmar have yet to do so. Similarly, HCFs are not practicing source separation and proper labelling of HCW to an adequate standard. These challenges are further complicated by the wide unavailability of instruments and technologies to collect, treat, and dispose of HCW in an environmentally sound manner across much of the country, with open burning and dumping continue to be widely employed. These capacity constraints, together with a lack of early preparation, have both contributed to increasing the country's vulnerability to the spread of COVID-19 while also posing additional challenges for HCWM. These include:

- *Unanticipated growth in HCW generation:* HCW volumes have increased rapidly, creating an additional burden on HCFs that are already coping with increasing numbers of COVID-19 patients. The use of single-use plastics (SUP) has also grown due to the use of plastic-derived personal protective equipment (PPE), including gloves, masks, disinfectant bottles, as well as packaging material.

- *Diminished capacity of waste collection, treatment, and disposal systems:* HCW collection and treatment systems in all CDCs and Townships are overloaded, leading to stockpiling of HCW and potentially inadequate disposal. The growing reliance on temporary COVID-19 clinics and quarantine centres resulting from the unavailability of beds in many existing hospitals pose challenges for assigning specific workers and vehicles for the effective collection and treatment of HCW. The limited utilisation of source separation, reduction techniques, and recycling practices has further compounded these challenges. Although the immediate factor contributing to the slowdown of waste activities is the perceived risk of COVID-19 transmission, the majority of waste collected is currently directed to incineration, landfills or temporary dumpsites, which is countering progress on recycling efforts.
- *Worker safety issues:* The COVID-19 pandemic is endangering the safety of frontline healthcare workers involved in the handling, sorting, collection, transport, treatment, and disposal of HCW, many of whom already operate in challenging conditions. These workers are in the need of adequate PPE and other provisions, yet many HCFs lack the necessary inventory resulting from an increased reliance on volunteers, which has led to budgetary pressures and the wider unavailability of supply. Moreover, the rising number of COVID-19 infections presents additional challenges for continued safety monitoring, as many patients must self-quarantine at home. Another compounding issue concerns the management of household waste, which is managed through domestic collection systems: many of Myanmar's sanitation workers are not equipped with necessary PPE.
- *Livelihood concerns for the informal sector and small and medium enterprises (SMEs):* In Myanmar, the informal sector and SMEs are largely responsible for carrying out waste recycling. Many individuals are unable to conduct this critical service due to safety issues, who also face reduce incomes and interruptions of livelihoods due to COVID-19 lockdowns and the associated decline of recycling activities.

3.2 Planning HCWM Based on Triple R Framework (Response, Recovery, and Redesign)

Much like other countries, the Government of Myanmar has introduced some immediate actions to respond to the COVID-19 pandemic including with the support of WHO and other international and development partners. For instance, Myanmar has implemented country wide lockdowns, including closing its borders, suspending visas on arrival, and issuing public recommendations to avoid crowds and organising mass events (i.e., restricting groups to no more than five). Government agencies have also passed restrictions leading to the cancellation or postponement of large meetings, such as workshops, events, and other ceremonies. With support from WHO, Myanmar's MOHS developed a flash appeal aimed at mobilising immediate financial resources for responding to the COVID-19 pandemic. The MOHS also prepared the Myanmar Health Sector Contingency Plan on COVID-19 and Other Emerging Respiratory Illnesses (the Contingency Plan), aligned with WHO recommendations and requirements. Several other development partners, including World Bank and the ADB, have provided direct loans and related financial assistance with a view towards strengthening the country's response to the pandemic. Myanmar also convened a National-Level Central Committee on Prevention, Control and Treatment of Coronavirus Disease in 2019 under Presidential Order (45/2020) in order to spearhead national response efforts, which included coordinating with public authorities, the private sector as well civil society to control and manage the COVID-19 outbreak.

While the abovementioned efforts and funding strategies have included measures for addressing HCWM to a limited extent, Myanmar still needs to prioritise HCWM as an urgent and essential public service in order to minimise potential secondary impacts on health and the environment as part near term response and recovery measures. Furthermore, in view of the continued challenges Myanmar faces in addressing the COVID-19 crisis, the country also needs to redesign its HCWM system to be more resilient over the long term. In this context, this section discusses a range of policy-oriented and practical measures to achieve these objectives, based on the Institute for Global Environmental Strategies (IGES) Triple R Framework. The Triple R Framework—short for COVID-19 Response, Recovery, and Redesign— aims to advance integrated, evidence-based policy solutions for managing the pandemic while guiding more transformative societal changes (Zusmn et al., 2020; Mori et al., 2020). This calls for policymakers in Myanmar to link together (1) immediate responses, (2) broader recovery policies (including subsidy reforms) backed by stimulus funding, and (3) forward-looking redesigns of infrastructures and institutions aimed at aligning interests for mitigating the impacts of COVID-19 and supporting sustainable HCWM.

3.2.1 Response

Response refers to interventions that address immediate emergency needs while also helping to enhance the well-being of those adversely impacted by the health crisis. The following key measures provide examples of potential areas for tackling the rapid increase of HCW resulting from the COVID-19 pandemic:

Ensure the Health, Safety, and Implement Necessary Welfare Measures for Waste Workers All sanitation workers, both formal and informal, must be protected from health risks. The enforcement of basic hygiene measures (e.g. regularly washing hands) as well as the provision of regular health examinations (e.g. daily body temperature check) need to be carried out. Safe work practices (e.g. rotating work shifts) and the dissemination of adequate PPE must be made available. In addition, all employers and workers need to be informed and educated about the risks and hazards associated with exposure to COVID-19, which includes conducting training on workplace health protocols and the appropriate use of PPE. In addition to physical protection measures, it is also critical to ensure waste workers are protected from stigma, which includes social safety guarantees aimed at preventing disruptions to livelihoods and health. For example, the UK Government has granted sanitation employees "key worker" status, which entitles their children to in-school childcare throughout the COVID-19 crisis so that they may continue to provide critical waste services (Government of UK, 2021).

Prepare Contingency Plans for HCWM All cities and their respective HCFs need to have a contingency plans for HCWM in place. These can be prepared independently, appended to existing city waste management strategies or developed as a more general contingency plan. The contingency plan should outline measures ensuring that HCWM can continue uninterrupted and do not pose additional health risks or exacerbate negative impacts of the pandemic. In so doing, the plans should also assess existing capacities and outline immediate measures and potential alternative solutions for managing HCW during emergency operations, such as appointing staff, allocating available vehicles, and defining collection, treatment, and disposal methods. In addition, sufficient budget should be made available for the implementation of HCWM, specifically for the delivery of immediate actions, which includes purchasing safety equipment for workers, procuring additional storage bins, hiring/ renting collection vehicles, and establishing temporary treatment and disposal sites, inter alia. These interventions should be aligned with national contingency plans, and central and state budgets should be regularly reviewed to identify potential areas for earmarking funds from existing programmes in order to carry out activities related to COVID-19.

Establish Responsible HCWM Committees and Coordination Mechanisms Following the example of Myanmar's national-level Central Committee, all cities should establish respective HCWM committees to coordinate the planning and implementation of contingency plans aimed at preventing, controlling, and ensuring the proper treatment of HCW. Higher rank political and administrative staff from relevant departments should be tasked with chairing these committees, which in turn should coordinate activities with other external stakeholders.

Protect the Livelihood of the Informal Sector and Small and Medium Enterprises (SME) Given that many actors engaged in waste collection and recycling activities are informal workers or SMEs, actions must be taken to provide livelihood support including by coordinating with local waste associations, NGOs, and other relevant groups (Kojima et al., 2020). Supportive activities may include stipends for purchasing critical necessities, such as hygiene kits and food supplies. Moreover, urgent stimulus and safeguard measures should be introduced to ensure that informal workers and SMEs are capable of maintaining their business continuity, thereby minimising the negative effects of COVID-19 on livelihoods. Such measures should concentrate on protecting jobs and supporting business operation, such as the provision no-or-low-interest bridging loans, trade finance, working capital loans with flexible and deferred terms, tax relief, loan payment relief, rent/lease relief, grants, and wage subsidies in order to help keep workers on the payroll (ICC, 2020).

3.2.2 Recovery

Recovery measures refer to broader policy and spending decisions aimed at "building back better" from exogenous shocks. From a perspective of environmental sustainability, a *green and just* recovery is essential. The following measures offer the examples of potential recovery policy actions and programmes for HCWM in Myanmar:

Develop a National Policy and Regulatory Framework for HCWM A national HCWM policy and regulatory system needs to be established with a view to foster a successful and sustainable recovery that encompasses effective collection, treatment, and disposal of HCW. This proposed policy framework should drive decision making at the highest political level while mobilising government efforts and resources in supporting respective HCFs with pooling efforts to implement effective HCWM. The framework should be designed in line with national needs, priorities, and capacities, while also taking into account various international agreements and conventions that have been nationally adopted concerning public health, sustainable development, environment, climate change, and safe management of hazardous waste. In order for the effective implementation of the framework, national regulatory actions should clearly outline the respective roles and responsibilities of each actor involved in HCWM, ranging national healthcare providers to HCFs. The envisaged framework should also articulate main compliance mechanisms and methods of enforcement.

Improve HCW Treatment and Disposal Capacities In order to build upon and strengthen HCW treatment and disposal systems, priority should be given at the recovery stage to enhance the capacities of national and local waste management authorities with a view towards minimising potential hazards posed by HCWM and endeavouring to protect the environment. Identified treatment and disposal methods should be evaluated in the context of the waste management hierarchy with selected technologies adapted to the waste characteristics of HCFs, noting capabilities and requirements, environment and safety factors, as well as associated costs. Treatment technologies should employ thermal, chemical, irradiative, biological or mechanical processes with the most common types of treatment based on autoclaved, integrated or hybrid steam-based treatment systems, microwave treatment technologies, dry-heat treatment technologies, chemical treatment technologies, and incineration (Chartier et al., 2014). The technologies should be supplemented by allocating an annual budget for periodic maintenance and repair.

Restore, Protect, and Expand Recycling Services Downward pressure on oil prices during the COVID-19 pandemic have contributed to price volatility in recycling markets, further underlining the need for waste prevention and avoidance policies. Cities should consider additional measures for supporting waste recovery activities including by promoting recycling markets while resisting any industry push for new plastic manufacturing.

3.2.3   Re-Design

Redesign refers to the restructuring of infrastructure, institutions, and wider socio-economic systems with a view towards advancing a sustainable and resilient post-COVID-19 future. Strengthening public health systems as part of building long-term resilience to pandemics and environmental crises is a critical goal of redesign, which includes a range of potential measures:

Shift towards Circular Economy Models by Optimising Local Recovery of Resources circular economy (CE) strategies and solutions need to be fully implemented to ensure sustainable HCWM operations. The adoption of CE business models will not only facilitate the diversion of HCW from final disposal sites, redirecting these waste streams to recycling plants, it will also assist in supporting waste prevention and overall avoidance. Such models will also help in recovering local resources and materials, adding value to continued development efforts. For instance, many CE activities, including waste recycling and repair, are highly labour intensive: targeted investments in related sub-sectors can stimulate rapid service-sector led job creation, which is critical for rebounding from the health crisis. In addition, source separation, including the removal of hazardous from non-hazardous waste and recyclable from non-recyclable materials is also important for the successful realisation of CE. Lastly, developing local markets for resource recovery, including by supporting reuse, and recycling activities will help cities withstand negative global pressures and build resilience against future economic shocks. Cities can take steps to develop local markets for refurbished products, secondary materials and the reuse of goods, as well as support "right to repair" legislation.

Integrate Waste Management into Disaster Management Planning and Preparation waste management, including HCWM, should be a key consideration in disaster management planning and preparation, which currently only focuses on debris removal. Disaster management planning response measures and associated guidelines need to be reviewed and adjusted to better account for the dynamics of waste generation during the current as well as future pandemics. In addition to such provisions, disaster waste management planning guidelines should also outline relevant processes for convening state and local actors, including areas for participatory engagement of communities. Concerned actors should be well trained on the management of potentially hazardous HCW, including through developing national and international knowledge-sharing platforms defining best practices on HCWM.

4 Conclusions

This paper highlighted some of the urgent issues and challenges surrounding HCWM in Myanmar in context of the COVID-19 pandemic. Developing countries such as Myanmar face important challenges including limited capacities, knowledge gaps, deficient infrastructure, and lack of resources, and in consequence are highly susceptible to external shocks such as COVID-19. As the unprecedented increase in HCW volumes continue to exert pressure on waste collection, transport, treatment, and disposal systems in Myanmar, efforts should be made to protect the safety of frontline healthcare workers as well as the livelihoods of formal and informal waste workers involved in HCWM. Several identified issues were found to be common across major Myanmar cities and local HCFs, giving rise to some general recommendations. First, potentially contaminated waste from indiscriminately discarded facemasks, gloves, and other HCW mixed domestic waste streams pose health risks to sanitation workers. HCW collected from quarantined households, centres, and hospital wards needs to be managed separately from conventional waste management operations. While plastic usage is expected to increase due to the increased demand for PPE and associated healthcare items, ensuring proper HCWM, such as by conducting source separation (segregating hazardous and non-hazardous waste), implementing proper labelling, and monitoring the safe handling, collection/ transport, treatment and recycling of plastic waste, can reduce and prevent adverse impacts on public health and the environment. Policymakers should pursue a more inclusive and integrated approach to HCWM by way of linking (1) immediate responses, (2) broader recovery policies, and (3) a forward-looking redesign of infrastructure and institutions, aligning interests around bold and transformative change for mitigating the impacts of COVID-19 and supporting sustainable HCWM in Myanmar.

References

ADB. (2020). *Managing infectious medical waste during the COVID-19 pandemic*. ADB. Retrieved 21 August, 2021, from https://www.adb.org/publications/managing-medical-waste-covid19

Agamuthu, P., & Barasarathi, J. (2020). Clinical waste management under COVID-19 scenario in Malaysia. *Waste Management & Research, 39*, 1–9.

Aung, T. S., Luan, S., & Xu, Q. (2019). Application of multi-criteria-decision approach for the analysis of medical waste management systems in Myanmar. *Journal of Cleaner Production, 222*, 733–745.

Central Statistical Organisation (CSO). (2019). *Myanmar statistical yearbook 2019, Ministry of Planning and Finance*. The Government of the Republic of the Union of Myanmar.

Chartier, Y., Emmanuel, J., Pieper, U., Pruss, A., Rushbrook, P., Stringer, R., Townend, W., Wilburn, S., & Zghondi, R. (2014). *Safe management of wastes from health-care activities* (2nd ed.). WHO.

Frontier. (2021). *As COVID-19 surges, prices of medical goods follow suit, 16 July 2021*. Frontier. Retrieved 11 August, 2021, from https://www.frontiermyanmar.net/en/as-covid-19-surges-prices-of-medical-goods-follow-suit/

Gerin, R. (2020). *Myanmar fights COVID-19 in biggest city and in war refugee camps*. RFA – Radio Free Asia. Retrieved 15 March, 2021, from https://www.rfa.org/english/news/myanmar/yangon-covid-19-09282020220512.html

Government of UK. (2021). *Guidance: Children of critical workers and vulnerable children who can access schools or educational settings*. Retrieved 20 March, 2021, from https://www.gov.uk/government/publications/coronavirus-covid-19-maintaining-educational-provision/guidance-for-schools-colleges-and-local-authorities-on-maintaining-educational-provision

Htut, M. (2019). *A Study on Medical Waste Management System and Practices among the Health Care Personal of No. (2) Military Hospital in Yangon. A thesis submitted in partial fulfilment of the requirements for the Master of Public Administration (MPA) Degree*. Yangon University of Economics.

ICC. (2020). *ICC COVID-19 response: A call to action to save our SMEs*. Retrieved 15 March, 2021, from https://iccwbo.org/content/uploads/sites/3/2020/03/2020-sos-call-to-action.pdf

Jang, Y. C., Lee, C., & Yoon, O. S. (2006). Medical waste management in Korea. *Journal of Environmental Management, 80*, 107–115.

Kapoor, K., Calisesi, F., Kapp, N., Takeuchi, N., & Jinno, S. (2020). *Strategy guidance: Solid waste management response to COVID-19, UN-Habitat*. Retrieved 15 February, 2021, from https://unhabitat.org/sites/default/files/2020/05/un-habitat_strategy_guidance_swm_reponse_to_covid19.pdf

Karstensen, K. H., Saha, P. K., Win, T. A., Oo, K. T., & Mar, Y. Y. (2017). *Generation and management of hazardous wastes in Myanmar*. ECD, MONREC.

Khaing, A. T. (2020). *Mandalay COVID-19 centers produce 10 tonnes of waste daily*. Myanmar Times. Retrieved 03 March, 2021, from https://www.mmtimes.com/news/mandalay-covid-19-centres-produce-10-tonnes-waste-daily.html

Kojima, M., Iwasaki, F., Johannes, H. P., & Edita, E. P. (2020). *Strengthening Waste Management Policies to Mitigate the COVID-19 Pandemic. Policy Brief, No. 2020-05*. ERIA.

Lie, Y., Yu, X., Wu, X., Wang, J., Yan, X., Jiang, S., & Chen, Z. (2021). Emergency response to the explosive growth of health care wastes during COVID-19 pandemic in Wuhan, China. *Resources, Conservation & Recycling, 164*, 105074.

MOHS. (2014). *Myanmar essential health service access project: Environmental management plan*. The Republic of Union of Myanmar.

MOHS. (2016). *Greater Mekong subregion health security project*. Initial Environmental Examination.

MOHS. (2019). *Environmental management plan: Essential health services Access Project & Additional Financing*. The Republic of Union of Myanmar.

MOHS. (2021). *Coronavirus Disease 2019 (COVID-10) Surveillance Dashboard of Myanmar*. Retrieved 09 March, 2021, from https://mohs.gov.mm/Main/content/publication/2019-ncov

Mori, H., Zusman, E., Kojima, S., Waters, N. A., Mader, A., Onogawa, K., Kawazu, E., Janardhanan, N. K., Hengesbaugh, M., Otsuka, T., & Takahashi, Y. (2020). *Implications of COVID-19 for the environment and sustainability (version 2): Perspectives from the triple R (response, Recovery and Redesign) Framework*. IGES. Retrieved 12 March, 2021, from https://www.iges.or.jp/en/pub/covid19-ver2-e/en

Nikkei Asia. (2021). *Myanmar faces COVID surge amid lack of medics and jabs, 29 June 2021*. Nikkei. Retrieved 11 August, 2021, from https://asia.nikkei.com/Spotlight/Myanmar-Crisis/Myanmar-faces-COVID-surge-amid-lack-of-medics-and-jabs

Nor Faiza, M. T., Noor Artika, H., & Yusof, M. Z. (2019). Health care waste management and sustainable development goals in Malaysia. *Journal of Wastes and Biomass Management (JEBM), 1*(1), 18–20.

OCHA. (2020). *Myanmar: COVID-19: Situation report No. 01*. Myanmar.

Premakumara, D. G. J., Hengesbaugh, M., & Hlaing, O. M. T. (2016). *Quick Study on Waste Management in Myanmar: Current Situation and Key Challenges*. IGES.

Premakumara, D. G. J., Hlaing, O. M. T., Maw, A. M., & Hengesbaugh, M. (2019). *Status of solid waste management in Myanmar: Challenges and opportunities, in sustainable waste management challenges in developing countries* (pp. 223–247). IGI Global.

Price, A. D., Cui, Y., Liao, L., Xiao, W., Yu, X., Wang, H., Zhao, M., Wang, Q., Chu, S., & Chu, L. F. (2020). *Is the fit of N95 facial masks effected by disinfection? A study of heat and UV disinfection methods using the OSHA protocol fit test. medRxiv* 2020.

Regan, H. (2021). *Delta variant is ravaging the world but it's pushing Southeast Asia to breaking point, 5 August 2021*. CNN. Retrieved 11 August 2021, from https://edition.cnn.com/2021/0 8/04/asia/southeast-asia-delta-covid-explainer-intl-hnk/index.html

Saha, P. K., Karstensen, K. H., Maw, A. M., & Hlaing, O. M. T. (2020). *COVID-19 situation in Myanmar and its impact on waste management*. SINTIF Community.

Scheinberg, A., Woolridge, A., Humez, N., Mavropoulos, A., Filho, C. S., Savino, A., & Ramola, A. (2020). *Waste management during the COVID-19 pandemic*. ISWA. Retrieved 21 January 2021, from https://www.iswa.org/fileadmin/galleries/0001_COVID/ISWA_Waste_Manage ment_During_COVID-19.pdf

Sharma, H. B., Vanapali, K. R., Cheela, V. R. S., Ranjan, V. P., Jaglan, A. K., Dubey, B., Goel, S., & Bhattacharya, J. (2020). Challenges, opportunities and innovations for effective solid waste management during and post COVID-19 pandemic. *Resources, Conservation and Recycling, 162*, 105052.

Singh, N., Tang, Y., & Ogunseitan, O. A. (2020). Environmentally sustainable Management of Used Personal Protective Equipment. *Environmental Science & Technology, 54*(14), 8500–8502.

Sinha, R., Michelsen, J. D., Akcura, E., & Nijie, L. (2020). *COVID-19's impact on the waste sector*. IFC.

Taung, G. S. (2010). *Study on environmental health situation in healthcare setting (general practitioner clinics) at Lashio*. University of Public Health.

The Global Fund. (2020). *Technical brief: Sustainable health care waste management*. TG Fund.

The Republic of the Union of Myanmar. (2008). *Constitution of the republic of the Union of Myanmar*. Ministry of Information.

Theresa, N. (2021). *Myanmar plunges into deadly third COVID-19 wave, 19 July 2021*. The Diplomat. Retrieved 11 August, 2021, from https://thediplomat.com/2021/07/myanmar-plunges-into-deadly-third-covid-19-wave/

Thien, H. M., Maw, M., Thin, N. N., Latt, H., Onogawa, K., Premakumara, D. G. J., Hengesbaugh, M., & Hlaing, O. M. T. (2020). *National Waste Management Strtaegy and master plan for Myanmar (2018–2030)*. ECD, MONREC. Retrieved 15 March, 2021, from https://www.iges.or. jp/en/pub/nwmsamp-myanmar-2018-2030/en

Tsukiji, M., Premakumara, D. G. J., Pratomo, I. S. Y., Onogawa, K., Alverson, K., Honda, S., Ternald, D., Dilley, M., Fujioka, J., & Condrorini, D. (2020). *Waste management during the COVID-10 pandemic: From response to recovery*. UNEP and IGES. Retrieved 21 January 2021, from https://www.iges.or.jp/en/pub/waste-management-during-covid-19-pandemic-response-recovery/en

WHO. (2014). *Safe management of wastes from health care activities*. World Health Organization. Retrieved 11 August, 2021, from https://apps.who.int/iris/bitstream/handle/10665/85349/ 9789241548564_eng.pdf?sequence=1

WHO. (2017). *Report on health-care waste management (HCWM) status in countries of the South-East Asia region (SEA region)*. World Health Organization.

WHO. (2019). *Overview of technologies for the treatment of infectious and sharp waste from health care facilities*. World Health Organization.

WHO. (2020). *Prevention, identification and management of health worker infection in the context of COVID-19*. Retrieved from https://apps.who.int/iris/bitstream/handle/10665/336265/WHO-2019-nCoV-HW_infection-2020.1-eng.pdf?sequence=1&isAllowed=y

WHO and UNICEF. (2019). *WASH in health care facilities: Global baseline report 2019*. World Health Organization.

WHO and UNICEF. (2020). *Water, sanitation, hygiene and waste management for SARS-CoV-2, the virus that causes COVID-19*. Retrieved 15 February, 2021, from https://www.who.int/publications/i/item/WHO-2019-nCoV-IPC-WASH-2020.4.WHO/2019-nCoV/IPC_WASH/2020.4

Win, A. (2020). Rapid rise of COVID-19 second wave in Myanmar and implications for the Western Pacific region. *QJM: An International Journal of Medicine, 113*(12), 856–857. https://doi.org/10.1093/qjmed/hcaa290

Win, E. M., Saw, Y. M., Oo, K. L., Than, T. M., Cho, S. M., Kariya, T., Yamamoto, E., & Hamajima, N. (2019). Healthcare waste management at primary health centers in Mon State, Myanmar: The comparison between hospital and non-hospital type primary health centers. *Nagoya Journal of Medical Science, 81*, 81–91.

Windfeld, E. S., & Brooks, M. S. (2015). Medical waste management – A review. *Journal of Environmental Management, 163*, 98–108.

World Bank. (2020). *Myanmar's economy hit hard by second wave of COVID-19: Report. 15 December 2020*. Retrieved 11 August, 2021, from https://www.worldbank.org/en/news/press-release/2020/12/15/myanmars-economy-hit-hard-by-second-wave-of-covid-19-report

You, S., Sonne, C., & Ok, Y. S. (2020). COVID-19's unsustainable waste management. *Science, 368*(6498), 1438.

Zusmn, E., Kawazu, E., Mader, A., Watabe, A., Takeda, T., Lee, S. Y., Hengesbaugh, M., Otuka, T., Singh, R. K., Premakumara, D. G. J., Onogawa, K., Elder, M., Zhou, X., Moinuddin, M., Kojima, S., Kumar, P., Nugroho, S. B., Janardhanan, N. K., Waters, N. A., ... Takahashi, Y. (2020). *A sustainable COVID-19 response, Recovery and Redesign: Principles and Applications of the Triple R Framework*. IGES. Retrieved 15 March, 2021, from https://www.iges.or.jp/en/pub/tripler/en

Part V
Healthcare Waste Management and Impact of COVID-19

Healthcare Waste Management Practices in Nigeria: A Review

David O. Olukanni, Justin D. Lazarus, and Emmanuel Fagbenle

1 Introduction

In most low-income economies, healthcare waste (HCW) management, a major component of solid waste presents a serious concern. HCW being categorized as an hazardous or infectious solid wastes are unwanted products of healthcare activities that are generated during diagnosis or treatment of patients, immunization or vaccination of human beings or animals mostly during production or biological testing (Idowu et al., 2013; Olukanni et al., 2014; Chukwunonye, 2015; Awodele et al., 2016; Afolabi et al., 2018; Akpan & Olukanni, 2020). Hierarchical structure and complexity characterize HCW and healthcare facilities. The categories of HCW range from infectious, pathogenic, and sharp to genotoxic, chemical, and radioactive waste, each with a complement of various health risks. They include basically hazard prone materials such as blades, needles, and syringes, known as sharps, then non-sharps including bandages and swabs; body/tissue/organs or blood parts and various unfriendly chemicals and solvents such as mercury and disinfectants, radioactives, and pharmaceuticals (Longe, 2012; Oyekale & Oyekale, 2017). Not only that, but some other major healthcare by-products also include unused and expired drugs, pharmaceutical products, and vaccines that need to be disposed of immediately and appropriately.

These wastes may also include thrashed items contaminated in the process of pharmaceuticals usage and handling ranging from boxes, bottles, and related materials containing residues, drugs, surgical instruments, and workers' protective equipment (Uwa, 2014). Thus, medical waste is an unavoidable concern to healthcare workers and government at all levels because it can be both infectious and hazardous. As a result, poor management/improper disposal of these HCW may lead to

D. O. Olukanni (✉) · J. D. Lazarus · E. Fagbenle
Department of Civil Engineering, Covenant University, Ota, Nigeria
e-mail: david.olukanni@covenantuniversity.edu.ng

© The Author(s), under exclusive license to Springer Nature Singapore Pte Ltd. 2022
S. K. Ghosh, P. Agamuthu (eds.), *Health Care Waste Management and COVID 19 Pandemic*, https://doi.org/10.1007/978-981-16-9336-6_9

diseases and injuries to healthcare workers and endanger the environment. Several healthcare services are in the center of the cities, therefore littering of any wastes that is not adequately handled will cause harmful diseases and pose a possible risk to the local community.

Nigeria, one of the developing nations, has health challenges competing for scarce resources; this is not shocking that HCW treatment attracts little consideration. Therefore, in developed nations, where there are no formal provisions for healthcare waste treatment, there is a significant problem. In addition to municipal waste, clinical waste is freely disposed of in the dumpsite, and the practice encourages the people in the neighborhood to do the same. A report from the world health organization (WHO, 2018), 50% of the overall waste produced in the medical facility is hazardous and should be carefully separated at the source of generation to escape the danger of 100% of all hospital waste from being hazardous. The vulnerability of healthcare waste (HCW) and its management has gained considerable interest from health practitioners and non-professionals around the world. If healthcare facilities are aware of the types and quantities of clinical waste generated, the planning and budgeting of adequate revenue for the maintenance of hazardous waste would be facilitated.

Established gaps, such as the absence of bags with color codes at the point of use for HCW segregation, lack of segregation, and recycling guidelines for health workers, contribute to inadequate handling of waste generated in hospitals. Mishandling of HCW by healthcare facilities does not only cause medical problems to patients and hospital staff, but to the guests of patients and the environment where they are poorly discarded. Medicare programs are structured to maintain the health of individuals in their society and not to create possible health hazards. While healthcare waste management continues to present several problems, much research on the disposal of healthcare waste has continued to evolve. However, limited study has been performed on HCW segregation, which is a critical feature of HCW management. The division of waste in healthcare facilities aims to reduce the volume of radioactive waste. Once healthcare waste is separated, it would be easier to gather, adequately preserve, and dispose of infectious waste in a manner that would not affect the health of staff, patients, and the community (WHO, 2018). Proper healthcare waste management relies on effective organization, proper resources, and the active presence of qualified professionals.

The management of HCW must be taken as a fundamental part of healthcare facility hygiene and contamination control and management. This would help healthcare wastes generated within and around the healthcare facility follow a well-defined and appropriate track from the area of generation to the point of final discharge. Standard healthcare waste management should involve segregation of the waste, collection, storage and treatment, transportation and safe disposal, and follow-up monitoring of the process (Ezirim & Agbo, 2018; Olukanni & Oresanya, 2018). However, information deficiency and economic weakness could be accrued to the increasingly HCW management inadequacies thereby compounding the adverse environmental and public health situations in most of the developing world nations. The biggest challenge lies in the fact that as the rate of healthcare service delivery activities are increasing, they are usually accompanied some waste materials, which then eventually accumulates and generates some health and environmental hazards to the society, but mechanisms to minimize these effects are mostly ignored or unaware of.

Therefore, it was agreed by the healthcare professionals and policy makers that efficient process of managing Healthcare waste would make a formidable ground for ensuring quality healthcare service delivery (Oyekale & Oyekale, 2017). This cannot be ignored in Nigeria, because although there are guidelines in Nigeria on healthcare waste disposal to strengthen track of waste disposal system, but little to no effort is made for effective implementation designated monitoring bodies or the healthcare facilities managers (Oyekale & Oyekale, 2017).

Research shows that in most Nigerian cities, the different categories of solid wastes are neither defined nor properly managed, and thus, healthcare waste management toady remains a challenge in Nigerian cities as well as in many developing countries of the globe (Olukanni et al., 2020). Various researches have been made to investigate the situation of healthcare waste management in different states of the federation with corresponding significant results such that of (Oli et al., 2016; Awodele et al., 2016), etc. However, there is a need to have a broad picture of the general state of the healthcare management in Nigeria. To this end, this review report intends to bring on board the various activities individuals, organizations, and the government of Nigeria have and are putting in place to ensure proper healthcare waste management for human and environmental protection. This study will focus mainly on secondary data from literatures, which would include published article documents and web pages on subject matter.

1.1 Classification of Healthcare Wastes

Hospitals, animal testing centers, and labs produce healthcare waste, all of which can be reflected in hospitals. The most agreed and remembered by UNEP/WHO is that. Table 1 displays the Healthcare Waste (HCW) rating of the United Nations Environmental Program/World Health Organization (UNEP/WHO, 2005).

From Table 1, a detailed description of these types of healthcare waste is as follows:

1. Pathological waste: It consists of parts of the body or organs, tissues, or fluids such as blood. Even if healthy body parts can contain pathological waste, it must be considered as hazardous wastes for safety reasons.
2. Non-hazardous waste: It includes wool, kitchen waste, etc. that does not pose any handling issue, health, or environmental hazard. In patients' ward areas, out-patient department (OPD), kitchens, offices, etc., non-hazardous waste is generated.
3. Hazardous waste: Pathological, infectious, sharp, and chemical waste are included in the portion (specific definitions of these are given below). In labor

Table 1 UNEP/WHO Classification of Healthcare Waste (HCW)

Main group	Sub-group I	Sub-group II
A: Non-risk HCW	A: Recycle	
	A2: Biodegradable waste	
	A3: Other non-risk waste	
B: HCW requiring special attention	B1: Human anatomical waste	
	B2: Sharps	
	B3: Pharmaceutical waste	B31: Non-hazardous Pharmaceutical waste
		B32: Potentially hazardous pharmaceutical waste
		B33: Hazardous Pharmaceutical waste
	B4: Cytotoxic pharmaceutical waste	
	B5: Blood and body fluids	
C: Infectious and highly infectious waste	C1: Infectious waste	
	C2: Highly infectious waste	
D: Other hazardous waste		
E: Radioactive waste		

Source: UNEP/WHO (2005)

wards, operation theaters, laboratories, etc., hazardous waste are usually generated.

4. Infection waste: It is defined as waste containing sufficient building or quantity of pathogens that can lead to diseases when exposed to it, e.g., waste from operations with infectious diseases, contaminated plastic items, etc.
5. Sharps: These are described as anything that could cause a wound-inducing cut or puncture. The sharp waste comprises items such as needles, syringes, scalpels, knives, broken glass, etc.
6. Chemical waste: Substances consisting of discarded chemicals are usually cleaned and disinfected.
7. Pressurized containers: Complete or emptied containers or aerosol containers containing pressurized liquids, gas, or powdered materials.
8. Pharmaceutical waste: It includes pharmaceutical products, drugs, and vaccines, expired, unused, spilt, and contaminated. Discarded items used in handling pharmaceuticals such as bottles, vials, and linkage tubing are also included in this category.
9. Waste with a high heavy metal content: and derivatives, potentially highly toxic (e.g., cadmium or mercury from thermometers or manometers). They are regarded as a chemical waste sub-group, but they should be specifically treated.

10. Highly infectious waste includes microbial cultures and stocks from Medical Analysis Laboratories of highly infectious agents. They also have the body fluids of highly infectious disease patients.
11. Genetoxic waste: This is derived from medicinal products and commonly used in oncology or radiotherapy units with a highly dangerous mutagenic or cyto-toxic effect. Feces, vomiting or urine should be considered as genotoxic in patients treated with cytotoxic drugs or chemicals. Their proper treatment or disposal raises serious safety problems in specialist cancer hospitals.
12. Radioactive waste: comprises radionuclide-contaminated liquids, gases, and solids whose ionizing radiation has a genotoxic effect.

1.1.1 Biomedical Waste in Nigeria and Their Sources

Biomedical waste combined with domestic waste at Nigerian dump sites is the most common of all the issues associated with HCW in Nigeria. In rural areas, the people go to the waste dump to search for objects and things that could be useful for them and has become a habit for youngsters, and much less so in urban areas. The connection between unsafe handling of HCW and practice of collecting from dumpsites places by scavengers and children could make them sick and lead to the spread of diseases. The sources of biomedical waste must be addressed to tackle the problems associated with HCW management. Nigeria's sources of biological waste include medical facilities, mortuary, and autopsy centers, labs and research insti-tutes, laboratories for animal testing, blood donors, nursing homes, etc. Even though HCW requires careful management, it is unfortunate that only a few hospitals in most developing countries deem this a priority; it has become a significant problem and a major source of outbreaks of communicable diseases. The lifetime of certain infecting species is very long; anthrax will famously stay alive in the soil for many years. In certain Nigerian hospitals, such a situation has been identified. Yet many waste treatments are not handled properly and among surgical patients, staphylo-coccal and pseudomonas infections are prevalent.

1.1.2 Biomedical Waste Control Laws and Policies in Nigeria

It is considered that biological waste has significant hazards and the potential to pose a danger to public health. To avoid this danger to the environment and human health, many developed nations and worldwide bodies have adopted and implemented norms, legislation, rules, and recommendations for the treatment of biomedical waste. It was possible to enforce these rules and guidelines at the foreign, national, local, regional, and private levels. Numerous studies have said that there is no fixed standard for biomedical waste treatment in Nigeria.

Lagos State, one of Nigeria's most industrialized states, stands out as the only state with a set standard waste management scheme for the state government's healthcare. The Lagos State Waste Management Authority is the organization responsible for this (LAWMA). In 2006, after realizing the inappropriate disposal of waste from hospitals and labs. LAWMA delivered services to 2900 clinical centers in 22 local districts in June 2012. Latest studies have shown that in Lagos State, 394,625 kg of biological waste received pre-disposal treatment. Free distribution of colored containers for waste segregation and protective boxes to healthcare facilities was described as the most successful way to support the healthcare waste program. To ensure the success of the program, other measures were put into motion. This covers healthcare staff awareness workshops, adequate supervision and execution, annual stakeholder well-being meeting, and stringent laws on infectious waste management.

It is normal, though, to see a particular hospital with a distinctive set of waste management policies for the hospital. The Department of Public Health has set standards for appropriately dressing waste handlers, waste segregation, and waste disposal. Several techniques have been developed inside the hospital premises to reduce waste. But the question is, does the traditional waste management scheme follow these strategies in developing countries? Are the issues of disease prevalence and air degradation addressed by these policies? Obviously, the answer to these questions is "NO." Therefore, the government is responsible for setting a general model, regulations, and laws for all certified hospitals in the world, or adapting the same policy for individual states as the LAWMA (Olukanni & Oresanya, 2018).

1.2   Major Sources of Healthcare Wastes (HCW) in Nigeria

Since the beginning of epidemics such as Ebola, Lassa fever, and Corona virus, the world is striving to improve healthcare services and Nigeria is not left out of this effort. This has culminated to equivalent increase in medical care, laboratory analysis, treatment procedures, and processes that are generating more healthcare waste that needs to be safely managed. In 2018, through probability sampling Ezirim and Agbo (2018), were able to locate a total of 7667 private and public healthcare facilities in the country that could carry out HIV/AIDS services, out of which they selected a representative sample from each of the six (6) geopolitical zones to show the distribution of healthcare facilities in the country (Table 1). While the above authors did not specify the type of healthcare facilities, the World Health Organization has established that healthcare facilities includes all those institutions involved in the production or testing of biological materials such as clinics and hospitals, medical laboratories, and doctor's offices which includes dental and veterinary services (WHO, 1999, 2005). However, in Nigeria, most of the established findings on healthcare waste were seen to be conducted in the most organized or coordinated healthcare facility, the hospitals. These healthcare facilities are broadly categorized into primary, secondary, and tertiary healthcare facilities depending on whether it is governed and financed by the local, state, or federal government (Ezirim & Agbo, 2018). Treatment and disposal also vary according to the category.

Table 2 Representative distribution of healthcare facilities in Nigeria

Geopolitical Zone in Nigeria	Number of health facilities
North East	121
North West	171
North Central	426
South East	451
South West	226
South-South	526

Source: Ezirim and Agbo (2018)

On the side of healthcare facilities, the institutions range from teaching, specialist, general, and maternity hospitals; others are clinics, psychiatric, and orthopedic hospitals with different sizes, ownership, specialty, and competence (HCWM TWG, 2004). They are complemented by internal actors in the medical and paramedical fields, and their external counterparts in government and other non-medical specialists and professionals. This classes of healthcare facilities can either be private or public health institutions, giving precedence to the ownership by either the government or private individuals. As expected, the larger percentage of healthcare facilities are government-owned, estimated to be about 80%, while the remaining 20% are private healthcare waste producing facilities. However, the primary and secondary healthcare facilities owned by local and state government, out numbers the tertiary facilities owned by the federal government. While on distribution basis, the south-south zone of Nigeria has the greatest number of healthcare facilities and accordingly should be generating more healthcare wastes.

It was reported a few years back, it was cited by Idowu et al. (2013) that in 2006 on estimate, healthcare waste generation rate in Nigeria ranged between 0.562 kg/bed and 0.670 kg/bed daily but has gone as high as 1.68 kg/bed in a day by 2009. Although, the researchers admitted there may be some variations in the mode of waste generation in the different healthcare facilities and the techniques of their management in Nigeria. One of the papers consulted argued that tertiary healthcare facilities produced less or insignificant wastes because they are having modern equipment and most of their specialists are well trained in dealing with variety of health issues, hence, they mainly serve as referral hospitals to the primary and secondary hospitals in the country (Table 2). Like earlier highlighted in the introduction, healthcare waste is a concern in any society since it could be infectious/hazardous. These category of solid wastes are capable of transmitting infectious diseases when they come in contact with and contaminate pathogenic organisms and this makes hazardous healthcare wastes a source of momentous public health problems (Idowu et al., 2013) (Tables 3 and 4). According to the analysis of Nigerian guidelines on managing solid hazardous wastes, FEPA (1991) described healthcare wastes, especially the infectious ones as those arising from:

1. Residues and effects of culturing media and stored infectious agents.
2. pathological waste resulting from medical, surgical, pathological, and radiological processes.

Table 3 Types of healthcare waste generated from different categories of healthcare facilities

HCW generation rate by each facility (kg/day)				
Healthcare waste components	Tertiary	Primary	Secondary	Private facilities
Plastic and paper wastes	–	30.20	08.51	12.53
Sharp objects (needles and blades)	–	11.40	04.83	07.1
Swabs, solvents, or absorbents	–	12.90	02.04	03.01
Used drips and other pharmaceuticals	–	09.40	02.17	03.2
Infectious /contaminated wastes	–	03.30	00.56	00.8
Total daily waste generation rate (kg/d)	–	67.00	18.11	26.64
Average generation rate (kg/day)	–	00.45	00.31	00.39

Source: Olukanni et al. (2014))

3. human blood and tissue wastes derived from emergency handling, surgical, and medical operations.
4. sharps, e.g., blades and syringes resulting during patient care and in process of medical and research and even industrial laboratory activities.
5. Careless handling of body parts and animal carcasses, etc.
6. A large number also come from pathology, pharmaceuticals, research, or commercial laboratories.

1.2.1 Healthcare Facilities Location in Nigeria

In Nigeria, 34,173 recognized health facilities are registered, including 21,463 main public health facilities, 8290 primary private health facilities, 969 secondary public health facilities, 3023 secondary private health facilities, 73 tertiary public health facilities, and a list of health facilities in the nation. The revised Nigeria Health Facility Registry shows that Nigeria's total health facilities are about 41,098. There are 29,102 major public health centers, 1172 secondary public health centers, and 106 tertiary public health centers. Seven thousand twenty-five private primary health facilities, 3596 private secondary health facilities, and 97 private tertiary health facilities are also open. In the north central geographic region, healthcare facilities are highest in Nigeria with 9101 and lowest in the south-south sector with 4869. At the same time, the population size of the two areas is close.

There are also many healthcare services in the north-west and south-west regions with the largest population in the world. There are 36,127 primary health facilities, a total of 4768 secondary health facilities, and a total of 203 tertiary health facilities in Nigeria, as seen in Table 2. Apparent discrepancies are disclosed as the number of healthcare services is measured against the population. The Kano, Jigawa, and Rivers States are the states with the most needs. There are health facilities above the threshold ratio of 3500 to 1 in some states, e.g. the FCT, Nasarawa, Benue, and Niger. Mustapha (2017) noted that the best fitted are tertiary healthcare facilities which carry the brunt of treatment. Communicable disease care services are mainly given by secondary and tertiary clinical facilities worldwide.

Table 4 Distribution of population and location of healthcare facilities in Nigeria

State	Population	Public Pri.	Sec.	Tert.	Total (A)	Private Pri.	Sec.	Tert.	Total (B)	Total (A + B)	Geopolitical Zone
ABIA	2,845,380	787	18	6	811	222	245	2	469	1280	SE
ADAMAWA	3,178,950	1079	25	11	1115	89	13	–	102	1217	NE
AKWA-IBOM	3,178,950	580	67	2	649	40	196	3	239	888	SS
ANAMBRA	4,177,828	876	49	8	933	498	125	5	628	1561	SE
BAUCHI	4,653,066	1085	24	2	1111	68	6	2	76	1187	NE
BAYELSA	1,704,515	223	41	2	266	36	20	–	56	322	SS
BENUE	4,253,641	1285	36	4	1325	390	66	1	457	1782	NC
BORNO	4,171,104	615	44	5	664	35	9	–	44	708	NE
CROSS RIVER	2,892,988	1106	27	5	1138	88	82	3	173	1311	SS
DELTA	4,112,445	389	73	2	464	302	40	2	344	808	SS
EBONYI	2,176,947	662	17	2	681	67	44	–	111	792	SE
EDO	3,233,366	541	35	4	580	382	30	8	420	1000	SS
EKITI	2,398,957	378	22	3	403	78	29	–	107	510	SW
ENUGU	3,267,837	56	–	–	56	635	392	7	1034	1090	SE
FCT	1,405,201	300	22	3	325	509	83	5	597	922	NC
GOMBE	2,365,040	595	31	–	626	33	9	–	42	668	NE
IMO	3,927,563	695	40	–	735	217	375	–	592	1327	SE
JIGAWA	4,361,002	690	18	–	708	8	2	–	10	718	NW
KADUNA	6,113,503	1162	26	11	1209	253	64	13	330	1539	NW
KANO	9,401,288	1227	39	2	1268	118	129	14	261	1529	NW
KATSINA	5,801,584	1725	20	–	1745	63	18	1	82	1827	NW
KEBBI	3,256,541	909	22	3	934	10	23	–	33	967	NW
KOGI	3,314,043	918	54	2	974	56	79	–	135	1109	NC
KWARA	2,365,353	607	43	1	651	122	116	2	240	891	NC

(continued)

Table 4 (continued)

State	Population	Public				Private				Total (A + B)	Geopolitical Zone
		Pri.	Sec.	Tert.	Total (A)	Pri.	Sec.	Tert.	Total (B)		
LAGOS	9,113,605	450	49	1	500	952	809	–	1761	2261	SW
NASARAWA	1,869,377	1127	27	3	1157	325	16	3	344	1501	NC
NIGER	3,954,772	1418	30	2	1450	173	41	1	215	1665	NC
OGUN	3,751,140	745	32	1	778	232	126	–	358	1136	SW
ONDO	3,460,877	611	28	3	642	61	64	1	126	768	SW
OSUN	3,416,959	834	30	5	869	144	13	7	164	1033	SW
OYO	5,580,894	854	44	4	902	339	198	4	541	1443	SW
PLATEAU	3,206,531	949	17	2	968	240	20	3	263	1231	NC
RIVERS	5,198,605	417	31	–	448	36	52	4	92	540	SS
SOKOTO	3,702,676	787	17	3	807	23	14	3	40	847	NW
TARABA	2,294,800	1133	22	1	1156	160	24	–	184	1340	NE
YOBE	2,321,339	579	14	1	594	7	14	–	21	615	NE
ZAMFARA	3,278,873	708	28	2	738	14	10	3	27	765	NW
TOTAL	139,707,540	29,102	1172	106	30,380	7025	3596	97	10,718	41,098	

Source: Nigeria Health Facility Registry * FGN (2007), FMOH (2011, 2019)
FGN Federal Government of Nigeria, *FMOH* Federal Ministry of Health

1.3 Current Healthcare Wastes Collection and Handling Methods

The eruption of Lassa fever virus in Africa, which became a serious major health concern in Nigeria, posed a threat to so many citizens and healthcare workers mainly due to lack of proper handling of the wastes generated from the Lassa fever patients (Tobin et al., 2013). Similarly, Ebola outbreak in Nigeria in 2014 claimed people's lives and threatened the health of others and this attracted the attention of the World Health Organization that suggested the implementation of proper disposal of healthcare wastes, plus putting up measures to protect healthcare staff and their patients and other related personnel from contacting the deathly virus in the healthcare facilities (WHO, 2018). Recently, the COVID-19 pandemic also came with its challenges. Medical waste has increased owing to the mandated use of facemasks in public and other PPE used in healthcare facilities. In these COVID-19 times, the poor waste management practices raise new concerns for the country, given that recent research showed that the virus can last as long as 9 days on material surfaces (Kampf et al., 2020). The implication of this is that improper solid waste management would increase the risk of the spread of the virus in the country. Other issues are related to the availability of workers and staff for collection and disposal; co-mingling of general waste with COVID-19 waste, collecting waste from homes of individuals who would not go to isolation centers; capacity of Nigeria Center for Disease Control (NCDC) staff for handling post-COVID-19 Waste; Informal Workers and Their safety; How do we change behavior? Since then, various significant medical waste handling measures have been reported by researchers in different states of the federation. Some of such reports includes healthcare waste management studies conducted in south East states; including Owerri, and Enugu, in South-south states including Port Harcourt and Edo state, in the South west zone including Lagos and Ibadan, and in Northern zone of the country includes that of Zaria in Kaduna state and Jos, plateau state which was conducted earlier on in 2009 (Ogbonna, 2013; Joshua et al., 2014; Uwa, 2014; Adogu et al., 2015; Awodele et al., 2016). Cognizance of various factors such as level of awareness level of the health personnel to wastes management, types of waste generated and health policies, etc. are taken. The situation in most of the findings is not farfetched from what was expected as reported by the several authors.

1.3.1 Healthcare Waste Management Studies in Nigeria

Healthcare waste management (HCWM) issues in Nigeria have generated a lot of studies since the 1990s. An exploration to the literature depicts the approaches, themes, areas of concentration, arguments, and methodologies. A striking characteristic is an interest in the downstream aspect of HCWM founded on broad-based topics. For example, Akinwale and Sridhar (2010) examined non-segregation of HCW in rapidly increasing HCFs and the environmental problems caused by poor management of general waste, and the resurgence of sanitation-related diseases in Nigeria's urban centers. An identical urban environment quality stance, still from the downstream approach was adopted by Coker (2009) dealing with hospital waste in Ibadan with emphasis on the obstacle to and proposals for sustainable HCWM.

Abah and Ohimain (2011) discovered from their study of healthcare waste management in Nigeria that the level of healthcare waste management is zero and therefore, unsustainable. They also revealed no focal persons or waste managers responsible for healthcare waste management, preferably the orderlies in the various hospitals undertook to the sanitation of the hospital. According to their report, the personnel who understood the management in the provision of awareness programs were not in place. The research of Sridhar and Coker (2009) suggested that each facility's technical, human, and financial resources should be kept in mind while selecting the management method. However, the Coker (2009) study concluded that poor HCW management is a major cause of some illnesses, some of which are life-threatening. Health workers are more susceptible to non-communicable diseases, such as nurses, pharmacists, and waste handlers, including sweepers and other auxiliary staff, as well as patient relationships. Proper disposal at dump sites was suggested by the study of Coker (2009). The study found that at Aba-Ekuin Ibadan, which is indeed a bad practice, there is no sorting and segregation of these wastes at the source to be disposed of. To help the situation (Coker, 2009), part of which should be political will, a holistic approach should be adopted by integrating waste management with any HCF development from the outset.

As a representative of healthcare institutions in Port Harcourt city, Nigeria, the study by Ogbonna (2011) carried out in five different hospitals to investigate the category of medical waste and its management practices observed that major facilities, such as incinerator were not available in hospitals. In addition, there are no waste segregation or color coding in the different hospitals, nor do they keep records of wastes for the various waste streams. Furthermore, both hospital waste and domestic waste have been found to be treated equally. A common feature in the survey was a lack of relevant training and protective equipment for waste handlers. There was a lack of human and financial resources and institutional capacity necessary to manage medical waste effectively to improve the protection of human life and the environment from health hazards. It was further shown from the findings of Ogbonna et al. (2012) that all the hospitals surveyed fell below the recommended practices for waste management as maintained by the World Health Organization and other regulatory authorities. The study showed that waste for the various waste streams was not segregated into marked or color-coded containers/bins. No records of generated waste and disposal have been kept, no training of personnel on waste management and the provision of safety equipment and adequate education on reducing waste strategies have been provided.

Lessons from the above review of HCWM literature in Nigeria are of great relevance to the topic of this study. All the studies adjudged HCWM in Nigeria to be poorly managed. The studies highlight the precarious position of HCWM in Nigeria in several critical areas. Coker (2009) also underscored the existence of little knowledge of hospital waste management in Nigeria especially in the areas of record-keeping of the actual quantities of medical waste generated, and lack of any HCWM plan. They also held the inexistence of HCWM policy, plans, and the inadequacies of government regulatory body as the reason for the absence or guidelines. Other deficiencies identified include inadequate training of HCW handlers, staff training and awareness and education to influence the management of hazardous waste the situation in Nigeria has been found to exist in other developing countries characterized by the absence of national regulations for HCW management, low awareness about HCW management by personnel, and waste workers. Knowing the great importance of the health hazards posed by improper management of healthcare waste, it reassures that this study is critical.

Ezirim and Agbo (2018) researched 1921 health facilities selected from the six geopolitical zones in Nigeria for their study based on probability proportional to size. All the facilities selected provided HIV/AIDS services. 78% of the health facilities were government-owned, while 22% were private health facilities. 44.8% had medical waste management adopted and 41.4 had personnel or department monitoring the waste segregation and disposal. Medical staff training has enhanced a few waste management's sectors but will not address the price and cost of waste management materials and infrastructure. An example of Malaysia, where staff were trained but poor waste management continued as there were no facilities and instruments required for waste segregation. The findings of the current study confirm that implementing the 2013 national waste management policy for healthcare in Nigeria has improved waste management practices. By providing policies and resources, proper waste management can be encouraged by the government.

An analysis carried out by Olubukola at two general hospitals in Lagos showed that, due to the lack of quantification of healthcare waste, there was no waste control policy in hospitals. This lack of a plan for the treatment of healthcare waste ultimately contributes to incomplete segregation of waste at the point of use, processing, recycling, and final disposal. This ineffective way of handling hospital waste poses health problems for health staff, patients, and the community. Similarly, the management of healthcare wastes by healthcare facilities was assessed in Lagos metropolis by Idowu et al. (2013) with the help of thorough survey processes including questionnaires, focused group discussions, and interviews. Also, by application of representative facility sample analysis of the healthcare waste streams of two hospitals, they found that the facilities failed to adopt sound healthcare waste management procedures. They further highlighted that the healthcare waste handling practices and strategies in the state as at then were weak as it involved the co-mixing medical wastes with other types of waste. They were also characterized by lack of standard waste management policy/system that deterred proper identification of the kinds of healthcare wastes generated. Though little was mentioned about the level of awareness on healthcare waste management of the healthcare workers.

Elsewhere, within a period of 3 years, Awodele et al. (2016) took a broader picture view of the healthcare waste management situation of the state by assessing the HCW management practices of seven selected hospitals across the state metropolis. They also examined the influence of LAWMA (Lagos State Waste Management Authority) and its involvements. It was noted that virtually all the healthcare facilities were found to have started implementing a more systematic waste management process, waste segregation, waste collection and on-site and off-site transportation and on-site storage. The waste collection workers could be seen using hand gloves as personal protective equipment and majority of the facilities had employed the services provided by Lagos Waste Management Authority LAWMA for their final waste treatment and disposal. Although, there are still lapses such as lack of waste treatment with incinerators and inadequate policies and guidelines in almost all investigated health facilities in managing generated wastes. This agrees with report of David et al., (2014), where in a survey involving all classes of healthcare centers in Ota, also a south-west region as Lagos, it was observed that there was no pre-treatment of their medical waste before disposing them. Reasons for this ranged from poor enlightenment of the waste handlers and financial constraints in setting up pre-treatment facilities. On the waste transportation, treatment, and waste disposal methods/ procedures, it was observed in most of the research that the wastes collected were transported using trolleys, wheeled barrows and trucks etc., while in some cases waste handlers were seen conveying their waste on hand which is endangering.

In South East part of the country, the report of Uwa (2014), in Enugu metropolis, assessed the HCW management processes by hospital staff across four tertiary healthcare institutions in the state. The outcome of the research indicated significant disparity in HCW management activities and the adoption the 3Rs; sustainability factors of reduce, reuse recycle. In this case the dominant healthcare waste management was incineration and frequent waste disposal, omitting other new and improved waste handling techniques. It was concluded that though, healthcare institutions adopted slight process of reduce, recycle, and reuse and was not consistent.

2 Management of Healthcare Waste in Nigeria

The audit report on environmental and social protections showed more than 70%of facilities assessed in the world produce dangerous, and contagious waste (HCWM TWG, 2004). In the National Organization for the Management of AIDS, approximately 80% of all intermediate, tertiary, primary, and private clinical facilities researched created hazardous and infectious waste (Ezirim & Agbo, 2018; Idowu et al., 2013; HCWM TWG, 2004). It was clear that the National Health Management Plan in Nigeria provided prices for the generation of healthcare waste. It was clear from these reports, Nigeria's tertiary health facilities contain approximately 10.67 kg/bed/day of destruction. In comparison, primary and secondary health facilities produce approximately 2.79 kg/bed/day and 1.16 kg/bed/day, respectively. Tertiary hospitals operate predominantly in metropolitan areas and receive many patients, while secondary and primary health services are majorly in semi-urban to rural areas. Literatures suggest that tertiary care is responsible for the cost of services, which puts a burden on the administration of healthcare waste related to the required programs, personnel, and equipment (HCWM TWG, 2004). Nigeria has a 7,250,372.17 kg/year annual generation rate of healthcare waste, of which nearly 15%(1198.82 tons) is potentially contagious and harmful clinical waste. The quantity of clinical waste generated throughout the country in tertiary health facilities, and the same data for secondary and primary health facilities. The amount of waste produced in secondary and main healthcare facilities is clearly higher than what can be collected in tertiary healthcare facilities. However, because of the quality of treatment rendered at that level, a larger percent of waste created by tertiary healthcare facilities will contain dangerous and contagious clinical waste. At primary healthcare stage, the enormous amount of healthcare waste is produced.

2.1 Waste Incineration

Of the technology to be regarded in the management of HCW, a prominent medium is incineration, of which there are several kinds. Consequently, all facets of HCW incineration are an approach which in HCW literature has gained a great deal of coverage. This includes the detection of environmentally friendly technology, design and recording, autoclave forms of incinerators, and a recent invention known as hydroclave treatment, as well as a microwave and chemical disinfection. According to the authors, technology can aid, but it needs to be part of a broader solution. Instead, focus can be put on preparation for segregation, local housekeeping and disinfection, safety procedures from the viewpoint of an occupational hazard for nurses, ward boys and rag pickers, and waste minimization that will potentially help address the medical waste issue (Tables 5 and 6).

Table 5 Percentage distribution of health facilities based on waste generated

Characteristics	General hazardous waste	General universal waste	General infectious waste	General waste banned from landfill
Type of facility				
Tertiary	100	93.3	96.7	71.4
Secondary	85.9	82.3	87	48.3
Primary	83.7	81.6	86.9	43.3
Private	83.1	81.8	88.2	52.1
Geopolitical zone				
North East	71.2	52.1	65.5	41.7
North West	81.9	70.4	64.5	35.7
North Central	82.5	80.1	90	47.7
South East	85.7	97.3	95.9	50.7
South West	83.2	77.5	84.8	28.2
South South	88.1	82.2	90.6	51.8

Source: Environmental and social safeguards audit (NACA, 2015)

Table 6 Amount of healthcare waste generated in Nigeria

Geopolitical zones	Population per zone*	Healthcare facilities per zone **	Average waste Kg/bed/day**	Waste generation per zone (Kg/day)	Annual waste generation per zone (Kg/day*365)	Infectious and hazardous waste generated (15% of annual waste generated)
North Central	20,368,918.00	9101.00	0.59	5369.59	1,959,900.35	293,985.05
North East	18, 984,299.00	5735	0.71	4071.85	1,486,225.25	222,933.79
North West	35,915,467.00	8192.00	0.48	3932.16	1,435,238.40	215,285.76
South East	16,395,555.00	6050.00	0.43	2601.50	949,547.50	142,432.13
South South	20,320,869.00	4869.00	0.32	1558.08	568,699.20	85,304.88
South West	27,722,432.00	7151.00	0.37	2645.87	965,742.55	144,861.38
Total	139,707,540.00	41,098.00	0.48	19,864.03	7,250,372.17	1,087,555.83

Source: * FGN (2007), ** FMOH (2019), ***FMOH (2012)
FGN Federal Government of Nigeria, *FMOH* Federal Ministry of Health

2.2   HCW Treatment

Likewise, a significant number of studies have created the treatment component of HCW. The explanation is that the nature, toxicity, and possible risk of the components of HCWs are distinct, requiring different care and disposal procedures and choices. Various approaches have been developed to make biochemical waste chemically harmless and esthetically suitable. However, numerous experiments have demonstrated no way of handling or disposing of medical waste that entirely removes all threats to humans or the environment.

2.3   Sharps

Owing to the immense adverse health effects and vulnerability globally, the handling of sharps (i.e., syringes) in HCW has a high profile. For example, WHO reports that approximately 30,000 new HIV infections, eight million HBV infections, and 12 million HCV infections are caused by inappropriate injections per year worldwide. Two alternate methods were also explored by WHO/USAID. It is a cost-effective and waste control viewpoint of which the syringe may be reused in addition to the disposable syringe (after adequate washing and sterilization in a steam sterilizer); the other is a public welfare perspective that includes the usage of auto-destruct needles once. A good overview to base the preference of disposal facilities on the method of care and treatment choice to assess disposal facilities' choice is suggested.

2.4   HCW Segregation

Waste segregation is a crucial first step in achieving waste reduction, cost reduction, and sustainable waste management. It provides the health facility with the means to evaluate its waste composition more accurately and position the facility for practical strategies for HCW management. In HCW, the focus is on source isolation. The WHO advises the isolation of HCW ideally at the point of processing and provides advice in developed countries on the safe and sound disposal of medical waste. It is also essential to use instructive posters and color code bins to achieve successful waste segregation.

2.5 *Management*

As one of the key strategic variables that require consideration for the effectiveness of raising the standards of HCW management in HCFs, the management aspect of HCW depends on good governance. The others are administrative arrangements that involve the framing of relevant HCWM rules, legislation, laws, protocols, requirements, and tools for enabling them. Other considerations proposed by United Nations Environment Program (UNEP) include clear-cutting positions and capacity building obligations through knowledge exchange, allowing the policy system to resolve economic and market-based instruments, promoting, or supplying all stakeholders with reliable and timely information and funding, which is considered a key problem.

3 Legislative Framework for Healthcare Waste Management in Nigeria

According to Rampersaud (2008), the regulatory structure is assisted by a series of rules, legislation, and guidelines set up by governments to combat environmental problems by mitigating or avoiding emissions. Therefore, regulation is an essential instrument that underpins adequate and viable HCW management structures.

In Nigeria, the Constitution of the Federal Republic of Nigeria 1991 falls short of providing the necessary legal teeth favoring sustainable HCW management. Townend and Vallini (2008) noted this supreme legal order is robust in the clear definition of duties and responsibilities among the three constituent units of the federal entity in healthcare at central, state, and local government levels. However, the Environmental Law Research Institute (2011) considers the provision for HCW management issues in Nigeria in the constitution as marginal at best. In line with this contention, attention was called to Sections 33 and 34 of the Nigerian Constitution 1999 which guarantee fundamental human rights to life and human dignity respectively as seen as being argued to address HCW management via the need for healthy and safe environment to give these rights effect as being relevant to public health and safety.

Abah and Ohimain (2011) also noted the inadequacy of legislation in Nigeria for a proper and sustainable HCW management mechanism concerning the Public Health Act 1958 and various state decrees on environmental sanitation, providing solid waste management laws, especially non-hazardous general (municipal) waste. They faulted these laws for not adequately addressing the essential aspects of healthcare waste. With specific attention to the Draft Nigerian National HCW which is a national legislative policy, Abah and Ohimain (2011) went further to draw attention to its non-implementation at any level even though Nigeria is a signatory to several multilateral environmental agreements (MEAs) including the Basel Convention as well as Hazardous Chemicals and Waste Convention.

3.1 Challenges Faced by Nigerian Healthcare Centers

The health status of Nigerian citizens, especially in less developed areas, is still at risk. To better address public health, the Nigerian government has implemented legislative/regulatory measures. One of the key underlying reasons why the public health problem is the challenge faced by Nigerian hospitals is because of the lack of knowledge of health hazards related to healthcare waste, inadequate training in proper waste management, lack of waste management and recycling programs, insufficient financial and human capital, and the low priority provided to the issue (WHO, 2018). The misunderstanding of the "polluter pays" concept, which stresses the attribution of blame, is another issue. More precisely, the liability rests with the waste generators, typically the hospitals, or businesses engaged in similar operations.

It was found that the waste in neighborhoods was poorly disposed of by most hospitals and nursing homes. This was attributable to poor amenities for disposal, technological inadequacies, and lack of understanding. Another research undertaken in four separate clinics in Lagos examined procedures undertaken in treating medical waste. Around 0.562–0.67 kg of biomedical waste is estimated to be generated per bed regularly in private hospitals in Lagos State. In all hospitals examined, the observed medical waste activities reveal no adherence to the medical waste handling procedure as stated in the essential sections of the Nigerian environmental pollution management guidelines and norm.

3.2 Impacts of Poorly Managed Biomedical Waste on the Environment and Health

There are many threats associated with biological waste that is not adequately handled. There may be threats that are chemical, physical, or biological. More specifically, microorganism-causing infections, medical sharps, and hazardous biomedical waste chemicals can infect healthcare workers, patients, the general population, and the environment. In fact, radiation burns from hazardous waste, sharp accidents, poisoning and emissions from pharmaceutical goods, contaminated water, and hazardous substances such as mercury are human impacts and environmental consequences of poorly controlled biomedical waste (WHO, 2018). All those who have access to, or who are subjected to, infectious healthcare waste, and those who produce the waste, such as physicians and those responsible for treating the waste, may be at risk because of inadequate management and handling. People at high risk are healthcare workers, such as nurses, laboratory technicians and doctors, patients, waste handlers, and scavengers. This types of individuals at risk may be infected by punctures and other breaks in the skin, mucous membranes, inhalation, vector transmission, and swallowing. Antibiotic and chemical disinfectant resistant bacteria can also add to the risks posed by poorly handled waste. If we look at the environmental consequences of biomedical waste, excessive treatment will encourage soil biodegradation and threaten marine life, as well as underground water pollution. However, waste disposal has ecological impacts, such as the emission of toxic gases from incineration and combustion.

4 Conclusion

The disparities found in access to healthcare and the procedures and facilities of medical waste management could affect Nigeria's health system's sustainability. Efficient relational agreements will reinforce the network of healthcare services, establish organizational connections between different types of facilities, promote collaborations in public and private healthcare, and strengthen established capacity to manage and control healthcare waste. Provision of appropriate waste management activities for healthcare in Nigeria will entail proper waste management solutions in reasonable amounts and capacity building for healthcare professionals engaged in the use, storage, and disposal of medical waste materials. Nigeria's health policymakers should aim to ensure that health infrastructure and qualified medical professionals are equitably placed to increase equitable access to healthcare. In view of recent public health diseases of foreign importance, resolving healthcare management issues in the country has become a need of high significance. These guidelines will improve the stability and preparedness of health services in developed countries with respect to outbreaks of infectious health illnesses, the effects of climate action, and the accomplishment of health-related sustainable development objectives. From all these studies, it is imperative to state that most of the healthcare facilities are far behind in adopting new technologies and innovations of healthcare waste management. Thus, there is a need for improved efforts to ensure proper healthcare management practices and a consistent and intentional effort should be geared toward improving the efficiency of these actions.

References

Abah, S. O., & Ohimain, E. I. (2011). Healthcare waste management in Nigeria: A case study of Ibadan Teaching Hospital. *Journal of Public Health and Epidemiology, 3*(3), 99–110.

Adogu, P. O., Uwakwe, K., Egenti, B. N., Okwuoha, A. P., & Nkwocha, I. (2015). Assessment of waste management practices among residents of Owerri Municipal Imo State Nigeria. *Journal of Environmental Protection, 06*(05), 446–456. https://doi.org/10.4236/jep.2015.65043

Afolabi, O. T., Aluko, O. O., Kehinde, A. B., & Funmito, F. (2018). Healthcare waste management practices and risk perception of healthcare workers in private healthcare facilities in an urban community in Nigeria. *African Journal of Environmental Science and Technology, 12*(9), 305–311. https://doi.org/10.5897/ajest2018.2534

Akinwale, C., & Sridhar, M. K. C. (2010). Increase in Healthcare facilities and rapid environmental degradation: A technological paradox in Nigeria's urban centres. *African Journal of Environmental Science and Technology, 4*(9), 577–585.

Akpan, V. E., & Olukanni, D. O. (2020). Hazardous waste management: An African overview. *Recycling, 5*(3), 15.

Awodele, O., Adewoye, A. A., & Oparah, A. C. (2016). Assessment of medical waste management in seven hospitals in Lagos, Nigeria. *BMC Public Health, 16*(1), 1–11. https://doi.org/10.1186/s12889-016-2916-1

Chukwunonye, A. E. (2015). Healthcare waste management: What do the Health Workers in a Nigerian Tertiary Hospital know and practice. *Science Journal of Public Health, 3*(1), 114. https://doi.org/10.11648/j.sjph.20150301.30

Coker, A. (2009). Medical waste management in Ibadan, Nigeria: Obstacles and Prospects. *Waste Management Journal, 29*(2), 804–811.

Environmental Law Research Institute. (2011). Quaterly newsletter October 2011.

Ezirim, I., & Agbo, F. (2018). Role of national policy in improving health care waste management in Nigeria. *Journal of Health and Pollution, 8*(19). https://doi.org/10.5696/2156-9614-8.19.180913

Federal Government of Nigeria (FGN). (2007). Details of the breakdown of the national and state provisional population totals of 2006 census. *Federal Government of Nigeria (FGN) Official Gazette, 94*(24), B175–B198.

Federal Ministry of Health (FMOH). (2011). *A directory of health facilities in Nigeria*. Federal Ministry of Health. https://drive.google.com/file/d/0B1DAmtM1BcbMMGpSRFFwZGVGOTQ/view

Federal Ministry of Health (FMOH). (2012). *National healthcare waste management policy*. FMOH.

Federal Ministry of Health, Abuja, Nigeria Federal Ministry of Health. (2019). *Nigeria health facility registry*. Federal Ministry of Health (FMOH). https://hfr.health.gov.ng

HCWM TWG. (2004). *National Health-Care Waste Management Plan* (p. 108). Ministry of Health and Social Welfare.

Idowu, I., Alo, B., Atherton, W., & Al Khaddar, R. (2013). Profile of medical waste management in two healthcare facilities in Lagos, Nigeria: A case study. *Waste Management and Research, 31*(5), 494–501. https://doi.org/10.1177/0734242X13479429

Joshua, I. A., Mohammed, S., Makama, J. G., Joshua, W., Audu, O., Nmade, G., & Ogboi, J. S. (2014). Hospital Waste Management as a Potential Hazard in Selected Primary Healthcare Centres in Zaria, Nigeria. *Nigerian Journal of Technology, 33*(2), 215. https://doi.org/10.4314/njt.v33i2.11

Kampf, G., Todt, D., Pfaender, S., & Steinmann, E. (2020). Persistence of coronaviruses on inanimate surfaces and their inactivation with biocidal agents. *Journal of Hospital Infection, 104*(3), 246–251. https://doi.org/10.1016/j.jhin.2020.01.022

Longe, E. O. (2012). Healthcare waste management status in Lagos State, Nigeria: A case study from selected healthcare facilities in Ikorodu and Lagos metropolis. *Waste Management and Research, 30*(6), 562–571. https://doi.org/10.1177/0734242X11412109

Mustapha, A. (2017). Lassa fever: Unveiling the misery of the Nigerian health worker. *Annals of Nigerian Medicine, 11*(1), 1–5.

National Agency for the Control of AIDS (NACA). (2015). *Environmental and social safeguards audit report (World Bank Funded)*. National Agency for the Control of AIDS (NACA). http://documents.worldbank.org/curated/en/479211468097746610/pd f/SFG1506-EA-P102119-Audit-Report-Box393258B-PUBLICDisclosed- 11-24-2015.pdf

Ogbonna, D. N. (2011). Characteristics and waste management practices of medical wastes in healthcare institutions in Port Harcourt. *Nigeria Journal of Soil Science and Environmental Management, 2*(5), 132–141.

Ogbonna, D. N. (2013). Characteristics and waste management practices of medical wastes in healthcare institutions in Port. *Global Journal of Pollution and Hazardous Waste Management, 1*(1), 1–10.

Ogbonna, D. N., Chindah, A., & Ubani, N. (2012). Waste management options for health care wastes in Nigeria: A case study of Port Harcourt hospitals. *Journal of Public Health and Epidemiology, 4*(6), 156–169. https://doi.org/10.5897/jphe12.012

Oli, A. N., Ekejindu, C. C., Adje, D. U., Ezeobi, I., Ejiofor, S., Ibeh, C. C., & Ubajaka, C. F. (2016). Healthcare waste management in selected government and private hospitals in Southeast Nigeria. *Asian Pacific Journal of Tropical Biomedicine, 6*(1), 84–89. https://doi.org/10.1016/j.apjtb.2015.09.019

Olukanni, D. O., Azuh, D. E., Toogun, T. O., & Okorie, U. E. (2014). Medical waste management practices among selected health-care facilities in Nigeria: A case study. *The Scientific Research and Essays, 9*(10), 431–439. https://doi.org/10.5897/sre2014.5863

Olukanni, D. O., & Oresanya, O. O. (2018). Progression in waste management processes in Lagos State, Nigeria. *Journal of Engineering Research in Africa (JERA)., 35*, Pp11-23.

Olukanni, D. O., Pius-Imue, F. B., & Joseph, S. O. (2020). Public perception of solid waste management practices in Nigeria: Ogun state experience. *Recycling, 5*(2), 8.

Oyekale, A. S., & Oyekale, T. O. (2017). Healthcare waste management practices and safety indicators in Nigeria. *BMC Public Health, 17*(1), 1–13. https://doi.org/10.1186/s12889-017-4794-6

Rampersaud, P. (2008). *Systems Approach to Health-Care Management: Understanding System functionality using VSM – the case of Guyana*. M.Sc. Environmental Management and Policy.

Tobin, E. A., Ediagbonya, T., Asogun, D., & Aj, O. (2013). Assessment of Healthcare Waste Management practices in Primary Health Care Facilities in a Lassa Fever Endemic Local Government Area of Edo state Nigeria. *AFRIMEDIC Journal, 4*(2), 16–23.

Townend, B., & Vallini, I. (2008). Healthcare Waste Management – the global paradox. *Waste Management and Research, 26*, 215–216.

United Nations Environment Program (UNEP)/SBC and World Health Organization. (2005). *Preparation of National Healthcare Waste Management Plans in Sub-Saharan countries – Guidance Manual*. WHO Document Production Services.

Uwa, C. U. (2014). Assessment of Healthcare Waste Management Practices in Enugu Metropolis, Nigeria. *International Journal of Environmental Science and Development, 5*(4), 370–374. https://doi.org/10.7763/ijesd.2014.v5.512

WHO (World Health Organization). (1999). Unsafe injection practices and transmission of blood borne pathogens. *Bulletin of the World Health Organization, 77*, 787–819.

WHO (World Health Organization). (2005). *Understanding and simplifying bio-medical waste management. A training manual for trainers*. Available at: http://www.noharm.org/detailcfmID=1197andtype=documentwater_sanitation health/medical waste/decisionmguiderev221105.pdf (accessed January 2021).

World Health Organisation (WHO). (2018). *Fact sheet: Health care waste*. https://www.who.int/news-room/fact-sheets/detail/health-carewaste

Medical Waste Management in Lebanon and Impact of COVID-19

Amani Maalouf, Hani Maalouf, and P. Agamuthu

1 Introduction

The COVID-19 pandemic has triggered diseases and deaths, created immense confusion for the world, and modified solid waste management profiles (Klemeš, Fan, Tan, & Jiang, 2020; Penteado, 2021). This pandemic changed the behavioral pattern and consuming patterns of people, resulting in a sudden change in the generation amount, composition, and disposal rate (timing and frequency) of municipal solid waste (MSW), depending on location. However, the prediction is that the overall MSW generation would slightly decrease during COVID-19 (Fan, Jiang, Hemzal, & Klemeš, 2021; Klemeš et al., 2020). It also proposed that the use of plastic products for the prevention of the epidemic expanded during the COVID-19 pandemic (Zhou, 2021). A major rise in plastic use has been associated with packaging requirements and single-use products due to the demand for distribution and take-out of food/goods as well as for the demand for plastics for medicinal uses (Fan et al., 2021). In the opposite, because of the fear of inadequate food supply during the outbreak, people continued to conserve food and thereby decrease food waste (Jribi, Ben Ismail, Doggui, & Debbabi, 2020). In his evaluation of the expected trend of medical waste flow along with the epidemic/pandemic crisis (Klemeš et al., 2020) show that medical waste generation medical increased sharply, reaching up to 370% increase in some places such as Hubei Province, China. According to the latest published report by USEPA (United States Environmental

A. Maalouf (✉)
Earth Engineering Center, Columbia University, New York, NY, USA

H. Maalouf
Department of General Surgery, Saint Georges Hospital University, Medical Center, University of Balamand, Beirut, Lebanon

P. Agamuthu
Jeffrey Sachs Center on Sustainable Development, Sunway University, Sunway, Malaysia

© The Author(s), under exclusive license to Springer Nature Singapore Pte Ltd. 2022
S. K. Ghosh, P. Agamuthu (eds.), *Health Care Waste Management and COVID 19 Pandemic*, https://doi.org/10.1007/978-981-16-9336-6_10

Protection Agency) (2020), the rise in healthcare waste from COVID-19 associated healthcare facilities is also reported to be 3.4 kg per person per day worldwide. Approximately 2.5 kg/bed/day of COVID-19 healthcare waste is produced in developing countries (USEPA (United States Environmental Protection Agency), 2020). The COVID-19 pandemic has equally affected the management systems of MSW. In this context, recommendations were presented by different organizations such as (Basel Convention, 2020; CDC (Centers for Disease Control and Prevention), 2020; European Commission, 2020; ISWA (International Solid Waste Association), 2020; OSHA (Occupational Safety and Health Administration), 2020; USEPA (United States Environmental Protection Agency), 2020; World Health Organization (WHO), 2021; United Nations Environment Programme (UNEP), 2020) with various scopes at worldwide, regional or national levels to address different actors such as waste generators (citizens), waste collectors, employers (service providers), and local governments. Among them, ISWA (International Solid Waste Association) (2020) recommendations at worldwide level stand out as the most comprehensive and detailed. The main measures and recommendation for MSW waste handling and management included: providing adequate personal protective equipment (PPE) and ensure enhanced personal hygiene standards; discontinuing immediately manual sorting of mixed waste or commingled recyclables, including disabling and substituting the manual stages in mechanical-manual systems; in addition to providing other adequate information on COVID-19.

On average, high-income countries produce up to 0.5 kg of hazardous waste per day per hospital bed, while low-income countries generate an average of 0.2 kg (World Health Organization (WHO), 2018). However, in low-income countries, healthcare waste is also not classified into hazardous or non-hazardous waste, making the total volume of hazardous waste even larger (World Health Organization (WHO), 2018). Therefore, measures to ensure that healthcare waste is handled appropriately and environmentally soundly will avoid harmful health and environmental effects from such waste, including the accidental release into the atmosphere of chemical or biological contaminants, including drug-resistant microorganisms, thus protecting the health of patients, health workers and the public at large. In Lebanon, it was estimated that the infectious healthcare waste generation weighted mean is 1.14 per occupied kg per bed per day (Maamari, Brandam, Lteif, & Salameh, 2015). Nevertheless, epidemics or pandemics such as COVID-19 often result in higher generation of healthcare waste.

Globally, as of 23 July 2021, there have been 193.17 million, confirmed cases of COVID-19, including 4.14 million deaths, reported to WHO. In Lebanon, the confirmed cases on COVID-19 were 552,871 people, where 7889 people died as of July 23, 2021. The status on spread of COVID-19 with time in Lebanon is shown in Fig. 1 (MoI, 2021). As of July 23, 2021, the total number of vaccinated people against COVID-19 in Lebanon is 1.07 million out of which 711,334 (66.4% of total number vaccinated) are fully vaccinated and 359,699 (33.6%) are partly vaccinated. The share of the total population that received at least one dose of COVID-19 vaccine is 15.69% of the total population in Lebanon. Lebanon has a population of 6,007,000 in 2016, a life expectancy at birth of 74 for male and 79 for female in

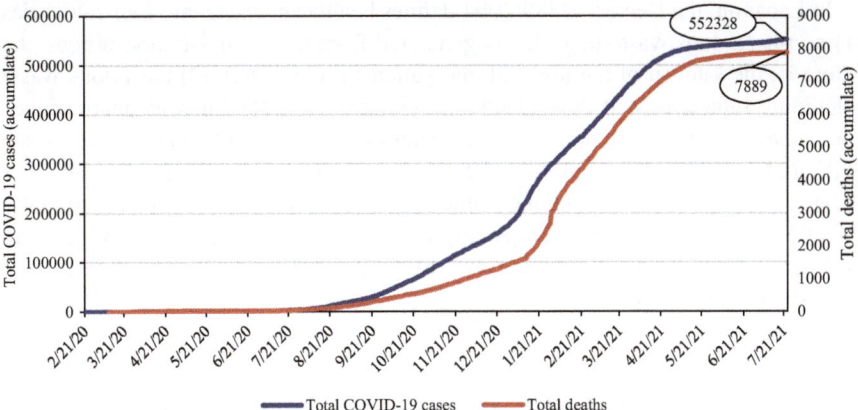

Fig. 1 Status on spread of COVID-19 in Lebanon. Source: Data extracted from Our World in Data (2021). Note: Numbers in oval represent the total COVID-19 cases and number of deaths (accumulated) as of 14 January 2021

2019 and a total expenditure on health as % of GDP of 6.39 in 2014 (World Health Organization (WHO), 2021).

With the rise in the number of COVID-19 infections, almost every region of the world has an increase in medical waste from healthcare facilities, such as hospitals, clinics, labs, temporary quarantine centers, and research laboratories. A similar rise in the production of medical waste is expected in Lebanon. This chapter clarifies the medical/healthcare waste management policies and regulations, reports on the generation and management of healthcare waste, and illuminates on the impact of COVID-19 on the existing challenges of medical waste management in Lebanon.

2   Classification of Medical Waste

Healthcare waste can also be known as medical waste, clinical waste, or scheduled waste (Department of Environment (DOE), 2009). It includes all the waste generated inside healthcare facilities, hospitals, research centers, and laboratories related to medical procedures, according to the World Health Organization (WHO) guideline reports. It also involves the waste generated in homes from handling and disposal of materials used in sick people (World Health Organization (WHO), 2014, 2017). The composition of healthcare waste can be heterogeneous including both hazardous and non-hazardous it can also be classified into the following groups: infectious waste, pathological waste, sharps waste, chemical waste, radioactive waste, cytotoxic waste, pharmaceuticals waste, and non-hazardous or general waste (World Health Organization (WHO), 2014). Of the overall amount of waste produced by healthcare operations, about 85% is non-hazardous general waste. The remaining 15% is known to be hazardous materials that may be infectious, toxic, or radioactive.

In Lebanon, the Decree 13389/2004 defines healthcare waste into four categories (1) non-hazardous waste (e.g. that is generated from the administration offices, the cafeteria, the kitchen of the hospital, the patients rooms, etc.), (2) hazardous waste (including infectious and non-infectious), (3) special waste (such as pharmaceuticals, chemical waste, cytotoxic, and pathological), and (4) radioactive waste (MOE/UNDP/ECODIT, 2011). Whereas infectious healthcare wastes are defined as: "any waste material that came in contact with blood and other potentially infectious fluids of the body, sharps, any waste produced by an isolated patient and biological fluids, small anatomic parts, tissues, cultures and stocks of infectious agents" (Maamari et al., 2015).

3 Policy/Regulations/Legal and Institutional Framework in Lebanon on Medical Waste

3.1 Existing Policy, Legal, and Institutional Framework

The healthcare sector is perceived to be one of the country's largest and most advanced service-providing sectors and is dominated mainly by the private sector. The private sector runs more than 85% of the hospitals in Lebanon (World Bank, 2020a, 2020b).

In the past decade, Lebanon has enacted key legislations related to healthcare waste management. Decree 8006 (dated 11/6/2002) is amended by Decree No. 13389 (dated 18/9/2004).[1] This decree classified the different healthcare waste categories and indicated the proper management and disposal of each category. In accordance to this degree, all hospitals and medical institutions/organizations are responsible for the management of their generated medical waste in an environmentally appropriate manner. In 2011, the government through the Ministry of Environment tried enforcing that decree with continuous follow-up with all these medical institutions (mainly hospitals). Moreover, hazardous waste is regulated under Law 64 dated back to 1988. Lebanon has also ratified two main international conventions dealing indirectly with the issue of healthcare waste management: (1) the Stockholm convention on persistent organic pollutants and (2) the Basel convention on the control of the transboundary movements of hazardous wastes and their disposal (Laceco/Ramboll, 2012). Table 1 summarizes the legal framework in Lebanon in relation to healthcare waste.

---

[1]Lebanese Decree 13389/2004, amends decree 8006/2002. 13389 regulates healthcare waste. It defines the type of healthcare wastes. It requires proper waste segregation and minimization. It sets guidelines for the collection and storage of waste. Finally, it requires an Environmental Impact Assessment for healthcare waste treatment facilities in order to get a license from Ministry of Environment (MoPH, 2018).

Table 1 Summary of the legal framework in relation to the healthcare waste management in Lebanon

Legal text	Year	Title
Law Decree 9826	1983	The law for establishing private hospitals implemented by Decree No. 9826 and amended by Legislative Decree No. 139 dated January 16 1983
Law 64	1988	The law of environmental preservation against harmful and hazardous waste pollution
Law 387	1994	Basel convention on the control of transboundary movements of hazardous wastes and their disposal
Law 444	2002	The framework law for the protection of the environment
Decree 8006	2002	Determination of the categories of healthcare waste and ways of disposal off
Law 432	2002	Stockholm convention on persistent organic pollutants
Decree 13389	2004	Amendments of the Decree 8006/2002 related to determining the types of waste in health institutions and how to dispose of them
Decision 2287/1	2014	Decision No. 2287/1 of 12/24/2014 regarding mandatory compliance by hospitals with legal texts regarding medical waste treatment
Decision 1/1229	2015	Decision No. 1/1229 of 7/1/2015 clarifying Decision No. 1/2287 of 12/24/2014
Decision 2824/1	2015	Decision No. 2824/1 of 12/29/2015 regarding the formation of a committee to treat medical waste
Memorandum 34	2019	Memorandum No. 34 dated 4/12/2019 related to the obligation to fulfill health and environmental conditions by hospitals

Source: Adapted from Sweep-net (2014)

Decree 13389 also established the permitting mechanism for infectious healthcare waste treatment facilities. Under this decree, healthcare waste should be treated at specialized treatment facilities licensed by the Ministry of Environment before being disposed of. It is important to highlight that environmental guideline for the establishment and operation of healthcare waste treatment facilities and the licensing of healthcare waste transportation vehicles are still lacking in the Lebanese legal framework. Recently, new decisions, memoranda and administrative circulars were circulated such as Decision No. 2824/1 in 2015 regarding the formation of a committee to treat medical waste and Memorandum No. 34 dated 4/12/2019 related to the obligation to fulfill health and environmental conditions by hospitals, in order to regulate the role of the Ministry of Public Health and promote a proper and comprehensive management of healthcare waste in Lebanon.

According to local legislations, the responsibility of the regulation and management of healthcare waste is distributed among a number of authorities namely, Ministry of Environment, Ministry of Public Health, Ministry of Interior and Municipalities, and Council for Development and Reconstruction. The various public authorities responsible for regulating the healthcare waste and their duties in this area are summarized in Table 2. Other stakeholders that might be involved with the healthcare waste management include the syndicate of private hospitals, United Nations Development Program, World Health Organization, Universities, the

Table 2 Summary list of the main stakeholders responsible of regulating the healthcare waste management sector in Lebanon

Stakeholder	Responsibility
Ministry of Environment	Responsible for legislation, licensing and monitoring
Ministry of Public Health	Responsible for studying the health impact of any waste management program Governs the hospital accreditation program
Ministry of Interior and Municipalities	Municipalities are responsible for providing waste collection and disposal facilities HCW treatment plants are granted a license from the Governorate based on the MoE license Internal Security Forces are responsible of Law enforcement
Council for Development and Reconstruction	Monitors the execution of major projects endorsed by the Government

Source: Adapted from Sweep-net (2014)

Order of Physicians, the Order of Dentists, the Syndicate of Medical Laboratories and the Syndicate of Dental Laboratories.

3.2 Policy and Regulations in Relation to COVID-19 Pandemic

During the COVID-19 pandemic, no new policies or regulations were introduced for the management of COVID-19 related waste since they are classified under the category of infectious healthcare waste. Therefore, the management of waste generated from COVID-19 related cases has been managed in compliance with existing regulations related to infectious healthcare waste. However, additional training and awareness program related to COVID-19 waste management in Lebanon were carried out mainly on the proper use of PPE, especially for the waste collectors. Moreover, during this pandemic, the non-governmental organization (NGO) "Arcenciel," a service provider for waste collection and treatment in Lebanon since 2003, is asking healthcare facilities to mark the COVID-19 waste for tracking and data purposes in terms of reporting to the Ministry of Environment and Ministry of Public Health.

4 Generation of Medical Waste

In Lebanon, the total amount of healthcare waste generated from all sources, particularly from clinics, is very difficult to estimate, as it is neither officially documented nor registered. In fact, most of the research carried out in relation to the healthcare waste concentrated on hospitals that provide an approximate

Table 3 Estimated healthcare waste generation in Lebanon

Year	Nb. of hospitals	Nb. of beds	Occupancy (%)	Generation rate per occupied bed (kg/day)	Healthcare waste generated (tons/day)
1998	160	13,493	56	1.5	10.81[a]
2000	–	–	–	–	11.32[a]
2005	–	–	–	–	12.46[a]
2010	174	13,668	60	1.0–1.5	8.2–12.3[a]
2012	164	15,342	60	1.0–1.5	9.2.5–13.8
2019	180	–	–	–	12.8[b]

Sources: Compiled by authors based on MOE/UNDP/ECODIT (2011); Laceco/Ramboll (2012); Sweep-net (2014)
[a]These studies mainly considered hospital risk waste (associated with categories 2, 3, and 4 of healthcare waste as defined earlier in this chapter) whereas hospital non-risk waste ranged between 43.2 and 55.2 tons/day, thus this result with a total (risk and non-risk) between 54.05 and 69.05 tons per day of healthcare waste during this period (1998–2010)
[b]Quantity was assessed by the use of provided monthly records of Arcenciel, a service provider for healthcare waste collection and treatment in Lebanon since 2003

estimation. Table 3 displays the estimated hospital healthcare waste generation in Lebanon based on previous studies.

As of 2012, assuming 60% occupancy and an average generation rate of 1.0–1.5 kg/bed/day, Lebanon generates about 9.2–13.8 tons of risk healthcare waste daily (about 3358–5037 tons/year) from 164 public and private hospitals (about 15,342 hospital beds) (MoE/UNDP/GEF, 2016). Approximately, Lebanon produced a total of 69 tons of healthcare per day (25,200 tons/year) divided into risk waste (14 tons/day) and non-risk waste (55 tons/day) (MOE/UNDP/ECODIT, 2011).

A study by Maamari et al. (2015) estimated that the total infectious healthcare waste generation rates in Lebanon is equal to 6,383,806 kg/year based on data records for the years 2009–2013 (assuming 15,342 hospitals beds). The study also showed that for large private hospitals (over 200 beds), the infectious healthcare waste generation average is 2.45 kg per occupied bed per day. For other hospitals, i.e. public hospitals and private hospitals under 200 beds, the rate is 0.94 kg per occupied bed. The weighted mean is estimated at 1.14 kg per occupied bed per day (assuming an average hospitals occupancy rate of 58%). These values are greater than the reference levels of WHO for the Middle East (0.54 kg per bed per day), but very close to reported infectious healthcare waste generation levels by Sanida et al. (2010) for Greece (0.85 kg per occupied bed per day).

We estimated the quantities of infectious healthcare generated in 2019 to be around 4,666,754 kg per year or 13 tons/day based on the monthly records of Arcenciel. The sample is formed of 150 hospitals (private) and 30 public, 80 Labs, 42 dispensaries, 4 associations, 4 universities, and around 200 primary healthcares (PHC) and isolation centers, from which infectious healthcare waste is collected on a regular basis. Therefore, based on a prepared collection route all of these institutions are served accordingly. Noting that Arcenciel requires from these institutions to have a cold storage room as pre-condition to sign a contract with them, in order to facilitate the collection scheme. Therefore, the sample represents 34.97% of total number of hospitals and 40.70% of the total number of beds in Lebanon.

The daily generation of infectious medical waste in year 2012 was 13.8 tons/day and 5037 tons annually, whereas currently it is estimated that the generation of clinical is 12.8 tons/day or 4668 tons annually. In this context, it is worth noting that Lebanon is facing a multi-faceted crisis, combining a financial and economic collapse, a political crisis, a health crisis and a waste crisis. Starting October 2019, the economic and financial crisis in Lebanon has greatly limited the ability of the health sector to deliver affordable and accessible services. This was attributed to many problems such as: (a) persistent delays in government payments to hospitals; (b) a lack of dollars and uncontrolled limits on the access of depositors to their accounts, hampering the importation of vital medical facilities, drugs and supplies; and (c) a rise in unemployment rates, contributing to a rise in the number of uninsured residents, especially among middle-income communities, needing government assistance. In addition, the drop in salaries' value has forced hundreds of doctors and nurses to leave the country altogether. The COVID-19 pandemic further aggravated the health sector's financial strains; the following is warranted as such (World Bank, 2020a).

The situation was exacerbated on August 4, 2020, whereby a massive explosion rocked the Port of Beirut, destroying much of the port and severely damaging dense residential and commercial areas within 5 km of the site of the explosion. Multiple public and private hospitals centrally located in Beirut reported extensive damage and were not able to welcome patients. Therefore, several facilities are still not operating at their pre-blast capacity. The explosion also caused damage to solid waste management facilities as well as damages to the healthcare waste storage waste facilities of three hospitals in Beirut, which was then collected and outsourced for treating to an external contractor (World Bank, 2020b). Lebanon was suffering since 2015 from a waste crisis after the premature closure of the main landfill (Naameh) that served as the primary largest sanitary landfill in the country serving more than half the total population (El-Fadel & Maalouf, 2020; Maalouf & El-Fadel, 2019). The problem was exacerbated following the Beirut explosion which generated large quantities of debris and demolition waste and affected severely damaged two key sorting, recycling, and composting facilities that were helping in reducing the amount of MSW that goes to landfills. It should be noted that in Lebanon estimates on amount of household infectious waste being generated related to COVID-19 is lacking as there is no MSW separation at source (Maalouf, Di Maria, & El-Fadel, 2019).

5 Management of Medical Waste

In the past, healthcare waste was not being adequately and properly managed thus resulting in significant environmental problems such as air, soil and water pollution as a direct implication of the haphazard burning and dumping of healthcare waste

(MOE/UNDP/ECODIT, 2011). Starting 2002, and after the implementation of Decree 8006 (dated 11 June 2002) on the proper handling of healthcare waste in Lebanon, many hospitals and institutions began to manage their medical waste in an environmentally sustainable manner. In 2003, a local Non-Governmental Organization (NGO) named Arcenciel started treating (autoclaving) and collecting potentially infectious healthcare and became active in several regions from Lebanon. As of 2010, 55–60% of the total healthcare waste stream (about 90% of Beirut's waste stream) collected from 81 public and private hospitals is being treated by Arcenciel. Therefore, autoclaving became the main method for treating waste generated by hospitals. The remaining portion (around 35–40%) of the healthcare waste is either being incinerated[2] at hospitals without permits or illegally dumped with MSW (MoE/UNDP/GEF, 2016). Starting 2003, autoclaving conducted by Arcenciel significantly reduced the amount of medical waste incinerated in hospitals or medical facilities from 1250 tons/year in 2003 to 60 tons/year in 2015 (MoE/UNDP/GEF, 2019). However, it should be highlighted that incineration carried out at the different medical establishments is without monitoring and contribute to a total of 1050 tons of CO_2 equivalent of greenhouse gas (GHG) emissions per year (MoE/UNDP/GEF, 2016). Uncontrolled and unlicensed incineration of healthcare waste in many hospitals were also associated with the release of persistent organic pollutants (POPs) and other pollutants (MoE/UNEP/GEF, 2017). Currently, it is estimated that around 80–85% of infectious healthcare waste in Lebanon is being managed by Arcenciel (NGO) in close collaboration with other stakeholders (MoPH, 2018).

Current technologies for medical waste treatment are not adequately monitored. There was no assessment of the efficacy of medical waste treatment using autoclaves. Moreover, the majority of treated medical waste is disposed of in landfill or dumpsite at the available site. All medical waste should, however, be adequately treated and disposed of, so the Ministry of Environment should continue to force institutions to comply with the related decree.

5.1 Privatization and Existing Medical Waste Treatment Facilities

The health sector in Lebanon is predominantly private, accounting for 85% of hospital beds, with NGOs running 80% of PHC centers. More than 90% of medicines are manufactured and 100% of medical devices and supplies are imported. Coverage for healthcare is low, with about 49% of Lebanese citizens being uninsured, resulting in high out-of-pocket health expenses (World Bank, 2020a, 2020b).

According to the Ministry of Environment, most of the medical waste generated is being collected and treated (mainly autoclaving) by the private sector. The role of

---

[2]Note that to this day, in Lebanon, MSW incineration is not practiced.

Table 4 Overview of healthcare waste treatment units in Lebanon

Location	Name of Healthcare center	Operator	Treatment type	Treated healthcare waste (kg/day)	Nb. of beds served
Beirut	Clemenceau Medical Center (CMC)	CMC and USM	Microwave (on-site)	315	94
	Hotel Dieu de France	Arcenciel	Autoclave (on-site)	385	343
	Jisr El Wati	Arcenciel	Autoclave (off-site)	3235	3371
North	Haykel Hospital	Arcenciel	Microwave and autoclave (on-site)	82	160
	Zgharta	Arcenciel	Autoclave (off-site)	783	1889
Bekaa	Zahle	Arcenciel	Autoclave (off-site)	332	929
South	Saida	Arcenciel	Autoclave (off-site)	1800	733
	Abassiyeh	Mirage	Autoclave (off-site)	450	325

Note: Most of the centers are operational. Zgharta, Zahle, and Saida centers' License pending for EIA Approval
Source: Compiled by authors based on MOE/UNDP/ECODIT (2011); Laceco/Ramboll (2012)

private sector participation is to centralize treatment, particularly in urban areas, where the majority of medical facilities are located, thereby reducing costs and enhancing service quality.

As of 2010, 33% of private hospitals, 20% of public hospitals, and only 2% of private labs/clinics treat their healthcare waste either on-site or in off-site units (inside and outside hospital premises, respectively). Other health centers/hospital located in Beirut, are reported to export small amount of their hazardous waste under the Basel Convention (Sweep-net, 2014). The total number of licenses issued by the Ministry of Environment for healthcare waste treatment facilities using autoclaves amount to 7. These treatment facilities are either built in within the hospitals premises and thus operated by the hospital itself or are off-site the hospitals premises and operated by Arcenciel (NGO). There are only two on-site treatment centers in two private hospitals and five off-site treatment centers operated by Arcenciel (Sweep-net, 2014). Table 4 presents the existing healthcare waste treatment units in different locations in Lebanon.

Overall, Arcenciel (NGO) is responsible for the management of around 80–85% of infectious healthcare waste in Lebanon in close collaboration with the Ministry of Environment, the Ministry of Public Health, the Syndicate of Hospitals and Healthcare Institutions, and municipalities (MoPH, 2018). The waste is treated by autoclaving in one of the five centers of Arcenciel covering all regions in Lebanon (in Jisr el Wati, Zahlé, Hotel Dieu de France, Saida, and Zgharta) (Table 4). The unit cost as charged by Arcenciel for the collection, treatment, training, and provision of bins: 0.64 USD/kg for hospitals and 1.72 USD/kg for labs, dispensaries, associations, universities, and PHC centers (Agamuthu & Barasarathi, 2020; MoPH, 2018).

5.2 Collection and Treatment of Medical Waste

Currently, in most hospitals in Lebanon, infectious healthcare waste sorting is conducted at the generation point. Approximately 80–85% of the generated healthcare waste is treated by shredding associated to autoclaving by a national infectious healthcare waste management network run by Arcenciel. This network includes healthcare waste sorting and management preparation and knowledge for administrative employees, cleaning workers, and healthcare staff. A collection frequency is calculated based on an assessment for the measurement of the generation of infectious healthcare waste and the availability and capacity of the refrigerated storage room within the hospital or other institutions. On a daily basis, Arcenciel collects infectious healthcare waste from hospitals or institutions. In each hospital or institution, the infectious healthcare waste is weighted directly before collection, in the presence of a hospital/institution representative who signs the waste tracking document. The infectious healthcare waste is then transported to the network's closest treatment center. Five treatment centers serving all regions of Lebanon are part of the network. Each treatment center contains one medical waste autoclave which will, when needed, be a backup system for the other centers. Arcenciel sends reports to the Ministry of Environment every trimester, including the amounts collected from each hospital. In Lebanon, limited, recent and accurate information is available on the amounts of the different types of waste produced in healthcare facilities (Maamari et al., 2015). Thus, in order to determine infectious healthcare waste sorting and management procedures, hospitals and governmental authorities lack reference records. For the assessment of environmental effects and the design of management strategies, site-specific data relating to the infectious healthcare waste generation rate is important (Graikos, Voudrias, Papazachariou, Iosifidis, & Kalpakidou, 2010).

An assessment of waste management in 213 primary healthcare centers, conducted by the Ministry of Public Health in 2017 (MoPH, 2018), in an attempt to assess the situation of infectious waste, revealed that 80.5% of surveyed centers (169 out of 210) sort their medical waste, yet, there is an uncertainty in the proper disposal of infectious waste. Moreover, only 39% (82 out of 210) have a contract with specialized companies (Arcenciel, Mirage and Safe) for proper disposal of their infectious waste and 27.6% (58 out of 210) hand them over to hospitals. About 1.5% (3 out of 210) have incinerators and the rest (32% or 67 out of 210) have their infectious waste dumped in landfills directly or through municipalities. The assessment also tackled 31 public hospitals. The results showed that 70% (21 out of 31) of the contacted hospitals hand them to specialized companies (Arcenciel or Safe),

20.3% (7 out of 31) did not answer, have an autoclave or are in transitory phase and looking for a solution. The remainder 9.7% (3 out of 31) dispose them in municipal dump sites or simply burn them.

6 Management of COVID-19 Related Medical Waste

In the face of the COVID-19 era, healthcare centers such as hospitals, clinics, laboratories, temporary quarantine centers, research laboratories hospitals, and public health centers, and even homes will produce COVID-19 related medical waste. As the source of medical waste relating to COVID-19 has been complex, the initial management of medical waste from the source control is important. The detailed composition of COVID-19 related infectious medical waste is not well documented in Lebanon. However, the COVID-19 related waste could be composed of swab, syringes, needles, sharps, blood or body fluid, excretions, mixed waste, laboratory waste, material or equipment contaminated with the virus, mask or disposable gloves, and PPE that are used for screening and treatment for COVID-19 infected patients (Agamuthu & Barasarathi, 2020).

Initially, medical waste was expected to increase greatly since the number of COVID-19 patients reported will probably increase. The comparison of medical waste generation before and after the COVID-19 in Lebanon is displayed in Fig. 2 from March to October between 2019 and 2020. The total amount of infectious medical waste generated monthly in 2020 (355,171 kg) was declined to 9% of 2019 (388,896 kg) and was up to 34% lower in April since hospital assessment was substantially reduced for most general mild patients due to concern of COVID-19 infection as shown in Fig. 1. In addition, the decrease of infectious healthcare waste

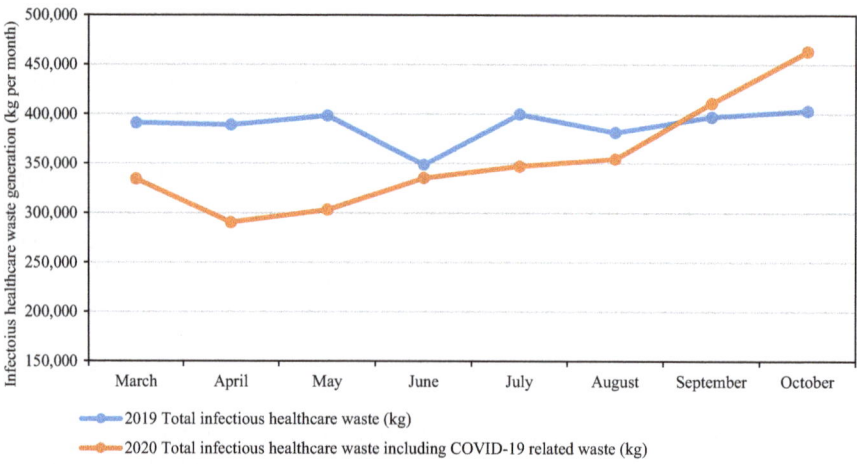

Fig. 2 Changes in pattern of medical waste generation by COVID-19 in Lebanon

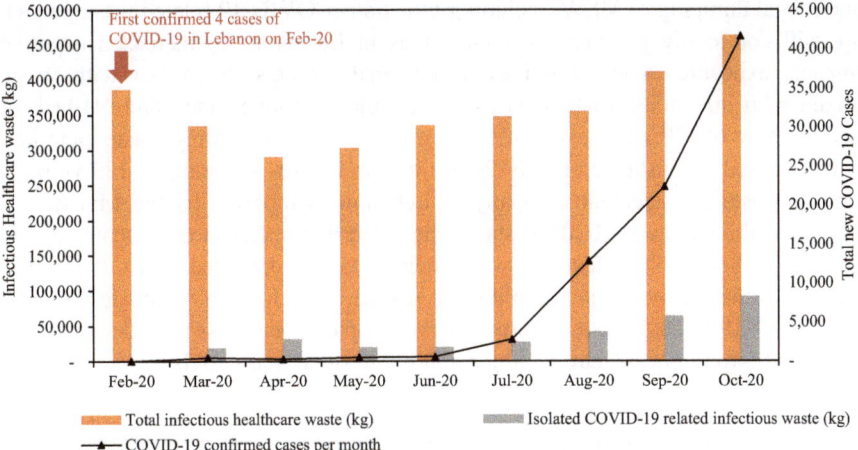

Fig. 3 Medical waste generation with spread of COVID-19 in Lebanon. Note: Sample was provided by Arcenciel and considered 150 hospitals (private) and 30 public—80 Labs—42 dispensaries—four associations and four universities—around 200 PHC's and isolation centers

generated from 2019 to 2020 could be attributed to the national economic and financial crisis that affected hospitals and other healthcare facilities which were doomed to reduce their staff members and work at half capacity which reflected on the number of patients they receive as described in Sect. 3. Another reason behind this decrease is the fear of the public of heading to hospitals because of COVID-19. Therefore, the large majority of cases in healthcare centers are COVID-19 related, except few urgent non COVID-19 cases which are allowed to be taken care of in hospitals.

However, the amount of isolated COVID-19 related medical waste generated in 2020 has increased with the spread of COVID-19 from February 2020. From March to April 2020, the amount of isolated infectious medical waste generated was about 71% higher and the generation amount of isolated medical waste has increased more than double after August 2020 as shown in Fig. 3. The correlation between COVID-19 patients and isolated COVID-19 related infectious medical waste from February 2020 was clearly shown in Fig. 3. The isolated COVID-19 related infectious medical generation in 2020 ranged between 18,351 and 91,352 kg/month, which constituted 5–20% of total infectious healthcare waste. The estimated average of COVID-19 related infectious healthcare waste per month is 39,035 kg or 1.3 tons/day.

The COVID-19 pandemic creates additional challenges in management of waste generated through both households (MSW) and healthcare facilities in Lebanon. Lebanon already lacks appropriate healthcare and MSW management practices due to technological, operational, economical and/or financial constraints is particularly vulnerable during the COVID-19 pandemic, with its new risks and challenges. The challenges associated with MSW management during the pandemic mainly include: (1) increased amount of mixed waste including infectious waste especially that in Lebanon there is no waste segregation at source; (2) increased littering, open burning and illegal dumping of MSW, including infectious COVID-19 related waste, which are still commonly practiced in many areas in Lebanon; (3) increased negative impacts exposure to transmission) to informal workers in contact with waste mixed with infectious waste such as gloves, masks, tissues, etc.; and (4) lack of daily supply of PPEs to waste collectors and awareness regarding waste handling, which caused for instance on July 2020 the Coronavirus infection of 133 workers from a waste management company in Lebanon as reported by the Ministry of Public Health (Reuters, 2020). It should be noted that in Lebanon estimates on the amount of household infectious waste being generated related to COVID-19 is lacking as there is no waste separation at source. Therefore, the estimated amount of COVID-19 related waste reported in this chapter is only associated with waste generated through healthcare facilities and not through households.

7 Conclusion

In Lebanon, medical waste management is under the responsibility of a number of authorities namely, Ministry of Environment, Ministry of Public Health, Ministry of Interior and Municipalities. Decree 13389/2004 amended through decree 8006/2002 regulates the healthcare waste by defining the different type of waste, setting guidelines for the collection and storage, and requiring proper waste segregation/minimization, management and disposal. The degree also requires an Environmental Impact Assessment for healthcare waste treatment facilities in order to get a license from Ministry of Environment. The estimated generation of healthcare waste is 1.0–1.5 kg/bed/day or about 9.2–13.8 tons of risk healthcare waste daily (about 3358–5037 tons/year). Almost 80–85% of the total healthcare waste stream is collected 150 hospitals (private) and 30 public, 80 Labs, 42 dispensaries, 4 associations, and 4 universities, and around 200 PHC's and isolation centers is being treated by Arcenciel (local NGO) through autoclaving. The remaining of the healthcare waste is either being incinerated at hospitals without permits or illegally dumped with MSW. The estimated average of COVID-19 related infectious healthcare waste per month in Lebanon is 39,035 kg or 1.3 tons/day, which constitute 5–20% of total infectious healthcare waste. The analysis showed that the isolated infectious medical waste has greatly increased with the increased number of confirmed COVID-19 cases, which more than doubled after August 2020. However, the total infectious medical waste in 2020 has decreased in comparison to previous years due to financial and economic crisis in Lebanon, which equally affected the implementation of an effective healthcare waste management plan. The management of COVID-19 related waste is managed according to existing legislation and regulations for the management of infectious waste.

Acknowledgments We appreciate and thank the Non-Governmental Organization "Arcenciel" and Mr. Mario Goraieb for providing us with the relevant information on the infectious healthcare waste management in Lebanon.

References

Agamuthu, P., & Barasarathi, J. (2020). Clinical waste management under COVID-19 scenario in Malaysia. *Waste Management & Research, 39*, 18–26.

Basel Convention. (2020). *Waste management an essential public service in the fight to beat COVID-19*. Retrieved 20 May, 2020, from http://www.basel.int/Implementation/PublicAwareness/PressReleases/-WastemanagementandCOVID19/tabid/8376/Default.aspx

CDC (Centers for Disease Control and Prevention). (2020). *What waste collectors and recyclers need to know about COVID-19*. Retrieved 25 May, 2020, from https://www.cdc.gov/coronavirus/2019-ncov/community/organizations/waste-collection-recycling-workers.html

Department of Environment (DOE). (2009). *Guideline on the handling and management of clinical wastes in Malaysia* (3rd ed.). Government Printers.

El-Fadel, M., & Maalouf, A. (2020). Challenges of waste management in a developing context: Lessons from Lebanon. In A. Pariatamby, F. Shahul Hamid, & M. Bhatti (Eds.), *Sustainable waste management challenges in developing countries* (pp. 166–185). IGI Global. https://doi.org/10.4018/978-1-7998-0198-6.ch007

European Commission. (2020). *Waste management in the context of coronavirus crisis*. Retrieved 25 May, 2020, from https://ec.europa.eu/info/sites/info/files/waste_management_guidance_dg-env.pdf

Fan, Y. V., Jiang, P., Hemzal, M., & Klemeš, J. J. (2021). An update of COVID-19 influence on waste management. *Science of the Total Environment, 754*, 142014. https://doi.org/10.1016/j.scitotenv.2020.142014

Graikos, A., Voudrias, E., Papazachariou, A., Iosifidis, N., & Kalpakidou, M. (2010). Composition and production rate of medical waste from a small producer in Greece. *Waste Management, 30*(8–9), 1683–1689.

ISWA (International Solid Waste Association). (2020). *Waste Management during the Covid-19 pandemic ISWA's recommendations*. Retrieved 20 May, 2020, from https://www.iswa.org/fileadmin/galleries/0001_COVID/ISWA_Waste_Management_During_COVID-19.pdf

Jribi, S., Ben Ismail, H., Doggui, D., & Debbabi, H. (2020). COVID-19 virus outbreak lockdown: What impacts on household food wastage? *Environment, Development and Sustainability, 22*(5), 3939–3955. https://doi.org/10.1007/s10668-020-00740-y

Klemeš, J. J., Fan, Y. V., Tan, R. R., & Jiang, P. (2020). Minimising the present and future plastic waste, energy and environmental footprints related to COVID-19. *Renewable and Sustainable Energy Reviews, 127*, 109883. https://doi.org/10.1016/j.rser.2020.109883

Laceco/Ramboll. (2012). *Preparation of pre-qualification documents and tender documents for solid waste management in Lebanon, sub report 1, baseline study*. CDR.

Maalouf, A., Di Maria, F., & El-Fadel, M. (2019). Economic and environmental consequences of implementing an EU model for collecting and separating wastes system in Lebanon. *Waste Management & Research, 37*(12), 1261–1270. https://doi.org/10.1177/0734242X19877677

Maalouf, A., & El-Fadel, M. (2019). Life cycle assessment for solid waste management in Lebanon: Economic implications of carbon credit. *Waste Management & Research, 37*(1), 14–26. https://doi.org/10.1177/0734242X18815951

Maamari, O., Brandam, C., Lteif, R., & Salameh, D. (2015). Health care waste generation rates and patterns: The case of Lebanon. *Waste Management, 43*, 550–554. https://doi.org/10.1016/j.wasman.2015.05.005

MOE/UNDP/ECODIT. (2011). *State and trends of the Lebanese environment*. Ministry of Environment, Lebanon. Retrieved October 2015, from https://www.lb.undp.org/content/lebanon/en/home/library/environment_energy/state-trends-of-the-lebanese-environment.html

MoE/UNDP/GEF. (2016). *Lebanon's third national communication to the UNFCCC*. Beirut, Lebanon. Retrieved July, 2017, from http://climatechange.moe.gov.lb/viewfile.aspx?id=239

MoE/UNDP/GEF. (2019). Lebanon's Third Biennial Update Report (BUR) to the UNFCCC. Beirut, Lebanon. Retrieved July, 2020, from https://unfccc.int/sites/default/files/resource/LEBANON-%20Third%20Biennial%20Update%20Report%202019.pdf

MoE/UNEP/GEF. (2017). *National implementation plan on persistent organic pollutants. Section 5- assessment of industrial POPs and unintentionally released POPs, Lebanon.*

MoI. (2021). *COVID-19 cases.* Ministry of Information, Lebanon. Retrieved January, 2021, from https://corona.ministryinfo.gov.lb/

MoPH. (2018). *Lebanon health resilience project.* Beirut, Lebanon: Ministry of Public Health. Retrieved April, 2020, from https://www.moph.gov.lb/userfiles/files/HealthCareSystem/PHC/20180523%20-%20LHRP%20DRAFT.pdf

OSHA (Occupational Safety and Health Administration). (2020). Worker exposure risk to COVID-19. Retrieved 20 May, 2020, from https://www.osha.gov/Publications/OSHA3993.pdf

Our World in Data. (2021). *Coronavirus pandemic data explorer.* Lebanon: University of Oxford. Retrieved January, 2021, from https://ourworldindata.org/coronavirus-data-explorer?zoomToSelection=true&time=2020-03-01..latest&country=~LBN&region=World&deathsMetric=true&interval=smoothed&perCapita=true&smoothing=7&pickerMetric=total_cases&pickerSort=desc

Penteado, C. S. G. (2021). *Covid-19 effects on municipal solid waste management what can effectively be done in the Brazilian scenario?* 9.

Reuters. (2020). *Lebanon records new coronavirus infection high with 166 cases.* Retrieved July, 2020, from https://www.reuters.com/article/us-health-coronavirus-lebanon/lebanon-records-new-coronavirus-infection-high-with-166-cases-idUSKCN24D0F5?il=0

Sanida, G., Karagiannidis, A., Mavidou, F., Vartzopoulos, D., Moussiopoulos, N., & Chatzopoulos, S. (2010). Assessing generated quantities of infectious medical wastes: A case study for a health region administration in Central Macedonia, Greece. *Waste Management, 30,* 532–538.

Sweep-net. (2014). *Country report on the solid waste management in Lebanon.* Sweep-net.

United Nations Environment Programme (UNEP). (2020). *Waste management during the COVID-19 pandemic: From response to recovery.* Retrieved 5 January, 2021, from https://www.unenvironment.org/resources/report/waste-management-during-covid-19-pandemic-response-recovery

USEPA (United States Environmental Protection Agency). (2020). *Recycling and sustainable management of food during the coronavirus (COVID-19) public health emergency.* Retrieved 25 May, 2020, from https://www.epa.gov/coronavirus/recycling-and-sustainablemanagement-food-during-coronavirus-covid-19-public-health

World Bank. (2020a). *Lebanon economic monitor: The deliberate depression.* Retrieved 5 January, 2021, from http://documents1.worldbank.org/curated/en/474551606779642981/pdf/Lebanon-Economic-Monitor-The-Deliberate-Depression.pdf

World Bank. (2020b). *Beirut rapid damage and needs assessment. Prepared by the World Bank Group in cooperation with the European Union and the United Nations.* Retrieved 5 January, 2021, from https://www.worldbank.org/en/country/lebanon/publication/beirut-rapid-damage-and-needs-assessment-rdna-august-2020

World Health Organization (WHO). (2014). *Safe management of wastes from healthcare activities.* World Health Organization.

World Health Organization (WHO). (2017). *Safe management of wastes from healthcare facilities: a summary.* World Health Organization.

World Health Organization (WHO). (2018). *Healthcare waste.* World Health Organization.

World Health Organization (WHO). (2021). *Statistics.* World Health Organization.

Zhou, C. (2021). The impact of the COVID-19 pandemic on waste-to-energy and waste-to-material industry in China. *Renewable and Sustainable Energy Reviews, 7,* 110693.

Global Picture of COVID-19 Pandemic with Emphasis on European Subcontinent

Abhijit Majumder, Debadatta Adak, Tapas Kumar Bala, and Nandadulal Bairagi

1 Introduction

While the whole world was busy welcoming the new year on December 31, 2019, China informed the World Health Organization (WHO) regarding a new coronavirus that causes unusual pneumonia (World Health Organization, 2020a). Least was known that the New Year 2020 will bring such distress, anxiety, misery, sadness, indebtedness in everyone's life to such an extreme level instead of success, happiness, prosperity and tranquillity. Life on earth has suddenly come to a standstill due to the invasion of this novel coronavirus. The question still remains unanswered that who transferred and how the human chain got affected with this virus in the beginning, but this coronavirus must be identified as one of the worst creations of God or the brainchild of the highly intellectual creation of the God, called mankind.

Anticipating the spread of the virus, WHO postulated and directed management policies for COVID-19 to every nation (World Health Organization, 2020b). Accordingly, most of the countries sealed their international borders. No inbound or outbound transportation was permitted by any means. As a next step, the interstate transportation was shut, followed by a total nationwide lockdown barring the emergency services. This led to sudden social chaos in the first stage, wherein people didn't know how to react to the situation or what to do. Businesses and industries

A. Majumder · N. Bairagi (✉)
Centre for Mathematical Biology and Ecology, Department of Mathematics, Jadavpur University, Kolkata, India
e-mail: nbairagi.math@jadavpuruniversity.in

D. Adak
Department of Applied Mathematics, Maharaja Bir Bikram University, Agartala, India

T. K. Bala
Department of Oral and Maxillofacial Surgery, KSDJ Dental College and Hospital, Kolkata, India

© The Author(s), under exclusive license to Springer Nature Singapore Pte Ltd. 2022   235
S. K. Ghosh, P. Agamuthu (eds.), *Health Care Waste Management and COVID 19 Pandemic*, https://doi.org/10.1007/978-981-16-9336-6_11

stopped functioning, leading to decreased or no production of goods. It initiated a cascade fall of the local economy to the national economy. Gradually people started losing jobs or asked to work from home those who could with a salary cut. Many labourers were asked to go back home since there was no work but then no transportation, hence again a chaos. The government realized that this novel coronavirus is on for a long haul but at the same time the economy is taking a toll, the stock market started tumbling, more people started losing jobs and had a psychological impact on society as well as at the individual level. Schools, examinations, transportation etc. were suspended to contain the spread of this pandemic. This whole situation was so sudden and new to every nation of the world that a lockdown was imposed to fabricate strategies and policies to tackle this pandemic which was spreading like a jungle fire. WHO's call of public health emergency of international concern (PHIC) on 30 January (World Health Organization, 2020b) for the sake of accelerated preparedness and better management to prevent the spread of infection to other countries did not work and by March 11, 2020 infection disseminated to 200 countries. As the outbreak was intensifying further, WHO had no option left but to declare COVID-19 a global pandemic on March 11, 2020 (Chen, Xiong, Bao, & Shi, 2020).

2   First Month of COVID-19 Infection

This highly infectious disease crossed the international border of China and reached Thailand on January 8, 2020 (World Health Organization, 2020a). On January 11, the first death was recorded in China due to coronavirus (Fig. 1). China gave a name to this new virus on January 12 and called it 2019-nCoV. It also shared the whole genome sequence with the WHO on the same day. The first case was detected in the USA on 20th January 2020, while the first case in Europe was detected in France on 24th January 2020. Gradually the disease had spread to many other European countries, including Italy, in no time and became the first epicentre of this disease outside China.

Table 1 compares the worldwide epidemiological data of COVID-19 till July 29, 2021 with the data of January 31, 2020. Just looking at the statistics, the severity

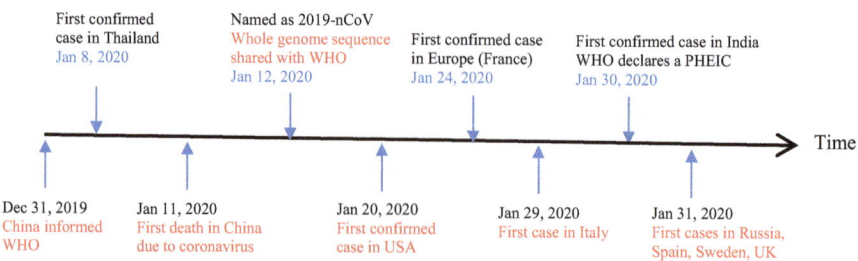

Fig. 1 Journey of coronavirus in the first month of 2020

Table 1 COVID-19 statistics comparison

Date	Affected countries	Confirmed cases	Death cases
By January 31, 2020	23	2,011	43
By July 29, 2021	220	198,034,753	4,224,421

Fig. 2 Model picture of coronavirus (courtesy internet)

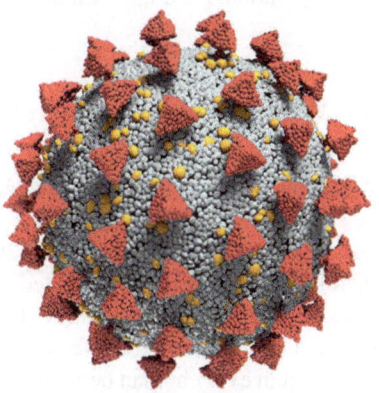

of the disease and its devastating effect on mankind was clearly evident. By January 31, 2020, the number of affected countries were 23 with 2011 infected cases and 43 death cases. WHO coronavirus disease (COVID-19) dashboard showed that 220 countries/territories were already affected by this dreaded disease with 198 million infected and more than 4.2 million death cases (Worldometers, 2020a) by July 29, 2021.

3   COVID-19 Disease and Coronavirus

The disease that has caused the pandemic is known as COVID-19 and its causative agent is called SARS-CoV-2 (Severe Acute Respiratory Syndrome Coronavirus 2). This name was given by the International Committee on Taxonomy of Viruses (ICTV) on February 11, 2020 (Lambert, 2010).

Coronavirus is a positive-stranded RNA virus. There are many spike proteins on the cell surface of a coronavirus (Fig. 2), which look like a crown (coronam is the Latin term for crown) hence the virus is called coronavirus. The virus uses this S-protein as a key to open the cellular door and enters into the cell. There could occur some errors in copying the genetic materials during the replication process of a virus in the host cell, giving rise to a new variant. Such changes may occur in most viruses, and the COVID-19 virus is not an exception. Occasionally, such evolution in the virus structure may bring some changes in the virus properties, e.g., infectivity and virulence, or the virus may not respond to the existing drug or vaccine. Depending on the changed properties of SARS CoV-2 variants and assessing their efficacy in posing threat to global public health, WHO has classified the existing variants of coronavirus into two categories, viz. variants of interest (VOI) and variants of concern (VOC). Currently, there are four VOC of COVID-19 virus, viz. alpha, beta, gamma, delta and four VOI, viz. eta, iota, kappa, lambda (World Health Organization, 2021). The future of the COVID-19 epidemic depends on these existing variants and the variants that will come up in the future.

4 Why Lungs Are Severely Affected?

ACE2 (angiotensin-converting enzyme 2) is a protein found on the surface of many cell types and SARS-CoV-2 spike protein has a strong binding affinity to ACE2 (Abdullahi et al., 2020; Lambert, 2010). In the lungs, ACE2 is in abundance and therefore lungs are mostly affected. ACE2 is also expressed in the epithelium of the nose, mouth, eye, heart, kidney, liver and digestive tract, implying that the other organ of our body may be affected by this virus (Xu et al., 2020). The receptor ACE2 is present in every human being but their abundance may vary from person to person. It is reported that individuals having hypertension, diabetes, and coronary heart disease may have a higher level of ACE2 and therefore be more prone to SARS-CoV-2 infection.

5 Other Coronavirus Diseases

This is the third attack of coronavirus in the twenty-first century. The first episode was recorded in 2002–2003, originated in Guangdong Province of China (Xie et al., 2020). The second one was MERS, which occurred in 2012–2020 and its origin was Saudi Arabia. The third one is SARS-CoV-2, which is trending now, originated from Hubei province of China. SARS CoV-1 and MERS both are mainly lower respiratory infections, whereas SARS-CoV-2 is an upper respiratory infection (Rockx et al., 2020; Worldometers, 2020b). Notably, SARS-CoV-2 has a relatively higher viral load than its other two counterparts (see Table 2). MERS-CoV has the highest case fatality ratio and SARS-CoV-2 has the lowest.

6 Who Can Be Infected?

SARS-CoV-2 is a novel coronavirus, meaning a new strain of virus previously not detected in the human race, and hence we have no immunity against this virus, making everyone susceptible to this disease. However, two categories of people are at a higher risk of getting severe COVID-19 infection: older people and those who have some specific systemic disease. In Italy, only one in four COVID-19 infected

Table 2 Comparison among between SARS, MERS, and SARS-CoV-2

SARS-CoV (2002–2003)	MERS-CoV (2012–2019)	SARS-CoV-2 (2019 till date)
Origin: Guangdong province, China	Origin: Saudi Arabia	Origin: Wuhan city, Hubei province, China
It is mainly a lower respiratory tract infection	Lower respiratory tract infection	It is mainly a upper respiratory tract infection
It takes 7–10 days after onset of symptoms until peak RNA concentrations	It also takes 7–10 days after onset of symptoms until peak RNA concentrations	Peak concentrations were reached during the first 5 days after symptoms onset
Average viral load 5×10^5 copies per ml	Average viral load 6.3×10^4 copies per ml	Average viral load 7×10^6 copies per ml
Spread over 29 countries; confirmed cases—8096; death—774; case fatality ratio—9.6	Spread over 27 countries; confirmed cases—2494; death—858; case fatality ratio—34.4	Spread over 220 countries; confirmed cases—198,034,753; death—4,224,421; case fatality ratio—2.13

individuals is in the age group 19–50 years, and the rest three are in the age group 50–90 years (Jin et al., 2020). A similar result is also observed in the case of the USA. Study shows that the death ratio of male to female is 2.4:1 while both have the same prevalence (Jordan, Adab, & Cheng, 2020). Individuals having other associated illnesses, e.g., coronary artery diseases, cancer, chronic obstructive pulmonary disease (COPD), Type 2 diabetes, and other immunocompromised conditions, are probably at a higher risk of severe illness due to SARS-CoV-2 (Statista, 2020).

7 Clinical Manifestation

COVID-19 infection may be classified into mild, severe and acute depending on the severity of the infection. About 81% of the COVID-19 cases are mild or asymptomatic. They do not require ICU intervention or management. The percentage of severe cases is about 14 and they need ICU intervention (their breathing rate per minute >30 (normal rate is 12–18) and oxygen saturation level are <93% (normal range is 95–100%). About 1% of acute cases show respiratory failure and MOD (multiple organ dysfunction) (An et al., 2020; Singhal, 2020).

8 Symptoms of COVID-19

The symptoms of COVID-19 are similar to the symptoms of the common flu. Infected individuals may have a fever, dry cough, shortness of breath, and tiredness. Other symptoms could be dyspnoea, muscular pain, chills, runny nose, nasal congestion, headache, sore throat, chest pain, nausea or vomiting, and diarrhoea. A large number of infected people are asymptomatic or may show mild symptoms. Infected individuals show radiological evidence of bilateral pneumonia and decreased peripheral lymphocyte counts. Some symptomatic patients rapidly develop acute respiratory infection symptoms or acute respiratory distress syndrome (ARDS) or even multi-organ dysfunction (Chen et al., 2020; Chu et al., 2020).

9   Laboratory Diagnosis

Respiratory tract swab specimen (nasopharyngeal and oropharyngeal swab) is collected by trained personnel and then tested for SARS-CoV-2 RNA in the authorized laboratory. The molecular test real-time reverse transcription-polymerase chain reaction (RT-PCR) assays are the CDC (Centre for Disease Control) recommended diagnostic test for the coronavirus disease. An antibody-based technique is also used as supplemental tool (Tang, Schmitz, Persing, & Stratton, 2020).

10   Transmission of COVID-19

Transmission of infection occurs through direct contact or inhaling the droplets (>5–10 μm in diameter) caused due to coughing and sneezing by an infected person. Infection can also spread if one touches a contaminated surface and then touches his/her mouth, nose, and eyes. Some recent studies show that an individual may be infected by inhaling aerosols (<5 μm in diameter) containing a virus during medical or dental procedures. For example, a jet nebulizer can produce virus-contaminated aerosols when a COVID-19 patient is given nebulizers under controlled laboratory conditions. Such contaminated aerosols can sustain 3–16 h in a closed environment (Fears et al., 2020; Van Doremalen et al., 2020). However, there is no such evidence of aerosols-based infection in the natural environment and normal cough conditions. Some studies report that SARS-CoV-2 RNA has been identified in faeces of COVID-19 patients and also from the urine (Sun et al., 2020; Xiao et al., 2020). There is no report on whether infection can spread through such biological samples.

11   Why Controlling of COVID-19 Is Very Challenging?

In epidemiology, the latent period is defined as the period between the occurrence of infection and the onset of infectiousness. It is approximately 4.6 days in the case of COVID-19. During this period an individual carries the virus but is not yet infectious. The incubation period represents the time period between the occurrence of infection (or transmission) and the onset of disease symptoms (Fig. 3). The mean incubation period for COVID-19 is about 5 days. There is a narrow interval during

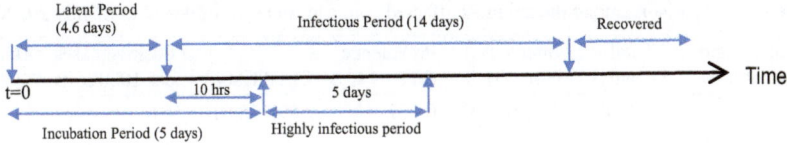

Fig. 3 Time line in COVID-19 infection

which an individual does not show any symptoms but can spread the disease. Such time for COVID-19 is approximately 10 h. The infectious period is the time during which an infected individual is capable of spreading the infection. In the case of COVID-19, it is about 14 days. Out of these 14 days, an infected individual remains highly infectious for an initial 5 and a half-day (Lauer et al., 2020).

Usually, an infected individual visits hospital on an average of 1–2 days after the appearance of the symptoms (post-onset) and it takes another 2 days to get the test report. Thus 3–4 days of the highly infectious period is over before the isolation takes place. These 3–4 days, the individual remains in the society as a common man and may spread the infection to others. This is one of the reasons why controlling COVID-19 is very challenging. This also explains why random testing is important for COVID-19. We have to identify and isolate COVID-19 positive patents as early as possible. The 3T principles: Trace, Test, and Treat are the best possible options in controlling the COVID-19 spreading. If we cannot trace the potentially infected individuals before 5 days, it will be difficult to break the chain. That is why large-scale testing and random testing are very important.

12 Comparison of Epidemiological Data

Table 3 compares the epidemiological status of the ten most COVID-19 affected countries in the world as of July 29, 2021, in terms of death (Worldometers, 2020a). The United States of America tops the list and Argentina being the last. The prevalence of the disease is defined by the ratio of infected cases to the susceptible population, which is highest in Argentina and lowest in Mexico. Positive cases per 1000 tests are highest in Argentina and lowest in Mexico. The case fatality rate, defined here by the ratio of deaths to total settled cases, is highest in Peru and lowest in India.

13 Prediction About Disease Dynamics

Mathematical models are vital in the understanding of the disease dynamics and can be used as a tool in predicting the COVID-19 cases along with the time frame for the same. The model can predict the probable time of the epidemic peak, incidence, and

Table 3 Comparison among the ten most affected countries in terms of death (as of July 29, 2021)

Countries and population (N)	Total deaths	Infected cases (I)	Prevalence (I/N)	Total tests	Positive cases per 1000 tests	CFR (%)
USA (333,093,143)	629,064	35,688,506	0.107	529,178,910	107	1.76
Brazil (214,186,368)	555,512	19,880,273	0.09	55,034,721	93	2.79
India (1,394,605,426)	423,842	31,613,993	0.023	466,427,038	23	1.34
Mexico (130,390,343)	240,456	2,829,443	0.021	8,421,353	21	8.5
Peru (33,465,698)	196,291	2,109,294	0.063	15,476,472	63	9.3
Russia (146,001,924)	157,771	6,242,066	0.043	164,700,000	43	2.53
UK (68,270,129)	129,583	5,830,774	0.085	244,114,700	85	2.2
Italy (60,365,948)	128,047	4,343,519	0.072	77,279,334	72	2.9
Colombia (51,466,439)	120,432	4,776,291	0.092	22,448,016	92	2.52
France (65,429,176)	111,824	6,103,548	0.093	104,074,220	93	1.8
Argentina (45,642,150)	105,586	4,919,408	0.108	19,334,462	108	2.15

duration (Paul et al., 2020, Adak et al. 2021). Such information is vital for the healthcare management authority. Based on the information, they can take necessary steps for designing effective control measures and facility development, so that the healthcare system does not collapse due to the surge of COVID-19 cases, and optimum treatment can be provided to the thousands of infected people. Policymakers have been increasingly relying on mathematical projections and taking various decisions on containment measures. Based on the mathematical prediction of Imperial College, London, the UK Government imposed strict restrictions on the movement to prevent the spread of infection during the first wave (Adam, 2020).

Lots of mathematical models have successfully predicted the COVID-19 epidemic burden on different countries (Fanelli & Piazza, 2020; Majumder et al., 2020). These models reasonably address the epidemic burden of an affected country to guide the policymakers and the healthcare providers on their preparedness. Such models usually extend the classic SEIR (susceptible, exposed, infected, recovered) models. The outcomes of these models, however, depends on the model assumptions based on which the models are constructed and the data they use for parameter estimation. Most of the models considered to study the COVID-19 epidemic are deterministic in nature which considers that the environment is unvarying and therefore, infection rate, recovery rate, death rate, etc. are constant. However, such assumptions may influence the model outcomes because there are significant variations and uncertainties in the estimation of rate parameters in case of a growing epidemic of unknown aetiology like COVID-19. Stochastic epidemic models can capture such uncertainty or randomness and give better predictions (Majumder, Adak, & Bairagi, 2021).

One of the most important metrics in epidemiology on which mathematicians rely is the basic reproduction number, usually denoted by R_0. It is the number of secondary cases produced by a single infected individual during his/her infectious period when introduced into a completely susceptible group of individuals (Ferguson, Donnelly, Woolhouse, & Anderson, 1999). For example, $R_0 = 3$ implies that one infected individual will infect an average of three individuals during his/her infectious period. It is therefore straightforward to understand that an infection will spread in a group if the basic reproduction number of the disease is greater than 1, otherwise the infection will be eradicated over time. The epidemic will grow in size and spread more rapidly in the community depending on the value of R_0. Thus, the severity of the epidemic outbreak depends on how large the basic reproduction number is compared to 1. This transmission potentiality measure, R_0, is used by public health officials as a community health indicator against infectious disease. The basic reproduction number depends on the other system parameters in a complex way and therefore its accuracy depends on the right estimation of the parameters and the accuracy of the data set in turn. The effectiveness of the intervention measures on containment or regression of an epidemic is reflected in the reproduction number. In the case of COVID-19, there exists a large variation in the basic reproduction number.

Epidemiologists are now using more advanced and sophisticated tools to measure the real-time reproduction number R_t, which gives the value of the reproduction number for every day on the basis of surveillance data (Cori, Ferguson, Fraser, & Cauchemez, 2013; Thompson et al., 2019). The aim of the health management officials is to keep R_t below 1. If the value is above 1 then the containment measures have to intensify to bring down R_t below unity. The first row of Fig. 4 represents the daily COVID-19 cases reported in the three European countries Italy, Spain, and France. The corresponding real-time reproduction numbers (R_t) using the R software package (EpiEstim 2.2) are presented in the second row. The time series of reported COVID-19 cases in Italy shows that the number of cases gradually increased and attained the first peak on March 21, 2020. The daily cases started to decrease but again started to increase in September and attained a second peak on November 13. A small third peak was observed at the end of the second week of March 2021. The corresponding R_t value shows that the initial reproduction number was above 3 and reduced to 1 on March 22, 2020 when the first epidemic peak was attained. R_t decreased after March 22 and then again raised above 1 and went below 1 after November 14, 2020. R_t value again crossed unity near the middle of March 2021 and went below unity in the subsequent time, indicating that the epidemic is under control. The disease will be eliminated once it becomes zero. In the case of Spain, the real-time basic reproduction number R_t crosses the critical value 1 as of March 26, 2020, just after the day of attaining the first epidemic peak. In the rush of the second wave of COVID-19 spreading, R_t went above 1 once more and crossed the

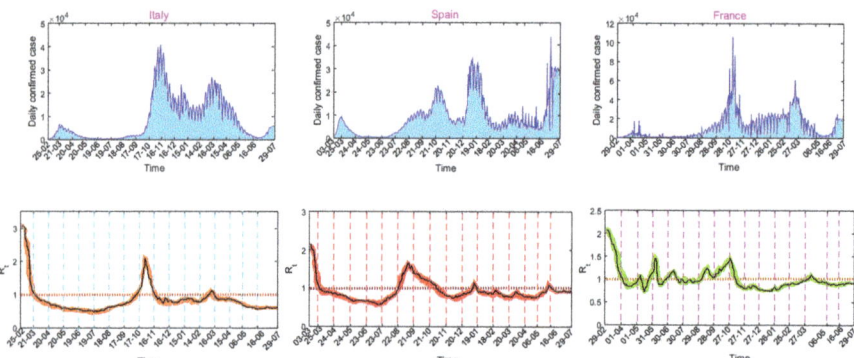

Fig. 4 Upper row: Per day COVID-19 cases in Italy, Spain, and France. Lower row: Real-time reproduction number (R_t) of the respective countries with 95% confidence interval. The R_t values as of July 29, 2021 are 0.724, 0.918, and 0.923 for Italy, Spain, and France, respectively

critical mark 1 on 29 October just after 2 days of attaining the second epidemic peak. It again crossed 1 and went below 1 on January 17, 2021. Due to the onset of the third wave, Spain's RTRN curve again crossed the threshold value 1 and went below 1 on 14 July. There are also significant fluctuations in the time series data of Spain, indicated by the wide confidence interval in the R_t plot. France attained its first epidemic peak on the last day of April 2020, the second peak on November 7 and the third peak in the second week of April 2021, and its R_t curve crossed the value unity on 1 April, 8 November 2020, and 9 April 2021, respectively. Noticeably, the R_t curve of France crossed the critical value 1 several times. The values of R_t as of July 29, 2021, are 0.724, 0.918, and 0.923 for Italy, Spain, and France, respectively, indicating that Italy was in a better position compared to the other two countries whose R_t values were close to 1. Although the epidemic is in decline mode ($R_t < 1$) in these countries, strict containment measures should be continued to avoid another wave.

14 Containment Measures

14.1 Non-pharmaceutical Interventions

The most effective way to reduce human-to-human transmission of this highly contagious disease is through non-pharmaceutical interventions (NPIs). For example, individual hygiene, cough etiquette, and safe distancing can reduce the transmission probability. Disease transmission can be protected at the individual level by using a face mask and extra protection can be added by protecting the eyes using spectacles and sunglasses (Chu et al., 2020). To reduce largescale contamination and the community transmission of COVID-19, lockdown is the most effective mechanism (Salzberger, Glück, & Ehrenstein, 2020; Wilder-Smith, Chiew, & Lee, 2020).

Many countries such as China, South Korea, and most European countries have restricted the transmission of coronavirus by implementing nationwide strict lockdown and contact tracing. Though transmission of the virus can be effectively reduced to tolerance level, but the eradication of SARS-CoV-2 is impossible in the absence of large-scale vaccination.

14.2   Drugs

There is no COVID-19 specific medication. Scientists are working relentlessly to find a suitable entry inhibitor to block the interaction between the spike protein (S-protein) and the ACE2 receptor (Roshanravan, Ghaffari, & Hedayati, 2020). If successful, an ACE2 blocker could be a promising therapeutic target. SARS-CoV-2 is a single-stranded RNA virus. After entering into the host cell, the RNA is transcripted into DNA with the help of a viral enzyme, called reverse transcriptase. People are also searching for a suitable drug called reverse transcriptase inhibitor (RTI) which can effectively block the transcription mechanism (Park, Thwaites, & Openshaw, 2020). In the last stage of a virus lifecycle, an enzyme called protease cuts the long chain of non-functional viral protein into a smaller functional protein. These smaller proteins mature and become a new virus. The drug protease inhibitors (PIs) may also be a therapeutic target to restrict the protease enzyme from its functioning and prevent virus replication (Chang et al., 2020).

Though specific drugs for SARS-CoV-2 is still unavailable, some repurposing drugs have been found to be helpful in treating COVID-19 patients (Goldman et al., 2020; Williamson et al., 2020). Existing drugs such as dexamethasone, favipiravir, and remdesivir, which are usually used for treating other diseases, have been given permission by the WHO and NIH (National Institute of Health) to treat COVID-19 patients (World Health Organization, 2020c). For instance, the immunosuppressive drug dexamethasone has been shown to save the lives of several COVID-19 patients and has remarkably reduced the mortality rate by one-third (National Institutes of Health, 2020). A recent study had shown that India has significantly reduced the COVID-19 burden through the combined application of repurposing drugs and NPIs during the first wave of the COVID-19 epidemic (Mondal, Majumder, Adak, & Bairagi, 2020). Passive immunization therapy, known as immunoglobulins therapy, is another potential treatment that can enhance the survival rate of individuals infected by a coronavirus (Chen et al., 2020). In this convalescent plasma therapy, plasma is collected from the COVID-19-recovered patients and is infused to a COVID-19 patient to boost the immune system of the recipient to fight against the virus.

14.3 *Vaccine*

The human immune system is unable to recognize the virus SARS-CoV-2 because it is a novel virus and there is no antibody in the body for this unknown pathogen. A weakened antigen collected from the pathogen is injected into the body to activate the immune system and to produce antigen-specific antibodies to counter the pathogen. In this process, a low level of antibody is produced in the body against this particular pathogen and which would protect any future invasion by this pathogen through acquired immunity (Goldman et al., 2020). Vaccine development is a lengthy and time-consuming multi-staged process. Pfizer/BioNTech-produced Comminate COVID-19 vaccine received the first approval of WHO for emergency use on December 31, 2020. It took around 12 months for the development and market availability of a COVID-19 vaccine since the date of the first information received by WHO about the outbreak. At present, 21 approved vaccines are being applied throughout the globe (COVID-19 Vaccine Tracker, 2021). These include the vaccines from AstraZeneca, Pfizer/BioNTech, Gamaleya Research Institute of Russia, Moderna, Bharat Biotech, Johnson & Johnson, Sinovac and Sinopharm of China, Chumakov Federal Scientific of Russia, Novavax, Valneva, the Israeli Institute for Biological Research, and others. There are another 89 promising vaccines that are in phase 2/3 trials (COVID-19 Vaccine Tracker, 2021).

15 Healthcare Waste Management

It is known that the virus SARS-CoV-2 spreads through respiratory droplets and contagious contacts. To create a barrier for the virus and to reduce the spreading of the disease, rational use of personal protective equipment (PPE) is a must. Healthcare providers and frontline workers have been advised to protect themselves from this deadly virus by using PPEs such as gowns, gloves, masks, head caps, shoe covers, etc. Common people have also been directed to wear at least a mask properly when they are outside. Hence billions of PPE are used per day throughout the world to stay safe from the coronavirus.

In the last one and half years, the healthcare system has been under tremendous pressure due to the surge of COVID-19 patients. Medical waste management has been jeopardized due to the excessive medical wastes generated from the COVID-19 associated treatment. It has become a big challenge to properly handle medical waste disposal during the COVID-19 pandemic since the used PPE is considered as infectious solid wastes. Medical and healthcare waste comprises the waste produced by healthcare centres, medical laboratories and medical research facilities. Medical waste management should be done properly to avoid infection to waste pickers, waste workers, health providers, patients, and the community. Since most of these medical wastes are infectious and contagious, especially with the COVID-19 virus, handling, disinfection, and disposal of COVID-19 wastes should be done following the national/international protocol to lower the dissemination of infection and for sustainable management of biomedical wastes (BMW) (Ilyas et al., 2020). Proper BMW management may add up to the containment and decrease in the number of COVID-19 cases in the health workers and the society. Many countries have adopted various safety and disposal norms for hazardous medical waste management. The advanced and many developing nations have moved from decentralization to centralization, from irregular to regular management, and from mostly incineration to non-incineration disposal technologies such as autoclave steam, dry heat, chemical disinfection or microwave (Singh et al., 2020). To avoid these sources of contamination and to inhibit the spread of COVID-19, information technology-based medical waste management methods like the Internet of Things (IoT), AI, and GIS, have been developed and used by many countries. Real-time tracking and controlling of the waste management process have been implemented in the city of Wuhan. This automated medical waste disposal method can decrease human contact and reduce the risk of infection. It is of utmost importance that every country should have a well-defined strategy for managing biomedical wastes during this ongoing pandemic. At the same time, the authority should have a strong monitoring system so that the stakeholders strictly follow the guidelines and help the administration for sustainable management of hazardous medical waste.

16 Conclusion

This COVID-19 pandemic has decelerated the world's economic, medical, and social factors which in turn have severely affected the entire globe through its devastating effects. The World Bank in its June report forecasted a 3% to 6% reduction in global economic growth in 2020. The World Health Organization is however uncertain about the eradication time and number of casualties in COVID-19 because of the underlying uncertainties prevailing in the ongoing pandemic. It is to be mentioned that life will never be the same as it was before this pandemic.

Many countries such as the USA, India, Brazil, France, Italy, and Spain have entered the third stage, where community spread is evident. In this stage, the spread of the disease is random, and contact tracing is not possible. Therefore, the only way to contain is by isolating himself from others and following the treatment protocol. Seeking medical help, early hospitalization will prevent the fatal sequelae of the disease for an individual. We have to learn, adapt, and improvise to tame and live the coronavirus till the definitive treatment is structured. One should make it a practice to wear a mask while going out of home even after the corona infection is under control and specific therapy is available. Some countries such as China, South Korea, New Zealand, and many European countries have controlled the coronavirus infection in this way. However, no country will remain safe, even if it has controlled the infection, because the infection can again arrive through international travels. It is to be remembered that infection supposedly started from an animal market in Wuhan and then has spread all over the world, so the uncertainty will always be hovering over mankind.

Social distancing and lockdown will reduce the spread of the virus for a short period, but lifting these measures might allow another wave of the pandemic. The bitter experience of the second wave had been tasted and experienced by many countries. The wave lasted for a longer time and the peak value was two to four times higher in comparison to the first wave. European countries such as Italy, Spain, and France have already witnessed the third wave. It may be difficult but not impossible for these countries to avoid the fourth wave. If 60% of the population is vaccinated and COVID-19 restrictions are maintained properly then another wave could be possibly avoided by these countries. Thus, there is a need for an all-around effort from all corners of the globe to fight against this virus and eradicate it for a better world.

It is worth mentioning that control and prevention of subsequent wave of COVID-19 are possible only through large-scale COVID-19 vaccination. The COVID-19 vaccination programme has been initiated throughout the world on a priority basis. Currently, about 21 approved vaccines are in circulation, however, there is a severe shortage of vaccines in developing and underdeveloped countries. It is, however, reported that people of many low-income countries, or even some developed countries and healthcare workers, are refusing to get vaccinated (Cerda & García, 2021; Dzieciolowska et al., 2021), which may cause a hurdle to prevent the next wave. Only 1.1% population of the low-income countries have received at least one dose of the vaccine, compared to 28% of the world population (Ritchie et al., 2020). European countries such as Italy, Spain, and France have crossed the 70% mark of at least one vaccine dose and around 59% of adult people have received full vaccination till the end of July 2021.

According to a recent report of the World Bank, if 60% of people of each country is vaccinated by March 2022 as a short-term target, then there will be an opportunity of getting herd immunity against COVID-19 (Agarwal & Reed, 2021). It is, however, a big challenge to vaccinate the majority of the world population by this timeline. The target may be difficult to achieve but not impossible. The foremost barrier is the inadequate production of COVID-19 vaccines. Maximum effort should be put in to enhance vaccine production. Perhaps, should be ensured that densely populated countries and low-income countries get access to these COVID-19 vaccines if we want to put an end to this pandemic. We have to keep patience and maintain COVID-19 hygiene protocols till the target is achieved. Mankind had always been experimental with nature and science, to which the history had been evident that we failed, then learned our lesson and eventually survived successfully through the adversaries.

Acknowledgment Research of A.M. is supported by CSIR, India, File No: 09/096(0874)/2017-EMR-I and Research of N.B. is supported by SERB MATRICS, File Number: MSC/2020/000020.

References

Abdullahi, I. N., Emeribe, A. U., Mustapha, J. O., Fasogbon, S. A., Ofor, I. B., Opeyemi, I. S., Obi-George, C., Sunday, A. O., & Nwofe, J. (2020). Exploring the genetics, ecology of SARS-COV-2 and climatic factors as possible control strategies against COVID-19. *Le Infezioni in Medicina, 28*(2), 166–173.

Adak, D., Majumder, A., & Bairagi, N. (2021). Mathematical perspective of Covid-19 pandemic: Disease extinction criteria in deterministic and stochastic models. *Chaos, Solitons & Fractals, 142, 110381*. https://doi.org/10.1016/j.chaos.2020.110381

Adam, D. (2020). Special report: The simulations driving the world's response to COVID-19. *Nature, 580*(7803), 316.

Agarwal, R., & Reed, T. (2021). *How to end the COVID-19 pandemic by March 2022*. World Bank Policy Research Working Paper.

An, P., Song, P., Wang, Y., & Liu, B. (2020). Asymptomatic patients with novel coronavirus disease (COVID-19). *Balkan Medical Journal, 37*(4), 229.

Cerda, A. A., & García, L. Y. (2021). Hesitation and refusal factors in individuals' decision-making processes regarding a coronavirus disease 2019 vaccination. *Frontiers in Public Health, 9*, 626852.

Chang, Y. C., Tung, Y. A., Lee, K. H., Chen, T. F., Hsiao, Y. C., Chang, H. C., Hsieh, T. T., Su, C. H., Wang, S. S., Yu, J. Y., & Shih, S. S. (2020). Potential therapeutic agents for COVID-19 based on the analysis of protease and RNA polymerase docking. https://doi.org/10.20944/preprints202002.0242.v1

Chen, H., Guo, J., Wang, C., Luo, F., Yu, X., Zhang, W., Li, J., Zhao, D., Xu, D., Gong, Q., & Liao, J. (2020). Clinical characteristics and intrauterine vertical transmission potential of COVID-19 infection in nine pregnant women: A retrospective review of medical records. *The Lancet, 395* (10226), 809–815.

Chen, L., Xiong, J., Bao, L., & Shi, Y. (2020). Convalescent plasma as a potential therapy for COVID-19. *The Lancet Infectious Diseases, 20*(4), 398–400.

Chu, D. K., Akl, E. A., Duda, S., Solo, K., Yaacoub, S., Schünemann, H. J., El-harakeh, A., Bognanni, A., Lotfi, T., Loeb, M., & Hajizadeh, A. (2020). Physical distancing, face masks, and eye protection to prevent person-to-person transmission of SARS-CoV-2 and COVID-19: A systematic review and meta-analysis. *The Lancet, 395*(10242), 1973–1987. https://doi.org/10.1016/S0140-6736(20)31142-9

Cori, A., Ferguson, N. M., Fraser, C., & Cauchemez, S. (2013). A new framework and software to estimate time-varying reproduction numbers during epidemics. *American Journal of Epidemiology, 178*(9), 1505–1512.

COVID-19 Vaccine Tracker. (2021). https://www.raps.org/news-and-articles/news-articles/2020/3/covid-19-vaccine-tracker

Dzieciolowska, S., Hamel, D., Gadio, S., Dionne, M., Gagnon, D., Robitaille, L., Cook, E., Caron, I., Talib, A., Parkes, L., & Dubé, È. (2021). Covid-19 vaccine acceptance, hesitancy, and refusal among Canadian healthcare workers: A multicenter survey. *American Journal of Infection Control, 49*(9), 1152–1157.

European Centre for Disease Prevention and Control. (n.d.). *COVID-19 vaccine tracker*. https://vaccinetracker.ecdc.europa.eu/public/extensions/COVID-19/vaccinetracker.html

Fanelli, D., & Piazza, F. (2020). Analysis and forecast of COVID-19 spreading in China, Italy and France. *Chaos, Solitons & Fractals, 134*, 109761.

Fears, A. C., Klimstra, W. B., Duprex, P., Hartman, A., Weaver, S. C., Plante, K. S., Mirchandani, D., Plante, J. A., Aguilar, P. V., Fernández, D., & Nalca, A. (2020). Persistence of severe acute respiratory syndrome coronavirus 2 in aerosol suspensions. *Emerging Infectious Diseases, 26* (9), 2168.

Ferguson, N. M., Donnelly, C. A., Woolhouse, M. E. J., & Anderson, R. M. (1999). Estimation of the basic reproduction number of BSE: the intensity of transmission in British cattle. *Proceedings of the Royal Society of London, 266*(1414), 23–32.

Goldman, J. D., Lye, D. C., Hui, D. S., Marks, K. M., Bruno, R., Montejano, R., Spinner, C. D., Galli, M., Ahn, M. Y., Nahass, R. G., & Chen, Y. S. (2020). Remdesivir for 5 or 10 days in patients with severe Covid-19. *New England Journal of Medicine, 383*(19), 1827–1837.

Ilyas, S., Srivastava, R. R., & Kim, H. (2020). Disinfection technology and strategies for COVID-19 hospital and bio-medical waste management. *Science of the Total Environment, 749*, 141652.

Jin, J. M., Bai, P., He, W., Wu, F., Liu, X. F., Han, D. M., Liu, S., & Yang, J. K. (2020). Gender differences in patients with COVID-19: Focus on severity and mortality. *Frontiers in Public Health, 8*, 152.

Jordan, R. E., Adab, P., & Cheng, K. K. (2020). Covid-19: risk factors for severe disease and death. *BMJ*. https://doi.org/10.1136/bmj.m1198

Lambert, D. W. (2010). The cell biology of the SARS coronavirus receptor, angiotensin-converting enzyme 2. In *Molecular biology of the SARS-coronavirus* (pp. 23–30). Springer.

Lauer, S. A., Grantz, K. H., Bi, Q., Jones, F. K., Zheng, Q., Meredith, H. R., Azman, A. S., Reich, N. G., & Lessler, J. (2020). The incubation period of coronavirus disease 2019 (COVID-19) from publicly reported confirmed cases: Estimation and application. *Annals of Internal Medicine, 172*(9), 577–582.

Majumder, A., Adak, D., & Bairagi, N. (2021). Persistence and extinction criteria of Covid-19 pandemic: India as a case study. *Stochastic Analysis and Applications*. https://doi.org/10.1080/07362994.2021.1894172

Majumder, A., Bala, T., Adak, D., N'Guerekata, G. M., & Bairagi, N. (2020). Evaluating the current epidemiological status of Italy: Insights from a stochastic epidemic model. *Nonlinear Studies, 27*(4), 1169–1177.

Mondal, C., Majumder, A., Adak, D., & Bairagi, N. (2020). Mitigating the transmission of infection and death due to SARS CoV-2 through non-pharmaceutical interventions and repurposing drugs. epidemic model. ISA Transaction. https://doi.org/10.1016/j.isatra.2020.09.015

National Institutes of Health. (2020). Dexamethasone: Coronavirus disease COVID-19. COVID-19 treatment guidelines. Retrieved 12 July 2020.

Paul, A., Chatterjee, S., & Bairagi, N. (2020). Prediction on Covid-19 epidemic for different countries: Focusing on South Asia under various precautionary measures. *Medrxiv*.

Park, M., Thwaites, R. S., & Openshaw, P. J. (2020). COVID-19: lessons from SARS and MERS. *European Journal of Immunology, 50*(3), 308.

Rockx, B., Kuiken, T., Herfst, S., Bestebroer, T., Lamers, M. M., Oude Munnink, B. B., De Meulder, D., Van Amerongen, G., Van Den Brand, J., Okba, N. M., & Schipper, D. (2020). Comparative pathogenesis of COVID-19, MERS, and SARS in a nonhuman primate model. *Science, 368*(6494), 1012–1015.

Roshanravan, N., Ghaffari, S., & Hedayati, M. (2020). Angiotensin converting enzyme-2 as therapeutic target in COVID-19. *Diabetes and Metabolic Syndrome: Clinical Research and Reviews, 14*(4), 637–639.

Ritchie, H., Mathieu, E., Rodés-Guirao, L., Appel, C., Giattino, C., Ortiz-Ospina, E., Hasell, J., Macdonald, B., Beltekian, D., & Roser, M. (2020). Coronavirus pandemic (COVID-19). Our World in Data.

Salzberger, B., Glück, T., & Ehrenstein, B. (2020). Successful containment of COVID-19: the WHO-Report on the COVID-19 outbreak in China. *Infection, 48*, 151–153.

Singh, N., Tang, Y., Zhang, Z., & Zheng, C. (2020). COVID-19 waste management: Effective and successful measures in Wuhan, China. *Resources, Conservation and Recycling, 163*, 105071.

Singhal, T. (2020). A review of coronavirus disease-2019 (COVID-19). *The Indian Journal of Pediatrics, 13*, 1–6.

Statista. (2020). https://www.statista.com/statistics/1101680/coronavirus-cases-development-italy/

Sun, J., Zhu, A., Li, H., Zheng, K., Zhuang, Z., Chen, Z., Shi, Y., Zhang, Z., Chen, S. B., Liu, X., & Dai, J. (2020). Isolation of infectious SARS-CoV-2 from urine of a COVID-19 patient. *Emerging Microbes & Infections, 9*(1), 991–993.

Tang, Y. W., Schmitz, J. E., Persing, D. H., & Stratton, C. W. (2020). Laboratory diagnosis of COVID-19: current issues and challenges. *Journal of Clinical Microbiology, 58*, 6.

Thompson, R. N., Stockwin, J. E., van Gaalen, R. D., Polonsky, J. A., Kamvar, Z. N., Demarsh, P. A., Dahlqwist, E., Li, S., Miguel, E., Jombart, T., & Lessler, J. (2019). Improved inference of time-varying reproduction numbers during infectious disease outbreaks. *Epidemics, 29*, 100356.

Van Doremalen, N., Bushmaker, T., Morris, D. H., Holbrook, M. G., Gamble, A., Williamson, B. N., Tamin, A., Harcourt, J. L., Thornburg, N. J., Gerber, S. I., & Lloyd-Smith, J. O. (2020). Aerosol and surface stability of SARS-CoV-2 as compared with SARS-CoV-1. *New England Journal of Medicine, 382*(16), 1564–1567.

Wilder-Smith, A., Chiew, C. J., & Lee, V. J. (2020). Can we contain the COVID-19 outbreak with the same measures as for SARS? *The Lancet Infectious Diseases*. https://doi.org/10.1016/S1473-3099(20)30129-8

Williamson, B. N., Feldmann, F., Schwarz, B., Meade-White, K., Porter, D. P., Schulz, J., Van Doremalen, N., Leighton, I., Yinda, C. K., Pérez-Pérez, L., & Okumura, A. (2020). Clinical benefit of remdesivir in rhesus macaques infected with SARS-CoV-2. *Nature, 585*(7824), 273–276. https://doi.org/10.1038/s41586-020-2423-5

World Health Organization. (2020a). Coronavirus disease 2019 (COVID-19) situation report – 92.

World Health Organization. (2020b). COVID-19 Public Health Emergency of International Concern (PHEIC) Global research and innovation forum.

World Health Organization. (2020c). *Dexamethasone and COVID-19*. Retrieved from https://www.who.int/news-room/q-a-detail/q-a-dexamethasone-and-covid-19

World Health Organization. (2021). *Tracking SARS-CoV-2 variants*. Retrieved from https://www.who.int/en/activities/tracking-SARS-CoV-2-variants/

Worldometers. (2020a). *Coronavirus updates*. Retrieved 1 August, 2020, from https://www.worldometers.info/coronavirus/

Worldometers. (2020b). https://www.worldometers.info/coronavirus/coronavirus-age-sex-demographics/

Xiao, F., Sun, J., Xu, Y., Li, F., Huang, X., Li, H., Zhao, J., Huang, J., & Zhao, J. (2020). Infectious SARS-CoV-2 in feces of patient with severe COVID-19. *Emerging Infectious Diseases, 26*(8), 1920.

Xie, C., Jiang, L., Huang, G., Pu, H., Gong, B., Lin, H., Ma, S., Chen, X., Long, B., Si, G., & Yu, H. (2020). Comparison of different samples for 2019 novel coronavirus detection by nucleic acid amplification tests. *International Journal of Infectious Diseases, 93*, 264–267.

Xu, J., Zhao, S., Teng, T., Abdalla, A. E., Zhu, W., Xie, L., Wang, Y., & Guo, X. (2020). Systematic comparison of two animal-to-human transmitted human coronaviruses: SARS-CoV-2 and SARS-CoV. *Viruses, 12*(2), 244.

Part VI
Preventive Measures to Indirectly Reduce COVID 19 Waste Generation: Vaccination and Rehabilitation

COVID-19 Vaccination in India

Sadhan Kumar Ghosh, Anjan Adhikari, Anirudhha Mukhopadhyay,
Samprikta Bose, and Komal Sharma

1 Introduction

As 2019 ended, the world was still unaware of the extent of the threat posed by the
novel coronavirus then circulating in Wuhan, China. In the following months, it has
become clear that the interruption of regular personal and economic activity in all
corners of the world will have an overwhelming effect on the lives of people
everywhere. The world was facing a novel pathogen, the scope and spread of
which was unknown in the beginning of 2020 subsequently WHO declared a Public
Health Emergency of International Concern on 30 January and a Global Pandemic
on 11 March 2020 (World Health Organization, 2020). In December 2019, COVID-
19, caused by the SARS-CoV-2 coronavirus which is a novel coronavirus, was
recognised (Fauci, Lane, & Redfield, 2020). The world is facing challenges as novel
as the coronavirus that ushered in the most severe contraction of the economy since
the Second World War. Terrible unprecedented impact of the pandemic will have on
the poorest and emerging economies. People worldwide are at a loss! What is the
course of actions! Possible vaccination is the best options with all other precautions
associated.

Outbreaks of vaccine-preventable diseases, such as measles, are a sign of weakness in routine immunisation programmes—often exacerbated by rapid urbanisation;
crowded, unsanitary living conditions; climate change; and population growth and
movement. When vaccination rates drop below the herd immunity threshold, outbreaks can spread—infecting and killing. In the poorest countries, nearly 25% of the

S. K. Ghosh (✉)
Mechanical Engineering, Jadavpur University, & ISWMAW, Kolkata, India

A. Adhikari
Department of Pharmacology, MJN Medical College, Cooch Behar, India

A. Mukhopadhyay · S. Bose · K. Sharma
Department of Environmental Science, University of Calcutta, Kolkata, India

deaths of children under the age of five are caused by pneumonia or diarrhoea. These diseases were even more of a death sentence for young children before new vaccines were made available. Pneumococcal vaccine prevents the most common bacterial cause of pneumonia. Rotavirus vaccine protects against the deadliest type of diarrhoeal disease.

Coronaviruses were named from their morphology, which resembled a solar corona with spherical virions with a core shell and surface projections. The four subfamilies of coronaviruses are alpha, beta, gamma, and delta. While alpha and beta coronaviruses are thought to have originated in mammals, especially bats. The other two stains gamma and delta viruses are thought to have originated in pigs and birds (Lipsitch, Swerdlow, & Finelli, 2020).

COVID-19 vaccinations help to prevent the virus that causes COVID-19 from spreading. While the quick development of COVID-19 vaccines is a remarkable feat, vaccinating the entire world has numerous hurdles, from production to distribution, deployment, and, most crucially, acceptability (Enhancing Public Trust in COVID-19 Vaccination, 2021). The first mass vaccination programme started in early December 2020 and the number of vaccination doses administered is updated on a daily basis by national regulators as well as the WHO. At least 13 different vaccines (distributed across four platforms) have been given out (World Health Organization, 2020). To maintain public trust in vaccines, governments must be able to convey the benefits of vaccination and distribute immunisations in a safe and effective manner.

Never has the world's attention been so focused on the critical role vaccines play in protecting lives, livelihoods, and economies. The international Alliance groups and the national government agencies are working hand in hand with countries to help immunisation services adapt; to mitigate the adverse effects on domestic financing of immunisation programmes; and to procure COVID-19 vaccines. While the pandemic has demonstrated just how devastating the economic impact of infectious disease can be, equally the reverse is true: through health, comes wealth. Investing more in health should be a fundamental part of economic development recovery, with countries mobilising and improving the resource allocation for health (Annual Progress Report, 2019).

The ability of governments to explain the benefits of vaccination and to provide vaccinations safely and effectively is crucial to public confidence in vaccines (Enhancing Public Trust in COVID-19 Vaccination, 2021). In 206 economies, campaigns have already begun. Various countries, including India, are concerned about concerns and obstacles associated to vaccination, such as storage, transportation, distribution, and availability, cost, vaccine administration awareness, and so on. One of the major issues that scientists face is the rapid discovery of vaccines (Ledford, 2021; Saad-Roy, 2020). Worldwide, there is a dynamism observed in all the affected countries including India to absolutely minimise the spread of COVID-19 through vaccination (Fang et al., 2021). However, some questions remain, like, what are the steps to develop vaccines? Are all the steps being adopted? What is the efficacy and clinical testing of vaccines developed in such a short period of time? What is the effectiveness of storing and distribution system? What are the methods by which awareness can be increased? What are the improvement scopes and how to improve? All of the aforementioned will be examined in this article and will be addressed. The study was based on secondary data gathered from literatures and reports through the use of search engines (Brann et al., 2020). The main objective of the study is to grow consciousness and awareness among common people about vaccination and present the proper review for researchers. It is to analyse and understand the supply chain and different methodologies used to develop COVID-19 vaccines especially in India.

The first attempt to prevent sickness in society is where the history of vaccines and immunisation begins (Malacrida et al., 2012). Smallpox (along with many other infectious diseases such as measles) has been known since ancient times and is thought to have originated over 3000 years ago in India or Egypt (Brickman, Bazin, & Bazin, 2002; Brimnes, 2004). In India, inoculation was common which was "the process of injecting an infective material into a healthy person, which causes typically mild sickness and protects that individual from future serious disease" (Driscoll & Weiss, 2008). Taking measures to protect the health and safety of workers and providing them assurance about how employers are addressing the risks related to the coronavirus crisis are essential for increasing workers' confidence and ensuring continuity of waste management services (Enhancing Public Trust in COVID-19 Vaccination, 2021). Vaccination of the focused groups with priority builds confidence of the groups.

At the UN General Assembly in September 2019, when heads of state and government met for their first-ever High-Level Meeting on Universal Health Coverage, the outcome declaration strongly prioritised the role of immunisation and primary healthcare in achieving Universal Health Coverage and, by extension, the Sustainable Development Goals 3, "achieve good health and well-being". Many countries and regions around the world, public health, and healthcare capacities and capabilities vary. Even in the most developed countries planning and implementation of emergency response plans for large-scale public health emergencies are a significant challenge, which was evidenced in the pandemic COVID-19.

2 Resilience and the Risks to Global Health Security

COVID-19 has caused an unprecedented global crisis, including millions of lives lost, public health systems in shock and economic and social disruption, disproportionately affecting the most vulnerable. The pandemic has challenged local, national, regional, and global capacities to prepare and respond. The various national strategies taken to control viral transmission are widely debated (Glover et al., 2020; Baker et al., 2020). However, the relative success of these strategies depends largely on how an existing health system is organised, governed, and financed across all levels in a coordinated manner (Etienne et al., 2020). The pandemic has exposed the limitations of many health systems, including some that have been previously

Fig. 1 Determinants of
health systems resilience
framework

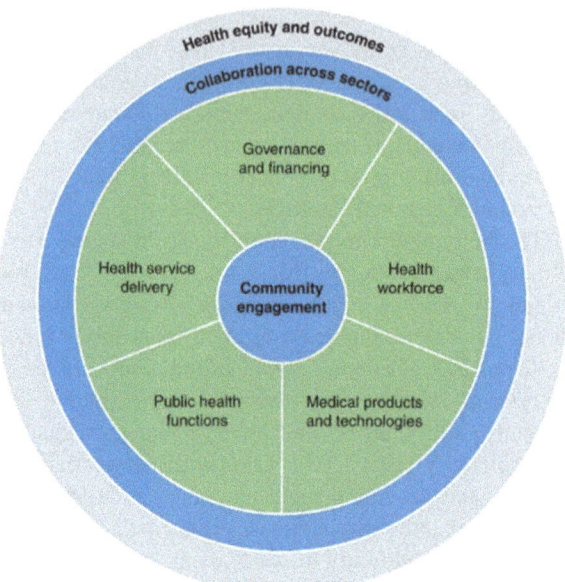

classified as high performing and resilient (El Bcheraoui et al., 2020). The pandemic
has fundamentally challenged health systems and the communities they serve
globally. While the evidence of system failures has come at a huge cost in human
and monetary terms, it has also pointed to what needs to change the health service
delivery, including non-COVID-19-related health services, has been directly threatened, and often compromised, at all levels by the demands of the pandemic, even in
traditionally high-performing health systems.

Pandemic needs actions to build resilient health systems globally. COVID-19
responses provide a clear illustration of the importance of governance supported by
scientific evidence and leadership willing to learn and adjust course for successful
health systems that protect health and well-being. Enhancing resilience to future
disease outbreaks requires longer-term work to create high-quality healthcare systems and build community trust (Haidari et al., 2016). Figure 1 illustrates the
components of the resilient health systems framework developed based on the
WHO's health systems building blocks framework. The five elements, governance
and financing, health workforce, medical products and technologies, public health
functions, and health service delivery need to collaborate centered around community engagement as core to all elements of health systems for resilience for health
equity and effective outcomes of resilient health systems (Haidari et al., 2016; World
Health Organization, 2020). Vaccination helps in bringing and sustaining resilient
health systems.

3   Immunisation Programmes in India

Immunisation programmes started in India way before independence and various milestones achieved are demonstrated in Table 1 including different vaccines, namely, DPT—Diphtheria-Pertussis-Tetanus vaccine, DT—diphtheria and tetanus, TT—tetanus toxoid, BCG—Bacille Calmette-Guerin, OPV—Oral polio vaccine. Until 1850, the vaccine was imported from the United Kingdom to India. However, transporting the vaccine to India presented significant logistical problems (Banthia, 2008). Increased demand in later years resulted in a shortage of vaccine or lymph in India, requiring the Indian government to develop new ways to increase vaccine supply on a long-term basis. There are traces of a few study efforts as early as 1832 in Bombay (Mumbai), and animal lymph trials began in Madras (now Chennai) in 1879, with initial results in 1880 (Basu, 2006; Roy, 2002). At the time of independence, the country was self-sufficient in producing smallpox vaccination. In 1948, Guindy established a BCG vaccine facility with the goal of producing enough BCG vaccine to meet the country's demand. Setting up vaccine manufacturing units in the private sector was also a significant breakthrough in Indian vaccine manufacturing initiatives. In addition to smallpox vaccines, these units were also active in the development of other vaccines (Lahariya, 2014; Shaji, Arun Thomas, & Sasidharan, 2019). A growing number of countries are using oral cholera vaccine (OCV), both to respond in emergencies and prevent outbreaks. Supply constraints and demand volatility limit the ability of countries to use OCV. Limited supplies are prioritised for emergency response efforts, often linked to outbreaks or humanitarian crises. This makes it harder to plan and deliver preventive campaigns. Alliance partners are working to improve demand predictability and support supply scale-up (Annual Progress Report, 2019).

4   Development Steps of COVID-19 Vaccine

As the rapidly evolving COVID-19 pandemic has made clear, infectious disease is proof of humankind's interconnectedness. The key to global health security, which shapes our interconnected economies and societies, is prevention. To reduce morbidity and mortality, the development of a safe and effective COVID-19 vaccination is critical. Vaccination benefits both the individual who receives it and the larger society by providing direct protection and group protection (also known as indirect or herd protection). A safe and effective vaccination takes several years to develop since it must go through clinical studies and receive tight regulatory permissions before it can be manufactured and distributed (Pronker, Weenen, Commandeur, Claassen, & Osterhaus, 2013). The development and testing of a new vaccine are being accelerated since the COVID-19 pandemic is having such a large worldwide impact.

Table 1 Significant breakthrough in the propagation and licensing of vaccines in India

Year	Milestone	Reference
1893	In Agra, cholera vaccine efficacy trials were undertaken	Lahariya (2014)
1897	Dr. Haffkine discovered Plague vaccine	Lahariya (2014)
1904/ 1905	First vaccine research institution was opened in Kasauli, Himachal Pradesh	Central Research Institute (n. d.)
1907	Coonoor's Pasteur Institute produced a neural tissue anti-rabies vaccination	India Science and Technology (n.d.)
1920–1939	The DPT, DT, and TT vaccines were made available in India	Driscoll and Weiss (2008)
1940	The drug and cosmetics act was passed	Ministry of Health and Family Welfare (1940)
1948	In Guindy, near Madras, a BCG vaccine laboratory was established	Lahariya (2014), Shaji et al. (2019)
1951	Liquid BGC vaccine was made available	Lahariya (2014)
1965	Live attenuated freeze-dried smallpox vaccine became available	Lahariya (2014)
1967	Freeze-dried BCG vaccine and OPV became available	Lahariya (2014)
1970	The indigenous Oral Polio Vaccine Trivalent (Sabin) was designed and produced for the first time in India	Lahariya (2014)
1980s	The production of indigenous measles vaccine has begun	Lahariya (2014)
1984	Inactivated polio vaccine production	Lahariya (2014)
1985/ 1988	The AEFI surveillance system has been built, and the first recommendations have been provided	Programme, Surveillance, and Aefi (2015)
1989	The public-private joint venture enterprises Indian Vaccine Company Limited (IVCOL) and Bharat Immunological and Biological Limited (BIBCOL) were established	Madhavi (2005)
1997	The very first recombinant DNA hepatitis B vaccine was developed in India	Chakma, Masum, Perampaladas, Heys, and Singer (2011)
2006	The Indian Council of Medical Research published clinical trial guidelines (ICMR)	Mathur and Swaminathan (2018)
2009	Three Indian companies have developed pandemic flu vaccines (Novel HINI: 2009)	Gatherer (2009)
2010	Meningitis was launched by India's National Pharmacovigilance Programme. A vaccine for African Meningitis Belt has been approved and is being used successfully in African campaigns. In the country, a bivalent oral cholera vaccine was produced and licenced based on indigenous research	Lahariya (2014)
2012	In the country, an indigenous "inactivated JE vaccine" has been approved. A firm in India has obtained the ability to generate inactivated polio vaccine	Lahariya (2014)
2020–2021	Vaccines against COVID-19 was developed	my gov (2021)

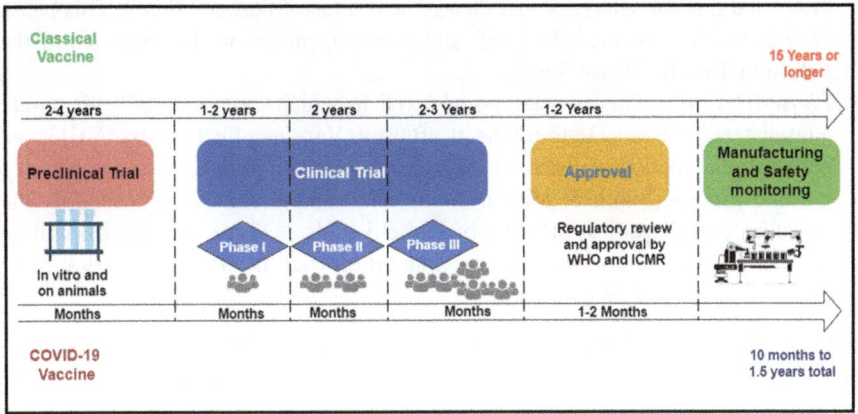

Fig. 2 The process of vaccine development for COVID-19 with comparison to classical vaccine production

Vaccines are often studied in a pre-clinical phase using in vitro human/mammalian cell cultures or appropriate animal models (in vivo). Because of the virus's novelty, the optimal animal model has yet to be found. However, in certain pre-clinical research, rhesus macaques, a non-human primate species that develops a COVID-19-like disease after infection with SARS-CoV-2, have been employed as an animal model (Gao et al., 2020; Yu et al., 2020). Pre-clinical testing of potential vaccines is required before clinical trials may begin. If positive pre-clinical results are obtained, vaccination candidates are tested in three clinical trial phases to evaluate whether they are safe and efficacious in humans. The stages of vaccine development has been shown in Fig. 2 (Sharma, Sultan, Ding, & Triggle, 2020).

Exploratory stage: The initial stage of vaccine development is an exploration phase that involves basic laboratory bench research and computational modelling to uncover natural or synthetic antigens that could be employed as a vaccine candidate to help prevent or treat disease.

- *Pre-clinical studies*: The potential vaccine's safety and immunogenicity, or ability to elicit an immune response, are tested in pre-clinical trials, which involve cell-culture or tissue-culture methods and animal experiments. Human clinical trials, which evaluate for safety and immunogenicity in small groups and then large groups across three phases, are carried out once animal safety, immunogenicity, and efficacy have been demonstrated.
- *Phase-1 Trials*: The vaccine is given to humans for the first time in this stage. A small number of healthy and immunocompetent people are given the vaccine to see if it's safe, get the right dose, and check for immunological response as a side effect.
- *Phase-II Trails*: Hundreds of people are administered the vaccination, who are divided into different groups based on their demographics (example: elderly vs. young). These examine for immunological response as a secondary

effect and test for safety, proper dosage, and interval between doses. This phase ensures that the vaccine is both safe and immunogenic, as well as determining the optimum dose for Phase 3 trials.

- *Phase-III Trails*: This is a large-scale trial in which thousands of patients are administered the vaccine to assess its efficacy. Vaccine effectiveness (VE) is the percentage reduction in disease incidence in vaccinated groups when compared to placebo groups (Singh & Mehta, 2016). The sample size is influenced by the prevalence of disease at the time of Phase 3 trials. A large sample size will be required to adequately establish vaccination efficacy in the case of a low incidence of disease in the population.

Once the human clinical trials are over and the vaccine's safety and clinical efficacy have been established, the vaccine will be moved to (Uttarilli et al., 2021):

- *Review and approval*: Vaccines should be approved by World Health organisation and then by the national regulatory bodies like Indian Council of Medical Research. Vaccines may be licensed for emergency use in a pandemic, as this procedure can take anywhere from 1–3 years.
- *Manufacturing and post-marketing surveillance*: This is done after the vaccine has been marketed to the general public and is checked for overall effectiveness. They also keep track of any negative side effects that may occur when the vaccine is widely used.

Participants or volunteers chosen for phase 1 and phase 2 clinical trials have to meet specified inclusion and exclusion criteria. The trials were conducted on participants ranging in age from 12 to 65 years old as shown in a guideline by ICMR.

Inclusion criteria for clinical trials (Worldwide COVID-19 Candidate Vaccines, n.d.)

1. Ability to provide permission and willingness and interest in meeting the study's requirements. Participants should permit to store and use biological samples in the future for study and they should stay in research region during investigation.
2. Good general health, as judged at the investigator's discretion (vital signs (heart rate 60–100 bpm; blood pressure systolic 90 mmHg and 140 mmHg; oral temperature 100.4 °F), medical history, and physical examination).
3. For a female participant of child-bearing potential, from the time of study enrolment until at least 4 weeks after the last vaccine, aim to prevent becoming pregnant (use of an effective means of contraception or abstention).
4. Male subjects with reproductive potential: From the first immunisation to 3 months after the last vaccine, use condoms to guarantee effective contraception with the female partner.
5. Male subjects agree not to donate sperm from the time of their first immunisation until three months after their last vaccine.
6. From the time of the first vaccine until 3 months following the last vaccination, participants must refrain from donating blood or plasma.
7. Consents to not take part in another clinical trial during the research term.

Exclusion criteria for clinical trials

1. Any other COVID-19 experimental vaccination history or employees in the medical field.
2. Confirmed SARS-CoV-2 infection using RT-PCR and/or ELISA at the time of screening or higher body temperature or in contact with COVID-19 positive patient.
3. A positive pregnancy test in the urine (within 24 h of administering each dose of vaccine) or any possible pregnancy, lactating or desire/intention to get pregnant while participating in the study.
4. Medical problems caused by the use of alcohol or illegal drugs in the previous 12 months.
5. Immunosuppression as a result of an underlying illness, immunosuppressive or cytotoxic medication treatment, or anticancer chemotherapy or radiation therapy within the previous 36 months.
6. Use of oral or parenteral glucocorticoids (glucocorticoids) for more than 2 weeks or high-dose inhaled steroids (>800 mcg/day of beclomethasone dipropionate or similar) in the previous 6 months (nasal and topical steroids are allowed).
7. Obese (BMI 35 kg/m^2) or underweight (BMI 18 kg/m^2) morbidly any chronic illness, cancer or any such co-morbidity.
8. Any other circumstance that, in the investigator's opinion, might compromise the safety or rights of a volunteer participant in the trial, or leave the subject unable to follow the protocol.

Leveraging policy dialogues and engagement opportunities with governments, as well as mobilising civil society at the global level, offer crucial levers for improving the ownership, strength, and sustainability of vaccine programmes at the national level.

5 Types of Vaccine Worldwide and Their Use

Many possible COVID-19 vaccines are being developed by scientists all around the world. These vaccines are all designed to train the body's immune system how to recognise and block the virus that causes COVID-19 in a safe and effective manner (World Health Organization, 2020)

- *Inactivated or weakened virus vaccines*: Inactivated or weakened virus vaccines that use a form of the virus that has been inactivated or weakened so that it does not cause disease but still triggers an immune response.
- *Protein-based vaccines*: Protein-based vaccines, which safely produce an immune response by using innocuous protein fragments or protein shells that resemble the COVID-19 virus.

- *Viral vector vaccines*: Vaccines based on viral vectors, which use a non-pathogenic virus as a platform for producing coronavirus proteins and eliciting an immune response.
- *RNA and DNA vaccines*: RNA and DNA vaccines are a cutting-edge method that generates a protein from genetically modified RNA or DNA, which then safely triggers an immune response.

COVID-19 fostered a great development in scientific technology in the development of vaccines. There are approximately 150 official vaccination projects as of December 2020, just 11 months after the SARS-CoV-2 genome was defined. About fifty of them have already been tested on humans, and a handful of them are currently being administered to certain segments of the general public (Thanh Le et al., n.d.). These anti-SARS-CoV-2 potential vaccines use a variety of methods to target the entire SARS-CoV-2 virus, as well as compounds or fragments of molecules expressed on the virus surface (Saad-Roy, 2020). An exhaustive list of COVID-19 vaccines available in presented in Table 2.

Moderna and Pfizer–COVID-19 BioNTech's vaccines, for example, are highly successful at averting symptomatic sickness, but it's yet unknown if they safeguard individuals from becoming affected or spreading the virus to others. The concept of Herd immunity is impaired as a result of this (Shinde et al., 2021).

Herd immunity is only useful if a transmission-blocking vaccination is available. If not, the only option to achieve herd immunity in a population is to vaccinate everyone (Sadoff et al., 2021). For vaccine to be transmission blocking, its efficacy may not be 100% but 70% can be considered as good enough (Aschwanden, 2021). Many vaccine data are convincing and encouraging towards this study and research is still on to achieve this.

6 Supply Chain of the COVID-19 Vaccine

The COVID-19 pandemic experiences a unique set of challenges for healthcare systems and for the immunisation programmes they support. Healthcare systems and the health workforce in the countries are being put under immense strain by the multiple demands: (a) caring for people sickened by the coronavirus, (b) trying to provide routine services, (c) protecting themselves from coronavirus, and (d) supply chain dependant inputs as well as the availability of vaccines. Immunisation services have been disrupted—with outreach suspended in many countries; many immunisation campaigns delayed and reduced service availability in some health facilities. In many countries, these challenges are multiplied by resource constraints and often weak underlying systems.

The supply chain of vaccine will be strengthened and sustainable with robust support of vaccine production, storing, data systems to supply and cold chains, transportation local, national, and international, availability of healthcare system integrated with trained manpower. Supply chains have been disrupted by the

Table 2 List of COVID-19 vaccines that are in use or under trial available in different countries

SL. No.	Vaccine name/manufacturer and country of origin	Storage temperature and ages in years, and dose	Technology used and approval of WHO EUA	Use in India	Announced efficacy and common adverse effect reported	References
1.	Comirnaty/BNT162b2/Tozinameran Pfizer-BioNTech, USA	−70 °C $\geq$12 2 jabs 21 days apart	mRNA Approved	No	95%; FDA issued warning of a likely association of heart inflammation among young adults	Teo (2021)
2.	m-RNA-1273 Moderna, USA	−20 °C $\geq$16 2 jabs 28 days apart	mRNA Approved	No	94.50%; FDA issued warning of a likely association of heart inflammation among young adults	Baden et al. (2021)
3.	Sputnik V The Gamaleya National Centre, Russia	−18 °C $\geq$18 2 jabs 21 days apart	Viral vector Approved	Yes	91.40%; Headache, Injection site pain/ swelling/redness or warmth and flu-like symptoms	Baraniuk (2021)
4.	Covishield Serum Institute of India (Oxford-AstraZeneca formulation)	2–8 °C $\geq$18 2 jabs 8–12 weeks apart	Viral vector Approved	Yes	90%; Rare occurrence of a blood clotting disorder with low platelet count	Knoll and Wonodi (2021)
5	AZD1222 Vaxzevria; Oxford-AstraZeneca, UK	2 doses, 4–12 weeks apart	Vector Approved	Yes	Reported occurrence of a rare blood clotting disorder with low platelet count	
6.	Janssen/Ad26.COV 2.S; Johnson & John-son; USA but developed in Netherlands	2–8 °C $\geq$18 2 doses at least 47 days apart	Vector Approved	Yes	66.0%; FDA warned of rare occurrence of Guillain-Barre Syndrome	Sadoff et al. (2021)
6.	Covaxin Bharat Biotech, India	2–8 °C $\geq$18 2 jabs 4 weeks apart	Whole virus (inactivated) Approved	Yes	93.4% and 63.6% against severe COVID-19 and asymptomatic disease; pain/swelling /redness/itching/headache/ fever/nausea/body ache	Sapkal et al. (2021), Bharat biotech
8.	NVX-CoV2373/Covovax; Novavax, USA	2–8 °C 2 jabs 3 weeks apart	Protein subunit Approved	No	Headache, lethargy, pain and/or swelling at injection site	Xia et al. (2021)

(continued)

Table 2 (continued)

SL No.	Vaccine name/manufacturer and country of origin	Storage temperature and ages in years, and dose	Technology used and approval of WHO EUA	Use in India	Announced efficacy and common adverse effect reported	References
9.	Covilo/BBIBP-CorV; Sinopharm/Beijing Institute of Biological Products (BIBP), China	2–8 °C 18–80 2 jabs 21–28 days apart	Whole virus (inactivated) Approved	No	86.0%; headache, lethargy, pain and/or swelling at injection site	
10.	Coronavac; Sinovac-Corona Vac, China	2–8 °C ≥18 2 jabs 2–4 weeks apart	Inactivated virus Approved	No	51% against symptomatic and 100% against severe COVID-19; Rare side effects include hypersensitivity, vomiting, nose bleed, fever, conjunctival congestion, muscle spasms among others	World Health Organization (2021)
11.	CVnCoV/CV07050101 Zorecimeran (INN); CureVac; Germany	– 2 jabs 4 weeks apart	mRNA Approved	No	Headache, lethargy, chills, muscle pain and less frequently fever	–
12.	Inactivated SARS-CoV-2 vaccine (Vero Cell); Sinopharm/Wuhan Institute of Biological Products (WIBP), China	2 jabs 3–4 weeks apart	Whole inactivated virus; pending	No	Headache, injection site pain/swelling/ redness or warmth	
13.	Ad5-nCoV; CanSinoBio, China	Single dose	Vector Approved	No	Fever, injection site pain, redness and swelling	
14	ZyCoV-D; Zydus Cadila Cadila Healthcare Ltd.	2–8 °C ≥12	DNA plasma vaccine Approved	Yes	66%	Zydus (2021)

Table 3 Cold storage requirements of mRNA and adenovirus-vectored COVID-19 vaccines (Antal, Cioara, Antal, & Anghel, 2021)

Manufacturer	Dose (mL)	Storage temperature (°C)	Refrigeration storage duration	Duration after reconstitution
Pfizer	0.3	−70	5 days	6 h room temp
Moderna	0.5	2–8	30 days	6 h room temp
J & J	0.5	2–8	3 months	6 h in refrigerator
AZ	0.25/0.5	2–8	6 months	6 h in refrigerator

reduction in global air travel (which has delayed shipments of vaccines and cold chain equipment), as well as by lockdowns and competing priorities. When a COVID-19 vaccine becomes available, it could put these systems under even greater pressure to deliver at speed.

Vaccine being a biological product, the active components in some vaccine formulations can decay and become less effective if it becomes too hot or too cold (some vaccine formulations must be protected against freezing). Rapid and convenient distribution chains combined with short end-user storage times require the liquid formulations with shelf lives of months and to be stored at 2–8 °C (Crommelin, Volkin, Hoogendoorn, Lubiniecki, & Jiskoot, 2021) (Table 3).

Once a vaccine dose has been created, it must be transferred to immunisation programmes, clinics, and health centres all over the world, where it must remain viable until it is needed (Lloyd & Cheyne, 2017). While the focus in the vaccine world has been on developing the required vaccines and measuring their effectiveness, the struggle to understand and properly address the issues of the Vaccine Supply Chain (VSC) greatly increases the impact of any vaccination programme (Lee & Haidari, 2017). COVID-19 vaccines are challenging to get to remote places, namely, several villages which are far from the storage facilities, in a timely and safe manner as vaccine transportation is far more complicated than that of conventional goods. The COVID-19 vaccinations are transported from the production plant to the final retailers in a complicated system that includes storage factories, cargo stations, aeroplanes, warehouses, and other facilities (Dai, Wu, & Si, 2021). Duijzer, van Jaarsveld, and Dekker (2018) conducted a study of the relevant literature and discovered that current vaccine supply chain research is mostly focused on four aspects: (a) vaccine quality, (b) demand in the locality, (c) allocation by the controlling agency, and (d) transportation from stores to destination. Using an unmanned aircraft system (UAS) to transport vaccinations could increase vaccine safety while also lowering costs (Haidari et al., 2016).

Covishield and Covaxin are the mostly used and available vaccine in India, whose storing and transportation need to maintain temperatures range from 2 to 8 °C. The vaccinations cold chain is maintained using active and passive cold chain equipment located at roughly 29,000 cold chain locations throughout India. Sputnik V storing temperature is very critically near cryogenic temperature of −180 °C or lower for which deep freezers are provided for storage across the country as part of the Universal Immunization Programme (My gov, 2021).

Despite concerted efforts to stop the spread of the disease, the pandemic continues to put a strain on healthcare systems around the world. To effectively contain the outbreak, a safe, highly effective, globally acceptable, and equitable vaccination programme, combined with pre-existing precautionary measures, are required.

7 Vaccine Nationalism

Vaccine nationalism is the priority of a country's own concerns over the needs of outsiders (Lagman, 2021). Some countries with high vaccination availability are currently resorting to "vaccine nationalism" due to significant public and political pressures and fears about fading immunity: stockpiling vaccinations in order to ensure that their citizens have access to them quickly (Fidl, 2020). According to the most recent data, the USA and the United Kingdom had each received 109 and 133 doses per 100 people, respectively, whereas India and Africa had received an average of 42 and 18 doses per 100 people, respectively (Josh Holder, n.d.). Vaccine nationalism is delaying the aim of whole vaccination programme even though it is one of the largest immunisation programmes in the world. This attitude of putting one's own interests first appears to improve viral defence and vulnerability. As a result, many wealthy countries purchase and stockpile vaccination supplies for future use (Chohan, 2021; Fidl, 2020). Vaccination, on the other hand, is not to protect one's country and does not help to solve the pandemic problem.

"Vaccine Nationalism would also lead to a prolonged epidemic since only a small number of nations would receive the majority of the supply," as said by World Health Organisation Director-General, Tedros Adhanom Ghebreyesus. "Vaccine nationalism is exclusively for the virus's benefit." Even if we simply vaccinate the wealthy world and a few developing countries, the virus will remain and spread in unvaccinated areas. Therefore, mutations are more likely to occur, allowing vaccines to elude the immunological response. As a result, due to the virus's mutation, all of the efforts put into developing and disseminating vaccines, as well as vaccination schedules, would be for naught.

8 Right to Vaccination

In the global context, just about 16% of the world's population has access to approximately 70% of the vaccines available. According to estimates, most high-income countries will finish their immunisation campaigns by the end of 2021 while that of poorer counties may take as long as 2024 to get vaccinated. In low-income countries, less than 0.1% of immunisations are administered (Kumar Kaustubh, 2021).

Human rights could be a powerful tool in the fight for a more equitable worldwide distribution of vaccines. Human rights experts invoke the right to the best achievable quality of health in Article 12 of the International Covenant on Economic, Social, and Cultural Rights (ICESCR), in a statement, asking for universal and inexpensive access to COVID-19 vaccines (The Human Right to Vaccines, n.d.).

The distribution of COVID-19 vaccinations must not enable sovereignty, which is inextricably tied to past power and resource disparities, to result in discrimination between affluent and poor countries. Poor countries must be assisted in obtaining COVID-19 vaccines by addressing the power dynamics that perpetuate and drive unfairness (Boehmer, 2020). Through participatory community involvement, equity, fairness, and transparency in the distribution of COVID-19 vaccinations must be assured. Development of public trust in COVID-19 vaccines is very important for ensuring the distribution and access to COVID-19 vaccines and it must be equal and should not be politicised (ScienceDirect, n.d.). Because wealthier nations have reserved more than they need and developers will not share their intellectual property, at least 90% of people in 67 low-income countries have limited chance of getting vaccinated against COVID-19 in 2021 (Covid-19, 2021).

9 Economics of Vaccination

The catastrophe caused by the COVID-19 pandemic in India as well as other countries of the world, has necessitated massive investments in both prevention, control, and treatment, as well as social welfare and livelihood programmes (Gupta & Baru, 2020). Several economic models regarding vaccination have been analysed. Persons do not internalise the positive externalities that their vaccination decisions may impact on other individuals, which is widely thought to be a market failure in the vaccine market (Francis, 1997). The socially optimal vaccination is contrasted to equilibrium vaccination. Diseases, vaccinations, and market characteristics that lead to inefficiency are found (Chen & Toxvaerd, 2014).

India has registered 25,467 new instances of COVID-19 in the previous 24 h, bringing the overall number of cases to 32,474,773. The death toll in the country has risen by 354–435,110 (as on 24 August 2021) (Coronavirus in India, 2021). Waiting for natural herd immunity to develop will be time consuming and costly, not only in terms of increased illness and mortality, but also in terms of the impact on livelihoods owing to the ongoing containment measures that would be required to prevent the virus from spreading faster (Randolph & Barreiro, 2020). For India, it is critical to develop a detailed plan that covers four domains: (1) production and financing, (2) the safety and efficacy of each candidate vaccine, (3) distributional elements such as availability and equity, and (4) accountability (Gupta & Baru, 2020). The health sector was not prepared for this unprecedented incident thus the economy faced a huge blow. Immediate economic and sustainable policies are required for free universal vaccination and combatting the disease.

10   Present Status of Vaccination in India

India on Thursday, October 21, 2021 reached the milestone of one billion or 100 crore COVID-19 shots administered in the country (Fig. 3). India reached the landmark in 278 days, having begun the vaccination drive on January 16. So far, over 70 crore people have received at least one dose of the COVID-19 vaccine, while over 29 crore people are fully vaccinated with both doses. The Narendra Modi government has been reiterating that all of India's adult population—around 94 crores—will vaccinated against COVID-19 by December (India's Journey to 1 billion COVID-19 Shots, 2021).

India has begun a widespread vaccination campaign across all the states. In India, 11.5%of the population is fully vaccinated, compared to 27.8%worldwide as of August 4 (Our World in Data, 2021). From the end of April to the beginning of May 2021, the dose administered curve began to decline, indicating an acute vaccine shortage. To alleviate Covidshield scarcity, the government has also suspended vaccination exports (Padma, 2021).

The number of days it took to reach the milestones of one billion mark of vaccination is India is given in Table 4. The Key vaccination drive dates in India in the year 2021 has been shown with expected outcome in Table 5.

Vaccine reluctance connected to vaccine efficacy is the COVID-19's main roadblock. Several health/frontline workers have been affected while being fully

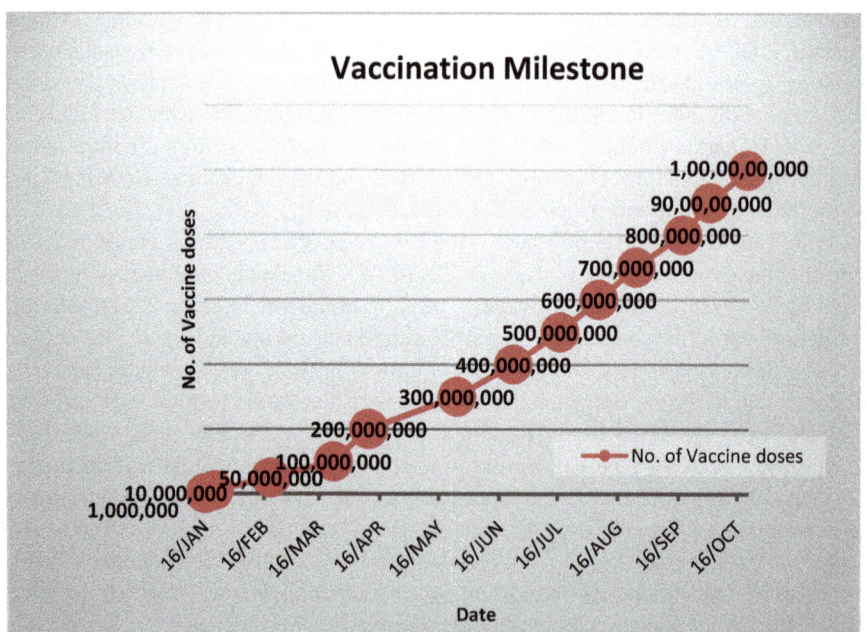

Fig. 3 Date wise data showing reaching One Billion marks of vaccination in India in 2021 (Modified from government of India data and NDTV.com)

COVID-19 Vaccination in India

Table 4 Days taken to reach the milestones of one billion (100 crore) mark of vaccination is India

Dates	Jan16–Jan 21	Jan 21–Feb 19	Feb 19–Mar 23	Mar 23–Apr 10	Apr 10–May 25	May 25–Jun 23	Jun 23–Jul 17	Jul 17–Aug 6	Aug 6–Aug 25	Aug 25–Sep 7	Sep 7–Sep 18	Sep 18–Oct 2	Oct 2–Oct 21
No. of days	5	29	32	18	45	29	24	20	19	13	11	14	19
No of vaccination (in crores)	0–0.01	0.90	4	5	10	10	10	10	10	10	10	10	10

Table 5 Key vaccination drive dates in India in the year 2021

Date	Description of the activities	Expected outcome
February 2, 2021	Vaccination of frontline workers, namely, doctors, nurses, other hospital workers, helpers, laboratory technicians, medical waste handlers	Vaccination to health workers who were in constant contact with COVID-19 patients brought protection from COVID-19 to them and so they could serve common people in a better way
March 1, 2021	Vaccination for 60+ age group and 45+ age group with co-morbidity began	Vaccination elder people and people with co-morbidity reduced the chance of infection and morbidity in them as they were more vulnerable to the disease
April 1, 2021	Vaccination for 45+ began	Vaccines to office going people and the one travelling and being in contact with many people allowed them to be safer
May 1, 2021	Vaccination drive for 18+ began	Vaccinating majority of population can help in controlling the spread of disease
August 27, 2021	India crossed 1 crore doses for the first time (10,353,290)	
August 31, 2021	India crossed 1 crore doses for the second time (13,318,718)	
September 6, 2021	India crossed 1 crore doses for the third time (11,353,571)	
September 17, 2021	India crossed 2 crore doses in a single day (22,619,002)	
September 27, 2021	India crossed 1 crore doses for the fifth time (10,222,525)	

vaccinated, raising public concerns about vaccination efficacy. The second barrier is vaccination illiteracy, which is particularly prevalent in rural areas. The third impediment is the vaccine's high price, which is prohibitive for low-income individuals (Chakraborty, Sharma, Bhattacharya, Agoramoorthy, & Lee, 2021).

The vaccine is distributed through both public and private hospitals and clinics. People can get free vaccinations at government centres by showing their identification cards. However, they must line for 10–12 h. The immunisation costs $7 for those who go to a private institution. Minimum immunisations are only provided on a daily basis in both locations, and consumers may return empty-handed after two or three visits, leading to apprehension. Although the government is taking steps to boost vaccine access, additional vaccines have also been licenced.

11 Discussion

The study deals with different aspects involved in COVID-19 vaccination. Firstly, starting off with the concept of vaccination and different types of vaccines that have been developed we have progressed towards the discussion on COVID-19. In this study we have talked about the vaccination procedure and management storage transportation and distribution of COVID-19 vaccine especially in India. The development of vaccines for COVID-19 in a short period of time is really challenging, next developing it on a large scale and distributing it throughout the world uniformly is which is still an important issue to be addressed.

The development of a safe and effective COVID-19 vaccine is vital for reducing morbidity and mortality. Vaccination benefits both the individual who receives it and society as a whole since it provides both direct and collective protection. Before clinical trials can begin, prospective vaccinations must undergo pre-clinical testing. If pre-clinical data are good, vaccine candidates are examined in three steps of clinical trials to see if they are safe. Once a vaccine dose has been developed, it must be distributed to immunisation projects, clinics, and health centres around the world, where it must survive and thrive until it is required. The vaccines also require special temperature and conditions for storage. Different vaccines have different storage technology and conditions. Besides storage of vaccines, transportation of vaccines is a complicated process. It can be done better and more economically through airplanes. This whole process of process of COVID-19 vaccination also involves huge costing.

Waiting for natural herd immunity to emerge will be time consuming and expensive, not just in terms of increased morbidity and death, but also in terms of the effect on economies due to continual containment measures that would be required to keep the diseases from spreading faster. It is vital for India to design a precise plan that addresses four domains: (1) production and financing, (2) the safety and efficacy of each candidate vaccine, (3) distributional components including availability and equity, and (4) accountability. Because the healthcare sector was unprepared for such a rare occurrence, the economy suffered a significant setback. According to the most recent data, the USA and UK received 109 and 133 doses per 100 people, respectively, while India and Africa received 42 and 18 doses per 100 people, respectively. This attitude of putting one's own interests first tends to increase viral defence and vulnerability. This is also delaying the process of immunisation and causing the poor countries invest much more. In spite of this economic setback, India was able to manufacture vaccines. It also exported vaccines to other countries in need.

Human rights could be an effective weapon in the fight for a more equitable global vaccination distribution. Article 12 of the International Covenant on Economic, Social, and Cultural Rights, according to human rights experts, guarantees the best possible health (ICESCR). COVID-19 immunisations should be available to everyone at a low cost, according to a statement. Poor countries must be helped to get COVID-19 vaccines by addressing the power dynamics that perpetuate and encourage injustice. Equitable, fair, and transparent distribution of COVID-19 vaccines must be ensured by community participation.

Misinformation, deliberate or not, is a feature of any emerging threat. COVID-19 is no different. An "infodemic" of rumours, fake news, and misinformation is spreading faster than our efforts to counter it. If unchecked, there is a risk of undermining confidence in routine immunisation and the introduction of future COVID-19 vaccines (Annual Progress Report, 2019).

12   Conclusion

A number of wealthy nations have decided to press ahead with plans to administer COVID-19 vaccine boosters in the coming months, yet with so many in under-resourced countries still without vaccines, we should be prioritising vaccinating all adults first.

Organisations are quickly responding to the pandemic by providing sanitation and public health education. Social distancing and lack of personal protective equipment have resulted in some programmes and organisations closing. As fundraising events are cancelled, organisations are finding their funding more constrained than ever before.

Such goals are highly ambitious, as global trends such as climate change, population growth, urbanisation, and human migration are changing the global health landscape in ways that made our work increasingly more difficult, even before the pandemic took hold. It is only by ensuring that everyone has access to vaccines, including COVID-19 vaccines, that our partnership can prevent pandemics and contribute to global prosperity, as one world, protected.

Real-time data on immunisation programmes and vaccine stocks are not available in many countries, and information systems are not as flexible as they need to be. From data systems to supply and cold chains, strong health systems are a vital part of the global response to COVID-19.

In this world of globalisation at one side science and technology are progressing but has a negative impact, as well. Rich countries are becoming richer and poor countries even poorer. Rich countries are the first one to enjoy the societal benefits while the third world countries have to struggle for facilities. In case of COVID-19, the picture is similar in access to the vaccines. More people are getting vaccinated in the first-world countries than that in the third-world countries. But to ensure the medication of the disease we have to make sure that there is equality and uniformity in the distribution and transportation of the COVID-19 vaccine. In spite of development of vaccines, mass campaigns, it is seen till now that the conception of eradication of COVID-19 is unknown. But, to reduce its effect, and lessen the spreading of the disease, uniform distribution proper administration dosage is absolutely essential.

COVID-19 demands dynamic systemic transformation. The pandemic has fundamentally challenged health systems and the communities they serve globally. The effect of a major shock represented by the pandemic is to manifest the points where the system is weakest, and to demonstrate the interdependencies of a range of health, social, and economic structures. While the evidence of system failures has come at a huge cost in human and monetary terms, it has also pointed to what needs to change. With over 3 million global deaths and pervasive social and economic costs, the pandemic must serve as a call for transformation and investment toward resilience and people centeredness, beginning with health systems. COVID-19 provides a renewed prospect for solidarity, both within and between countries. It also serves as a reminder that health is more than healthcare and that a whole-of-government approach to health and well-being is needed to create healthy populations able to collectively prevent and respond to crises, leaving no one behind. Pandemic needs actions to build resilient health systems globally.

References

Annual Progress Report. (2019). Retrieved February 27, 2022, from https://read.oecd-ilibrary.org/view/?ref=1060_1060300-enj5o5xnwj&title=Coronavirus-COVID-19-vaccines-for-developing-countries-An-equal-shot-atrecovery

Antal, C., Cioara, T., Antal, M., & Anghel, I. (2021). Blockchain platform for COVID-19 vaccine supply management. *IEEE Open Journal of the Computer Society, 2*, 164–178. https://doi.org/10.1109/OJCS.2021.3067450

Aschwanden, C. (2021). Five reasons why COVID herd immunity is probably impossible. *Nature, 2021*(591), 7851.

Baden, L. R., El Sahly, H. M., Essink, B., Kotloff, K., Frey, S., Novak, R., Diemert, D., Spector, S. A., Rouphael, N., Creech, C. B., McGettigan, J., Khetan, S., Segall, N., Solis, J., Brosz, A., Fierro, C., Schwartz, H., Neuzil, K., Corey, L., & Zaks, T. (2021). Efficacy and safety of the mRNA-1273 SARS-CoV-2 vaccine. *New England Journal of Medicine, 384*(5), 2035389. https://doi.org/10.1056/nejmoa2035389

Baker, M. G., Wilson, N., & Anglemyer, A. (2020). Successful elimination of Covid-19 transmission in New Zealand. *New England Journal of Medicine, 383*(8), e56. https://doi.org/10.1056/NEJMC2025203

Banthia, J. (2008). Fractured states: Smallpox, public health and vaccination policy in British India, 1800–1947 (review). *Bulletin of the History of Medicine, 82*(1), 13. https://doi.org/10.1353/bhm.2008.0013

Baraniuk, C. (2021). Covid-19: What do we know about Sputnik v and other Russian vaccines? *BMJ, 372*, 743. https://doi.org/10.1136/bmj.n743

Basu, R. N. (2006). Smallpox eradication: Lessons learnt from a success story. *National Medical Journal of India, 19*, 1.

Boehmer, T. K. (2020). Changing age distribution of the COVID-19 pandemic - United States, May-August 2020. *MMWR. Morbidity and Mortality Weekly Report, 69*(39), 1404–1409. https://doi.org/10.15585/mmwr.mm6939e1

Brann, D. H., Tsukahara, T., Weinreb, C., Lipovsek, M., Van Den Berge, K., Gong, B., Chance, R., Macaulay, I. C., Chou, H. J., Fletcher, R. B., Das, D., Street, K., De Bezieux, H. R., Choi, Y. G., Risso, D., Dudoit, S., Purdom, E., Mill, J., Hachem, R. A., & Datta, S. R. (2020). Non-neuronal expression of SARS-CoV-2 entry genes in the olfactory system suggests mechanisms underlying COVID-19-associated anosmia. *Science Advances, 6*, 31. https://doi.org/10.1126/SCIADV.ABC5801

Brickman, J. P., Bazin, H., & Bazin, H. (2002). The eradication of smallpox. Edward Jenner and the first and only eradication of a human infectious disease. *Journal of Public Health Policy, 23*, 4. https://doi.org/10.2307/3343253

Brimnes, N. (2004). Variolation, vaccination and popular resistance in early colonial South India. *Medical History, 48*, 2. https://doi.org/10.1017/S0025727300000107

Central Research Institute. (n.d.). Retrieved September 5, 2021, from https://crikasauli.nic.in/

Chakma, J., Masum, H., Perampaladas, K., Heys, J., & Singer, P. A. (2011). Indian vaccine innovation: The case of Shantha Biotechnics. *Globalization and Health, 7*, 9. https://doi.org/10.1186/1744-8603-7-9

Chakraborty, C., Sharma, A. R., Bhattacharya, M., Agoramoorthy, G., & Lee, S. S. (2021). The current second wave and COVID-19 vaccination status in India. *Brain, Behavior, and Immunity, 96*, 18. https://doi.org/10.1016/j.bbi.2021.05.018

Chen, F., & Toxvaerd, F. (2014). The economics of vaccination. *Journal of Theoretical Biology, 363*, 105–117. https://doi.org/10.1016/J.JTBI.2014.08.003

Chohan, U. W. (2021). Coronavirus vaccine nationalism. *SSRN Electronic Journal*. https://doi.org/10.2139/ssrn.3767610

Coronavirus in India. (2021). India reports 25,467 new COVID-19 cases, 354 deaths; lowest active case count in 156 days - The Economic Times.

Covid-19. (2021). Many poor countries will see almost no vaccine next year, aid groups warn - ProQuest.

Crommelin, D. J. A., Volkin, D. B., Hoogendoorn, K. H., Lubiniecki, A. S., & Jiskoot, W. (2021). The science is there: Key considerations for stabilizing viral vector-based Covid-19 vaccines. *Journal of Pharmaceutical Sciences, 110*(2), 627–634. https://doi.org/10.1016/J.XPHS.2020.11.015

Dai, D., Wu, X., & Si, F. (2021). Complexity analysis of cold chain transportation in a vaccine supply chain considering activity inspection and time-delay. *Adv. Difference Equ., 2021*(1), 73. https://doi.org/10.1186/s13662-020-03173-z

Driscoll, S., & Weiss, R. E. (2008). The history of vaccines. *Journal of Urology, 179*(4S), 60899. https://doi.org/10.1016/s0022-5347(08)60899-x

Duijzer, L. E., van Jaarsveld, W., & Dekker, R. (2018). Literature review: The vaccine supply chain. *European Journal of Operational Research, 268*(1), 15. https://doi.org/10.1016/j.ejor.2018.01.015

El Bcheraoui, C., Weishaar, H., Pozo-Martin, F., & Hanefeld, J. (2020). Assessing COVID-19 through the lens of health systems' preparedness: Time for a change. *Globalization and Health, 16*(1), 1–5. https://doi.org/10.1186/S12992-020-00645-5

Enhancing Public Trust in COVID-19 Vaccination. (2021). The role of governments.

Etienne, C. F., Fitzgerald, J., Almeida, G., Birmingham, M. E., Brana, M., Bascolo, E., Cid, C., & Pescetto, C. (2020). COVID-19: Transformative actions for more equitable, resilient, sustainable societies and health systems in the Americas Commentary. *BMJ Global Health, 5*, 3509. https://doi.org/10.1136/bmjgh-2020-003509

Fang, F. C., Benson, C. A., del Rio, C., Edwards, K. M., Fowler, V. G., Fredricks, D. N., Limaye, A. P., Murray, B. E., Naggie, S., Pappas, P. G., Patel, R., Paterson, D. L., Pegues, D. A., Petri, W. A., & Schooley, R. T. (2021). COVID-19—Lessons learned and questions remaining. *Clinical Infectious Diseases, 72*(12), 2225–2240. https://doi.org/10.1093/CID/CIAA1654

Fauci, A. S., Lane, H. C., & Redfield, R. R. (2020). Covid-19 — Navigating the uncharted. *New England Journal of Medicine, 382*(13), 1268–1269.

Fidl, D. P. (2020). Vaccine nationalism's politics. *Science, 369*, 6505.

Francis, P. J. (1997). Dynamic epidemiology and the market for vaccinations. *Journal of Public Economics, 63*(3), 383–406. https://doi.org/10.1016/S0047-2727(96)01586-1

Gao, Q., Bao, L., Mao, H., Wang, L., Xu, K., Yang, M., Li, Y., Zhu, L., Wang, N., Lv, Z., Gao, H., Ge, X., Kan, B., Hu, Y., Liu, J., Cai, F., Jiang, D., Yin, Y., Qin, C., & Qin, C. (2020). Development of an inactivated vaccine candidate for SARS-CoV-2. *Science, 369*, 6499. https://doi.org/10.1126/science.abc1932

Gatherer, D. (2009). The 2009 H1N1 influenza outbreak in its historical context. *Journal of Clinical Virology, 45*, 3. https://doi.org/10.1016/j.jcv.2009.06.004

Glover, R. E., van Schalkwyk, M. C. I., Akl, E. A., Kristjannson, E., Lotfi, T., Petkovic, J., Petticrew, M. P., Pottie, K., Tugwell, P., & Welch, V. (2020). A framework for identifying and mitigating the equity harms of COVID-19 policy interventions. *Journal of Clinical Epidemiology, 128*, 35–48. https://doi.org/10.1016/J.JCLINEPI.2020.06.004

Gupta, I., & Baru, R. (2020). Economics & ethics of the COVID-19 vaccine: How prepared are we? *The Indian Journal of Medical Research, 152*(1-2), 153. https://doi.org/10.4103/IJMR.IJMR_3581_20

Haidari, L. A., Brown, S. T., Ferguson, M., Bancroft, E., Spiker, M., Wilcox, A., Ambikapathi, R., Sampath, V., Connor, D. L., & Lee, B. Y. (2016). The economic and operational value of using drones to transport vaccines. *Vaccine, 34*, 34. https://doi.org/10.1016/j.vaccine.2016.06.022

India Science and Technology. (n.d.). *Indian contribution in vaccine development.* Retrieved September 5, 2021, from https://www.indiascienceandtechnology.gov.in/covid-19-vaccine/indian-contribution-vaccine-development

Josh Holder. (n.d.). Tracking coronavirus vaccinations around the world. *The New York Times.* Retrieved from https://www.nytimes.com/interactive/2021/world/covid-vaccinations-tracker.html

Knoll, M. D., & Wonodi, C. (2021). Oxford–AstraZeneca COVID-19 vaccine efficacy. *The Lancet, 397,* 10269. https://doi.org/10.1016/S0140-6736(20)32623-4

Kumar Kaustubh. (2021). Right to free vaccination amid current global scenario.

Lagman, J. D. N. (2021). Vaccine nationalism: A predicament in ending the COVID-19 pandemic. *Journal of Public Health, 43,* 2. https://doi.org/10.1093/pubmed/fdab088

Lahariya, C. (2014). A brief history of vaccines & vaccination in India. *Indian Journal of Medical Research, 139,* 491.

Ledford, H. (2021). Six months of COVID vaccines: What 1.7 billion doses have taught scientists. *Nature, 594*(7862), 164–167. https://doi.org/10.1038/D41586-021-01505-X

Lee, B. Y., & Haidari, L. A. (2017). The importance of vaccine supply chains to everyone in the vaccine world. *Vaccine, 35,* 35. https://doi.org/10.1016/j.vaccine.2017.05.096

Lipsitch, M., Swerdlow, D. L., & Finelli, L. (2020). Defining the epidemiology of Covid-19 – studies needed. *New England Journal of Medicine, 382*(13), 1194–1196. https://doi.org/10.1056/NEJMP2002125

Lloyd, J., & Cheyne, J. (2017). The origins of the vaccine cold chain and a glimpse of the future. *Vaccine, 35,* 17. https://doi.org/10.1016/j.vaccine.2016.11.097

Madhavi, Y. (2005). Vaccine policy in India. *PLoS Medicine, 2,* 5. https://doi.org/10.1371/journal.pmed.0020127

Malacrida, L. M. F., Fenner, D. A., Henderson, I., Arita, Z., Jezek, I., & Ladnyi, D. (2012). *Smallpox and its eradication.* World Health Organization.

Mathur, R., & Swaminathan, S. (2018). National ethical guidelines for biomedical and health research involving human participants, 2017: A commentary. *Indian Journal of Medical Ethics, 3*(3), 65. https://doi.org/10.20529/IJME.2018.065

Ministry of Health and Family Welfare. (1940). The drugs and cosmetics act and rules. The Gazaette of India.

My gov. (2021). FAQs on COVID-19 vaccines and vaccination program.

Our World in Data. (2021). Coronavirus (COVID-19) vaccinations - statistics and research. Coronavirus (COVID-19) vaccinations - statistics and research.

Padma, T. V. (2021). India's COVID-vaccine woes - by the numbers. *Nature, 592,* 7855. https://doi.org/10.1038/d41586-021-00996-y

Programme, U. I., Surveillance, A., & Aefi, T. (2015). Revised AEFI guidelines: Executive summary.

Pronker, E. S., Weenen, T. C., Commandeur, H., Claassen, E. H. J. H. M., & Osterhaus, A. D. M. E. (2013). Risk in vaccine research and development quantified. *PLoS ONE, 8*(3), e057755. https://doi.org/10.1371/journal.pone.0057755

Randolph, H. E., & Barreiro, L. B. (2020). Herd immunity: Understanding COVID-19. *Immunity, 52*(5), 737–741. https://doi.org/10.1016/J.IMMUNI.2020.04.012

Roy, S. (2002). Medicine and the Raj: British medical policy in India, 1835–1911. *Indian Historical Review, 29,* 1–2. https://doi.org/10.1177/037698360202900239

Saad-Roy, C. (2020). Immune life history, vaccination, and the dynamics of SARS-CoV-2 over the next 5 years. *Science, 370*(6518), 811–818. https://doi.org/10.1126/science.abd7343

Sadoff, J., Le Gars, M., Shukarev, G., Heerwegh, D., Truyers, C., de Groot, A. M., Stoop, J., Tete, S., Van Damme, W., Leroux-Roels, I., Berghmans, P.-J., Kimmel, M., Van Damme, P., de Hoon, J., Smith, W., Stephenson, K. E., De Rosa, S. C., Cohen, K. W., McElrath, M. J., & Schuitemaker, H. (2021). Interim results of a phase 1–2a trial of Ad26.COV2.S covid-19 vaccine. *New England Journal of Medicine, 384*(19), 2034201. https://doi.org/10.1056/nejmoa2034201

Sapkal, G. N., Yadav, P. D., Ella, R., Deshpande, G. R., Sahay, R. R., Gupta, N., Mohan, K., Abraham, P., Panda, S., & Bhargava, B. (2021). Neutralization of UK-variant VUI-202012/01 with COVAXIN vaccinated human serum. BioRxiv.

ScienceDirect. (n.d.). Uneven power dynamics must be levelled in COVID-19 vaccines access and distribution.

Shaji, B., Arun Thomas, E. T., & Sasidharan, P. K. (2019). Tuberculosis control in India: Refocus on nutrition. *The Indian Journal of Tuberculosis, 66*, 1. https://doi.org/10.1016/j.ijtb.2018. 10.001

Sharma, O., Sultan, A. A., Ding, H., & Triggle, C. R. (2020). A review of the progress and challenges of developing a vaccine for COVID-19. *Frontiers in Immunology, 11*, 585354. https://doi.org/10.3389/fimmu.2020.585354

Shinde, V., Bhikha, S., Hoosain, Z., Archary, M., Bhorat, Q., Fairlie, L., Lalloo, U., Masilela, M. S. L., Moodley, D., Hanley, S., Fouche, L., Louw, C., Tameris, M., Singh, N., Goga, A., Dheda, K., Grobbelaar, C., Kruger, G., Carrim-Ganey, N., & Madhi, S. A. (2021). Efficacy of NVX-CoV2373 Covid-19 vaccine against the B.1.351 variant. *New England Journal of Medicine, 384*(20), 1899–1909. https://doi.org/10.1056/NEJMOA2103055

Singh, K., & Mehta, S. (2016). The clinical development process for a novel preventive vaccine: An overview. *Journal of Postgraduate Medicine, 62*, 1. https://doi.org/10.4103/0022-3859.173187

Teo, S. P. (2021). Review of COVID-19 mRNA vaccines: BNT162b2 and mRNA-1273. *Journal of Pharmacy Practice*. https://doi.org/10.1177/08971900211009650

Thanh Le, T., Andreadakis, Z., Kumar, A., Gómez Román, R., Tollefsen, S., Saville, M., & Mayhew, S. (n.d.). The COVID-19 vaccine development landscape. *Nature Reviews Drug Discovery*. https://doi.org/10.1038/d41573-020-00073-5

The Human Right to Vaccines. (n.d.). Preventing discrimination against the unvaccinated. Health and Human Rights Journal.

Uttarilli, A., Amalakanti, S., Kommoju, P. R., Sharma, S., Goyal, P., Manjunath, G. K., Upadhayay, V., Parveen, A., Tandon, R., Prasad, K. S., Dakal, T. C., Ben Shlomo, I., Yousef, M., Neerathilingam, M., & Kumar, A. (2021). Super-rapid race for saving lives by developing COVID-19 vaccines. *Journal of Integrative Bioinformatics, 18*, 1. https://doi.org/10.1515/jib-2021-0002

World Health Organization. (2020). *Coronavirus disease (COVID-19): Vaccines*. World Health Organization.

World Health Organization. (2021). *The Sinovac COVID-19 vaccine what you need to know*. World Health Organization.

Worldwide COVID-19 Candidate Vaccines. (n.d.). http://ctri.nic.in/Clinicaltrials/showallp.php? mid1=45184&EncHid=&userName=BBV152

Xia, S., Zhang, Y., Wang, Y., Wang, H., Yang, Y., Gao, G. F., Tan, W., Wu, G., Xu, M., Lou, Z., Huang, W., Xu, W., Huang, B., Wang, H., Wang, W., Zhang, W., Li, N., Xie, Z., Ding, L., & Yang, X. (2021). Safety and immunogenicity of an inactivated SARS-CoV-2 vaccine, BBIBP-CorV: A randomised, double-blind, placebo-controlled, phase 1/2 trial. *The Lancet Infectious Diseases, 21*(1), 8. https://doi.org/10.1016/S1473-3099(20)30831-8

Yu, J., Tostanosk, L. H., Peter, L., Mercad, N. B., McMahan, K., Mahrokhia, S. H., Nkolol, J. P., Liu, J., Li, Z., Chandrashekar, A., Martine, D. R., Loos, C., Atyeo, C., Fischinger, S., Burk, J. S., Slei, M. D., Chen, Y., Zuiani, A., Lelis, F. J. N., & Barou, D. H. (2020). DNA vaccine protection against SARS-CoV-2 in rhesus macaques. *Science, 369*, 6505. https://doi.org/10.1126/science. abc6284

Zydus. (2021). *Zydus receives EUA from DCGI for ZyCoV-D, the only needle-free COVID vaccine in the world*. Retrieved from https://www.zyduscadila.com/public/pdf/pressrelease/Press/Release-Zydus-receives-EUA-from-DCGI-for-ZyCoV-D.pdf

Vaccination for the COVID-19 Pandemic

Kavitha Agamutu and P. Agamuthu

1 Introduction

A cluster of pneumonia cases was reported on the December 31, 2019 in the city of Wuhan, from the Hubei Province of China. A novel coronavirus, SARS-CoV-2, was identified as the causative microorganism, and on January 12, 2020, China shared the genetic sequence of what is currently known as the COVID-19 coronavirus. Subsequently, on March 11, 2020, the World Health Organization (WHO) declared this outbreak to be a pandemic after observing the alarming speed at which the disease spread, as well as the severity of the disease (WHO Guidance Document, 2021; World Health Organization, 2020a, 2020b, 2021a, 2021b). While several animals were initially suggested to be the intermediate host for the SARS-CoV-2, thorough research could not provide a definitive answer as to which animal or animals were indeed the intermediate host. As human-to-human transmission was established and caused the infection to spread rapidly, this novel coronavirus reached all around the globe at a speed never seen before, showing humanity its highly virulent nature. A year and a half later, this coronavirus proved to be a worthy opponent to the modern science of health, having taken 4.4 million lives thus far and infected 209 million people globally (WHO Guidance Document, 2021; World Health Organization, 2020a, 2020b).

In Malaysia, 1.5 million people had been infected by this virus till date, with more than 16,000 deaths. While the first wave swept over Malaysia with only 22 cases, the current and third wave has hit the country very hard, crippling both the economy as well as the healthcare system, with the states of Selangor and Kuala Lumpur

K. Agamutu
Public Health Development Division, Ministry of Health, Putrajaya, Malaysia

P. Agamuthu (✉)
Jeffrey Sachs Centre on Sustainable Development, Sunway University, Subang Jaya, Malaysia
e-mail: agamutup@sunway.edu.my

recording the highest number of cases (Ministry of Health Malaysia, 2021). Most people who test positive for the SARS-CoV-2, experience mild to moderate symptoms that is self-limiting and do not require hospitalization. Commonly, people present with fever, dry cough, and lethargy. Other symptoms may include headache, sore throat, generalized aches and pains, diarrhea, loss of taste or smell, conjunctivitis, and rashes. If the disease progresses further to a more severe form, the person could present with shortness of breath and chest pain as well and requires immediate medical attention (Ministry of Health Malaysia, 2021; World Health Organization, 2020a, 2020b).

1.1 SARS-CoV-2 Coronavirus Variants

The SARs-COV-2 coronavirus, just like any other virus, is subject to mutation over time, to enable it to better adapt to the environment. Each mutation that occurs produces a new variant or strain of the same virus, usually more virulent than the previous. The faster the virus spreads among the human population, the faster it is replicating and mutating. Generally, mutation of the virus would not impede its ability to cause an infection in the target host, but may produce a more virulent variant that could result in a more severe disease or a less virulent variant with less severe disease. More than a dozen variants of the SARS-CoV-2 coronavirus have been identified since the beginning of the pandemic. However, certain variants pose a higher threat to the population than others. The World Health Organization classifies each new variant as either a variant of concern (VOC) or a variant of interest (VOI). They further define VOC as a variant that evidently has a higher transmission rate, causes a more severe disease and reduced efficacy of treatment. Meanwhile, VOI is defined as a variant with genetic modification that can affect the characteristics of the virus and eventually also alter the transmission, disease severity as well as treatment efficacy (Centre for Disease Control and Prevention, 2021a, 2021b; World Health Organization, 2021a, 2021b). Table 1 presents the lists of current VOIs and VOCs globally as well as its presence in Malaysia.

In addition to data presented in Table 1, a total of nine (9) lesser-known variants (A, B, B.1, B.1.1, B.1.1.1, B.1.36, B.2, B.3, and B.6) were also detected in Malaysia. While the rare B.6 variant was predominantly detected following the Tablighi Jamaat religious gathering during early 2020, precipitating the second wave and significant local transmission, most of the other variants were associated with travel and limited local transmission (Chong et al., 2020). Towards the end of 2021, a new variant of concern, Omicron (B.1.1.529) was detected in samples from various countries and eventually in Malaysia as well (WHO, 2021a).

Table 1 Current VOIs and VOCs and its presence in Malaysia

Variant	WHO classification	Country of origin	Severity	Vaccine efficacy	Presence in Malaysia
Alpha B.1.1.7	VOC	United Kingdom	Potential to cause more infections and fatality	Currently approved vaccines are effective	✓
Beta B.1.351	VOC	South Africa	Currently no evidence of increased severity or fatality	Currently approved vaccines are effective	✓
Gamma P.1	VOC	Brazil	Currently no evidence of increased severity or fatality	Currently approved vaccines are effective	✗
Delta B.1.617.2	VOC	India	High transmissibility; potential to cause more severe disease	Current evidence suggests that fully vaccinated individuals have a less severe disease	✓
Lambda C.37	VOI	Peru	High transmissibility; potential to cause more severe disease	Insufficient data available	✗

Source: World Health Organization (2021a, 2021b), Centre for Disease Control and Prevention (2021a, 2021b), Chong et al. (2020)

1.2 Vaccinations

1.2.1 Brief Introduction to Vaccines

Most vaccines contain weakened or inactive components of a particular organism, acting as the antigen, to induce the body's immune system into producing antigen-specific antibodies to help fight the organism and subsequently offer protection to the person from future infections from the same or very similar organisms. This is not unlike the antibodies produced when a person is naturally infected with the organism. Hence, a vaccine simulates an infection but at a much-lowered severity, just enough for the person's immune system to respond with antibodies production. Therefore, some vaccines may require more than one dose to ensure adequate antibodies are produced against that particular organism or very similar ones.

Types of Vaccines

Over the course of time, and with the advancement of science in the field of vaccine-related research, there have been several types of vaccines available for most of the commonest communicable diseases. Traditionally, vaccines used to be either inactivated or live attenuated vaccines. Subsequently, toxoid-, recombinant-, conjugate-and recently, m-RNA-vaccines have also been developed to curb certain diseases (US Department of Health and Human Services (HHS), 2021; World Health Organization, 2021, 2021a, 2021b; British Society for Immunology, 2021). The type of vaccine that is developed for a specific disease usually is dependent on several factors, such as the type of organism causing the disease, the target population of the vaccine, cost of manufacturing, issues associated with mass production like storage space and temperature, etc.

Inactivated Vaccine

- These vaccines contain inactivated or killed organism that is causing a particular disease. Since the organisms are killed, these types of vaccines generally do not produce a very strong immune response and may need several doses or booster shots to achieve the required antibody titer that would protect the person against that disease.
- Example of inactivated vaccines include Polio (IPV), Rabies, and Hepatitis A vaccines.

Live Attenuated Vaccines

- These vaccines are made of weakened organisms that are able to mount a sizeable immune response from the person receiving it, but still not able to cause the potential harm of a full-blown infection. Generally, 1 or 2 shots are sufficient to provide lifetime protection from that particular disease.
- Examples of live attenuated vaccines include the Mumps Measles Rubella (MMR) combination vaccine, Smallpox, Rotavirus, Varicella (Chickenpox), and Yellow Fever.

Toxoid Vaccines

- Several bacteria that can cause diseases in the human population actually are not harmful by themselves but produce toxins that are the actual harmful component to humans. Hence, toxoid vaccines are made using the inactivated toxins from the bacteria to mount an immune response. The antibodies that are then produced will target the toxins specifically and not the organism, but still provide protection from the disease. Booster doses are usually required to achieve adequate protection.
- Examples of toxoid vaccines include diphtheria and tetanus.

Subunit and Conjugate Vaccines

- These types of vaccines contain only an isolated part of the causative organism to prompt an immune response.

- Examples of these vaccines include hepatitis B, pneumococcal, human papilloma virus (HPV), and pertussis (Whooping cough).

Messenger RNA (m-RNA) Vaccines

- This is a much recently developed type of vaccine, although research on m-RNA vaccines have been ongoing for decades. The messenger-RNA is a genetic component that is able to convey the message to the immune system on developing certain protein (for example, the spike proteins of the COVID-19 virus) and subsequently if the same person is infected by the actual virus, the body will be able to fight against the infection. This type of vaccine does not contain the organism or any of its component and can be developed in a short duration of time.
- Some COVID-19 vaccines are the best example of the m-RNA vaccine (Sources: US Department of Health and Human Services (HHS), 2021; World Health Organization, 2021a, 2021b).

1.3   Covid-19 Vaccines

Since the beginning of the pandemic, the world saw vaccine-related research and vaccine development being conducted at a speed never seen before. Following the onset of the pandemic, within a year, the first COVID-19 vaccine by Pfizer/ BioNTech which is an m-RNA vaccine, was granted the Emergency Use Authorization (EUA) by the Food and Drug Administration (FDA) (World Health Organization, 2020, 2020a, 2020b). The FDA is a regulatory body under the US Department of Health and Human Services and is responsible for the health of the human population by ensuring the safety, efficacy, and security of drugs, foods, cosmetics, all biological products and medical devices. Hence, in the event that a novel vaccine has been developed, the FDA approval is necessary before the vaccine can be used on the general population, in addition to the approval from WHO. Under the Emergency Use Authorization, the FDA is allowed to sanction the use of unapproved products in the event that it is a life saving measure (either as a treatment or a preventive measure) in an emergency situation as the pandemic is. As such, the Comirnaty vaccine, better known around the world as Pfizer vaccine was the first to obtain the EUA from FDA, and was subsequently followed by Moderna and Johnson & Johnson. Several other COVID-19 vaccines were granted the EUA for use in other countries, outside of USA, such as CoronaVac, Sputnik, Sinopharm, Sputnik Light, COVAXIN, and Oxford-AstraZeneca. This is similar to the vaccines Emergency Use Authorization (EUA) granted by the WHO for this pandemic (USFDA Covid-19 Vaccines, 2020; WHO Guidance Document, 2021; Centre for Disease Control and Prevention (2021a, 2021b); Yale Medicine, 2021; Covishield Vaccine Information, 2021; CanSinoBio Vaccine, 2021; Saeed et al. 2021; NPRA, 2021; Bharat Biotech Covaxin Fact Sheet, 2021; Montalti et al., 2021; Ministry of Health Malaysia, 2021, BCCDC, 2021) (Table 2).

Table 2 List of COVID-19 Vaccines Available Globally

Vaccine	Manufacturer	Country	WHO EUA	Type of vaccine	Vaccine schedule	Common/reported adverse effects
Comirnaty/ BNT162b2/ tozinameran	Pfizer/BioNTech	USA	✓	m-RNA	2 doses 21–28 days apart	Headache, chills, lethargy, pain and/or swelling at injections site FDA issued warning of a likely association of heart inflammation among young adults
m-RNA-1273	Moderna	USA	✓	m-RNA	2 doses 28 days apart	Headache, chills, lethargy, pain and/or swelling at injections site FDA issued warning of a likely association of heart inflammation among young adults
Ad26.COV2.5	Janssen (Johnson & Johnson)	USA but developed in Netherlands	✓	Vector	One dose	Headache, fever, lethargy, injection site pain and/or swelling FDA warned of rare occurrence of Guillain-Barre Syndrome
AZD1222 Vaxzevria	Oxford-AstraZeneca	UK	✓	Vector	2 doses, 4–12 weeks apart	Pain and/or swelling at injection site, lethargy, chills, fever Reported occurrence of a rare blood clotting disorder with low platelet count
CoronaVac	Sinovac	China	✓	While inactivated virus	2 doses, 14–28 days apart	Pain and/or swelling at injections site, lethargy, headache Rare side effects include hypersensitivity, dizziness, vomiting, fever, nosebleed, conjunctival congestion, muscle spasms among others
Covishield	Serum Institute of India	India	✓	Vector	2 doses 4–12 weeks apart	Pain, warmth, and/or swelling at injection site, lethargy, chills, fever Rare occurrence of a blood clotting disorder with low platelet count

Vaccine	Manufacturer	Country	Status	Type	Dosage	Side effects
Covilo/BBIBP-CorV	Sinopharm/Beijing Institute of Biological Products (BIBP)	China	✓	Whole inactivated virus	2 doses 21–28 days apart	Headache, lethargy, pain and/or swelling at injection site
Sputnik V	The Gamaleya National Centre	Russia	Pending	Vector	2 doses 3 weeks apart	Headache, Injection site pain/swelling/redness or warmth and flu-like symptoms
Inactivated SARS-CoV-2 Vaccine (Vero Cell)	Sinopharm/Wuhan Institute of Biological Products (WIBP)	China	Pending	Whole Inactivated Virus	2 doses 3–4 weeks apart	Headache, injection site pain/swelling/redness or warmth
Ad5-nCoV	CanSinoBio	China	✓	Vector	Single dose	Fever, injection site pain, redness and swelling
NVX-CoV2373/ Covovax	Novavax	USA	✓	Protein Subunit	2 doses 3 weeks apart	Headache, lethargy, pain and/or swelling at injection site
CVnCoV/ CV07050101 Zorecimeran (INN)	CureVac	Germany	✓	m-RNA	2 doses 4 weeks apart	Headache, lethargy, chills, muscle pain and less frequently fever
COVAXIN	Bharat Biotech	India	Approved	Whole Inactivated Virus	2 doses 4 weeks apart	Headache, fever, body ache, nausea, vomiting, and injections site pain/swelling/redness or itching

Sources: WHO Guidance Document (2021), Centre for Disease Control and Prevention (2021a, 2021b), USFDA Covid-19 Vaccines (2020), Yale Medicine (2021), Covishield Vaccine Information (2021), CanSinoBio Vaccine (2021), Saeed, Al-Shahrabi, Alhaj, and Adrees (2021), NPRA (2021), Bharat Biotech Covaxin Fact Sheet (2021), Montalti et al. (2021), Ministry of Health Malaysia (2021), BCCDC (2021)

1.3.1 The COVID-19 Vaccination Program

Since the very first dose of the Comirnaty vaccine was administered in the United Kingdom in December 2020, a total of almost 5 billion vaccine doses have been administered globally, with almost 1.9 billion people from around the world being fully vaccinated currently. As of August 25, 2021, the average daily vaccine dose administered stands at 33.8 million per day (CNN Tracking Covid-19 Vaccinations Worldwide, 2021; WHO Guidance Document, 2021). While every effort is being made to ensure that the vaccination program successfully inoculates as many people as possible to give or provide adequate protection against the SARS-CoV-2 virus and this pandemic, the efforts are still being met with some resistance from rather prevalent vaccine hesitancy and vaccine refusal.

Malaysia, through its National COVID-19 Immunization Program, records a total of 32 million doses administered, with 42.4% of the total population and 61% of the adult population being fully vaccinated. Meanwhile, 57.5% of the total population, and 80.3% of the adult population in Malaysia have received the first dose (JKJAV, 2021; Ministry of Health Malaysia, 2021). Five different vaccines were procured for the purpose of the vaccination program in Malaysia and they are Comirnaty (Pfizer/BioNTech), AstraZeneca vaccine, CoronaVac (Sinovac), CanSinoBio, and Sputnik (JKJAV, 2021). The National COVID-19 Immunization program aims to vaccinate every eligible adult in Malaysia through its three-phase plan. The first phase which was originally planned to be conducted between February and April, was aiming to vaccinate the front liners and those involved in all services deemed essential. The second phase was scheduled to be conducted between April and August and targeted the remaining front liners, disabled persons, those aged above 60 years and those in the high-risk category (had co-morbid diseases such as diabetes mellitus, hypertension, cardiovascular diseases, and obesity). The third and final phase of the program, scheduled to be conducted between May 2021 and February 2022, aimed to complete immunization for remaining Malaysian adults aged 18 years and above (JKJAV, 2021). The COVID-19 vaccination program in Malaysia is being coordinated largely using the MySejahtera application which allows the relevant ministries to be able to keep a close tab on the vaccination registration status of most individuals and enable them to swiftly allocate dates for those waiting for an appointment. However, a parallel and voluntary vaccination pathway was also established for any Malaysian above the age of 18 years, who was keen to be vaccinated with the AstraZeneca vaccine. With these efforts in place, the National COVID-19 Immunization Program rapidly picked up in pace. While off to a slow start after receiving the first batch of vaccine in February 2021, as of August 25, Malaysia recorded 423,380 vaccines doses being administered for the day in approximately 605 vaccine Administration Centres (PPVs) and the daily average lingers at about the same vicinity (Ministry of Health Malaysia, 2021). The National COVID-19 Immunization Program picked up its pace significantly, even recording the highest daily vaccine administration of 556,404 on July 29, 2021 (Ministry of Health Malaysia, 2021). Depending on the capacity of vaccination centers, staffing

Table 3 Cumulative vaccination status by states in Malaysia as of August 25, 2021

State	Dose 1	Dose 2	% of total population		% of adult population	
			Dose 1	Dose 2	Dose 1	Dose 2
Malaysia	18,792,979	13,842,928	57.5	42.4	80.3	59.1
Labuan Federal Territory	69,501	63,575	69.8	63.8	101.5*	92.8
Klang Valley	6,775,251	5,356,312	80.4	63.6	109.9*	86.9
Sarawak	1,840,963	1,754,803	65.4	62.3	90.1	85.9
Negeri Sembilan	748,658	606,482	66.3	53.7	91.9	74.5
Perlis	149,645	115,615	58.7	45.4	82.6	63.8
Malacca	516,872	343,118	55.4	36.8	76.3	50.7
Penang	1,092,150	690,486	61.6	38.9	79.9	50.5
Pahang	716,724	544,132	42.7	32.4	61.0	46.3
Terengganu	619,737	360,185	49.2	28.6	76.7	44.6
Perak	1,132,969	822,704	45.1	32.8	60.8	44.2
Johor	1,948,209	1,129,279	51.5	29.9	71.8	41.6
Kelantan	691,374	503,902	36.3	26.4	55.9	40.8
Kedah	987,048	619,577	45.2	28.4	64.1	40.2
Sabah	1,503,878	932,758	38.5	23.9	54.5	33.8

Source: Ministry of Health Malaysia (2021)

and severity of the pandemic in a particular location, several states have also offered "walk-in" vaccinations options to aid boost the number of Malaysian receiving the vaccine.

In the past decades, urban rural gaps and the discrepancies between West and East Malaysia has always been a troubling issue, even more so when the matter pertains to national health and safety. Thus, a program like the National COVID-19 Immunization Program also ensures that the vaccines were distributed accordingly and in a timely manner to allow the country to achieve mass protection from vaccination. This is apparent from data which shows the state of Sarawak ranking third, after fully vaccinating 62.3% of its total population and 85.9% of its adult population. Ranking first and second are Labuan Federal Territory and Klang Valley. However, the other East Malaysian state of Sabah has ranked at the bottom having only achieved fully vaccination status for 23.9% of its total population. Table 3 shows the cumulative vaccination achieved by states in Malaysia, as of August 25, 2021.

In line with global findings, the Ministry of Health Malaysia concurs that the deaths due to COVID-19 and Category 5 disease (very severe illness requiring assisted breathing) occurred mostly among those who were unvaccinated. On August 26, Malaysia recorded its highest daily new cases of 24,599, of which 118 cases (0.5%) were in Category 4 and 345 cases (1.4%) were in Category 5. Among those admitted in Category 4, 83.1% were unvaccinated meanwhile among those in Category 5, 89.6% were unvaccinated. In addition to this, it was observed that as the National COVID-19 Immunization Program picked up its pace rapidly, increasing number of new cases fell within Category 1 (asymptomatic) or

Category 2 (mild symptoms) (Ministry of Health Malaysia, 2021). This further demonstrated that while a fully vaccinated individual could still be infected, it was more likely for this individual to experience a mild disease or remain asymptomatic as opposed to being unvaccinated. This further proved the need for the program to succeed in vaccinating fully, every eligible Malaysian adult and be well on its way to win the battle against this pandemic.

2 Conclusion and Way Forward

As newer and increasingly potent variants keep emerging over time, it has become apparent that while mass vaccination to achieve herd immunity may not be sufficient to stop this pandemic in its track, it certainly plays a vital role in reducing progression to severe disease as well as deaths from SARS-CoV-2. As the virus evolves, so should the vaccines, to include protection against the newer variants that were not previously covered. This in turn necessitates further research and expedited development of newer vaccines as well as booster shots for current vaccines that may offer greater protection. In addition to being vaccinated, the human population must now get comfortable with wearing masks and social distancing, in what we are learning to be the new norm. In the wake of this pandemic, which has crippled many nations, we hope that the worst is over, and move forward with the knowledge and experience of facing emerging infectious diseases.

References

Bharat Biotech Covaxin Fact Sheet. (2021). https://www.bharatbiotech.com/images/covaxin/covaxin-fact-sheet.pdf

British Columbia Centre for Disease Control (BCCDC). (2021). *WHO emergency use authorization (EUA) qualified vaccines*. Retrieved from http://www.bccdc.ca/Health-Info-Site/Documents/COVID-19_vaccine/WHO-EUA-qualified-covid-vaccines.pdf

British Society for Immunology. (2021). *Types of vaccines for Covid-19*. Retrieved from https://www.immunology.org/coronavirus/connect-coronavirus-public-engagement-resources/types-vaccines-for-covid-19

CanSinoBio Vaccine. (2021). http://www.cansinotech.com/html/1//156/218/index.html

Centre for Disease Control and Prevention. (2021a). *SARS-CoV-2 variant classification and definitions*. Retrieved from https://www.cdc.gov/coronavirus/2019-ncov/variants/variant-info.html

Centre for Disease Control and Prevention. (2021b). *What you need to know about variants*. Retrieved from https://www.cdc.gov/coronavirus/2019-ncov/variants/variant.html

Chong, Y. M., Sam, I.-C., Chong, J., Kahar Bador, M., Ponnampalavanar, S., Syed Omar, S. F., et al. (2020). SARS-CoV-2 lineage B.6 was the major contributor to early pandemic transmission in Malaysia. *PLoS Neglected Tropical Diseases, 14*(11), e0008744. https://doi.org/10.1371/journal.pntd.0008744

CNN Tracking Covid-19 Vaccinations Worldwide. (2021). https://www.seruminstitute.com/product_covishield.php

Covishield Vaccine Information. (2021). Retrieved from https://www.seruminstitute.com/product_ covishield.php

JKJAV. (2021). *National Covid-19 immunization program Malaysia*. Retrieved from https://www. vaksincovid.gov.my/pdf/National_COVID-19_Immunisation_Programme.pdf

Ministry of Health Malaysia. (2021). *MOH Covid-19 pandemic*. Retrieved from https://covid-19. moh.gov.my/

Montalti, M., et al. (2021). ROCCA observational study: Early results on safety of Sputnik V vaccine (Gam-COVID-Vac) in the Republic of San Marino using active surveillance. *EClinicalMedicine*. https://doi.org/10.1016/j.eclinm.2021.101027

National Pharmaceutical Regulatory Agency (NPRA). (2021). *Frequently asked questions about coronavac vaccine*. Retrieved from https://www.npra.gov.my/easyarticles/images/users/1047/ FAQ-about-COVID-19-Vaccine-CORONAVAC.pdf

Saeed, B. Q., Al-Shahrabi, R., Alhaj, S. S., & Adrees, Z. M. A. A. O. (2021). Side effects and perceptions following sinopharm COVID-19 vaccination. *International Journal of Infectious Diseases*. https://doi.org/10.1016/j.ijid.2021.08.013

US Department of Health and Human Services (HHS). (2021). *USHHS vaccine types*. Retrieved from https://www.hhs.gov/immunization/basics/types/index.html

USFDA Covid-19 Vaccines. (2020). https://www.fda.gov/emergency-preparedness-and-response/ coronavirus-disease-2019-covid-19/covid-19-vaccines

WHO. (2020). *WHO news release: WHO issues its first emergency use validation for a COVID-19 vaccine and emphasizes need for equitable global access*. Retrieved from https://www.who.int/ news/item/31-12-2020-who-issues-its-first-emergency-use-validation-for-a-covid-19-vaccine- and-emphasizes-need-for-equitable-global-access

WHO. (2021). *The different types of Covid-19 vaccines*. Retrieved from https://www.who.int/news- room/feature-stories/detail/the-race-for-a-covid-19-vaccine-explained

WHO Guidance Document. (2021). *Status of Covid-19 vaccines within WHO EUL/PQ evaluation process*. Retrieved from https://extranet.who.int/pqweb/sites/default/files/documents/Status_ COVID_VAX_19August2021.pdf

World Health Organization. (2020a). *WHO coronavirus disease (Covid-19) pandemic*. Retrieved from https://www.who.int/emergencies/diseases/novel-coronavirus-2019

World Health Organization. (2020b). *Final report WHO-convened global study of origins of SARS- CoV-2: China part. Joint WHO-China Study*. Retrieved from https://www.who.int/publications/ i/item/who-convened-global-study-of-origins-of-sars-cov-2-china-part

World Health Organization. (2021a). *Tracking SARS-CoV-2 variants*. Retrieved from https://www. who.int/en/activities/tracking-SARS-CoV-2-variants/

World Health Organization. (2021b). *WHO Coronavirus (Covid-19) dashboard*. Retrieved from https://covid19.who.int/

Yale Medicine. (2021). *Comparing the Covid-19 vaccines: How are they different?* Retrieved from https://www.yalemedicine.org/news/covid-19-vaccine-comparison

Exercise in Immune Health Management and Rehabilitation Against COVID-19

Aparup Konar and Samiran Mondal

1 Introduction

The novel corona virus disease 2019(COVID-19) has made a serious threat to the humanity and enforced to adopt a "New Normal" of living for the matter of being continued to existence. Exposing of the COVID-19 infection, the immunity level gets weaken for which it is essentially concerned about the immune health to improve host defense of an individual.

Regular exercise possibly has the potential to enhance the natural immune defense that provides surveillance with covering measures of prevention, promotion, rehabilitation, and management healthcare to deal with this deadly virus. Resistance to specific infection is strongly associated with the effectiveness of immune function whereas moderate exercise may possibly influence the immune efficiency in protecting the host against the COVID-19 virus.

An effective and ideal immune enhancing exercise strategy is extremely significant that pre-requisites to adopt to fight against the deadly virus at the time when people apparently are becoming infected at large scale along with re-infection. As the recovered population has not achieved the herd immunity, there is uncertain around how long people might remain immune after exposure to COVID-19.

Present researchers intended to provide more specific exercise recommendations as that pertain to infection risk and control in both healthy and clinical populations. With admiration to this brief background, the specific objective of the present review

A. Konar (✉)
Jadavpur University, Kolkata, West Bengal, India
e-mail: dir.phy.ins@jadavpuruniversity.in

S. Mondal
Department of Yogic Art and Science, Visva-Bharati (Central University), Santiniketan, West Bengal, India

study was to observe the influence of regular moderate exercise on immune health in human body in the context of corona virus disease 2019 (COVID-19).

2 Exercise as Safeguard to Immune Health in the Perspective of COVID-19 Diseases

2.1 Role of Exercise on Immune Health

Health is determined not only by the absence of disease, but also resistance to infection. It depends on the effectiveness of specific homeostatic regulation and mechanism of general adaptation in functional economy and stability. Immunity is the most significant aspect of human body when it comes in health and active lifestyle. Regular exercise has a favorable impact over many systems throughout the body especially immune system and improves one's sense of well-being and general health. On the other hand, physical inactivity for longer period is associated with impaired immune cell activity and reduces immune health function. Physical exercise influences a variety of health changes, which are essential for increasing immune adaptability that is the most vital system of the human body. Regular moderate exercise can better able to formulate preventive, therapeutic, and rehabilitative measures by gaining immunity with its antiviral defense through specific and nonspecific mechanism to improve health and protect infection when there is no specific medicine or treatment for this deadly virus as of now. Regular exercise may affect the defense system and its antiviral protection (Walsh et al. 2011; Martin et al., 2009). Exercise therapies are advised as one of the most habitually prescribed remedies in both health and diseases (Vina et al., 2012).

2.2 Exercise Effect on SARC Cov2 Virus

The SARC Cov2 a new generation infectious viral disease resulted severe acute respiratory syndrome that leads to serious immune corrosion within a host. As a matter of severe public health risk, the COVID-19 virus mainly infects the lungs, respiratory tissues, and all major organs that lead to systemic wasting in the human body. The virus can damage directly to the respiratory system by compromising the immune system and finally leads to systematic failure and even death (Mousavizadeh & Ghasemi, 2020). To deal with the initial stage of infection of SARS CoV-2, regular exercise can boost up both innate immune cells (macrophages, neutrophils, dendritic cells, and natural killer cells or NK cells) and adaptive immune cells (T cells: CD4 T cells - Th1 and Th2 cell, CD8 T cells, B cells, antibodies, and cytokines). Exercise with different forms especially the cardiorespiratory exercise may produce huge number of immune cells that moving to attackable areas in the body mainly gut, URTI and epithelial cells of the lungs which are the prime target of SARS-CoV-2 to recognize and kill the virus-infected cells. Moderate intensity exercise has immune encouraging and enhancing impact against viral respiratory infections (Martin et al., 2009) and exercise reduces susceptibility to respiratory infections and make better antiviral immune function (Davis et al., 1997). The host defense mechanism is necessary to boost up by enhancing the body's natural immunity in the population as the number of asymptomatic persons with positive cases (active infection) is increasing day by day. Few data suggested that 74% of COVID-19 asymptomatic cases may be (Emery et al., 2020) with viral loads of similar percentage to those of positive COVID-19 patients (Zou et al., 2020).

2.3 Immune Response to Exercise against SARS-CoV2 Viral Infection

The reaction of immune system to viral infection is the matter of right balance of inflammatory immune response whether move away the virus from the body or the immune health damage caused by the virus. After infection, the SARS-CoV2 virus enters into the body and travels down to the alveoli in the lungs. When the cell becomes infected, macrophage encounters the antigen combining with the MHC and the antigenic peptide is presented on the surface of antigen presenting cells (APCs). The APCs moves to the draining lymph node to present viral antigen to T cells. Viral antigens are recognized by virus specific cytotoxic T cells (CD8). Soon after activating CD4 T cells, it helps to produce and recruit more number of T cells and also stimulates B cells. B cells activate plasma cells that lead to promote the production of virus specific antibody especially IgG and IgM. T cell responses are initiated by antigen presentation through dendritic (DC), natural killer (NK), and macrophage cells and CD8 T cells kill viral infected cells. Macrophage with NK cell releases cytokines, which modulates the humoral and cellular immune response by activating B and T cells. Macrophage is responsible to detect pathogen and to control the right balance of the inflammatory response by making up or down regulation as per requirement. SARS-CoV-2 bind with the DC specific intercellular adhesion molecule 3 non-integrin and related protein and it can phagocyte by DCs and macrophage. If the pro-inflammatory response becomes severe with an imbalanced immune response can lead to hyper-inflammation causing increase viral load and cytokine storm that results to damage the immune health. Inflammatory CD14, CD16 monocytes has high expression for IL6 that accelerate the progression of systemic inflammatory response. On the other hand, regular exercise has the potential to modulate the right immune balance against the normal homeostatic response to achieve the optimum immune health for viral (SARC-CoV2) protection as presented in Fig. 1.

Exercise in regular aerobic mode, probably has the ability of inhibiting chronic inflammation by stimulating anti-inflammatory cytokines and by reducing basal

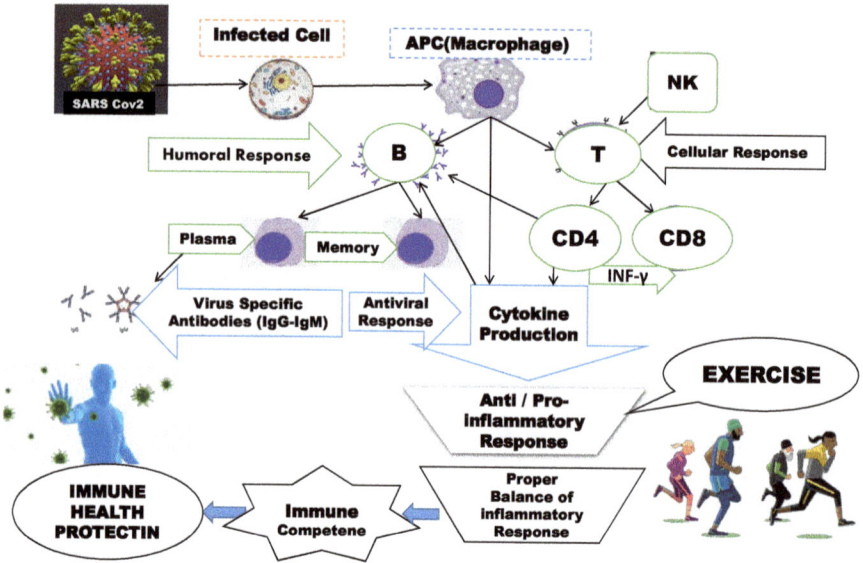

Fig. 1 Exercise induced immune health response to SARC-CoV2

concentration of inflammatory cytokines and the percentage of pro-inflammatory T effector memory CD45+ re-expressing T cells (T-EMRA cells) (Philippe et al., 2019). Exercise activates the autonomic nerves that stimulate the respiratory muscles to increase the rate of breathing. Exercise boosts the uptake amount of oxygen that the working muscles collect from the blood, which probably influence the increase in metabolic enzymes and byproducts. By the elevated CO, opens blood flow to more air sacs (alveoli) in the lungs that causes to increase the ventilation and allows more oxygen to enter into the blood.

2.4 Exercise and Host Immune Resistance

Host defense has a specific and nonspecific immune mechanism to recognize and eliminate the viral infection. In most viral infection, humoral (antibody mediated) and cellular (cell mediated) immune response can be developed specific to the virus over the course of week. Viruses cause infection of the host first by breaking natural protective mechanisms of the body, then evading the immune system of the host, and finally by killing off the host cells and triggering immune and inflammatory responses. The outcome of viral infection depends upon response of the host to infection by a virus and how effectively the host's defense mechanisms resist the offensive tactics of the virus? Exercise may be one of the best methods in sustaining a strong immune resistance to protect health and to prevent getting sick and help fight against infectious virus for recover faster. Physical activity offer protection from viral infections, structured exercise each day may be an effective art of war for optimizing the functional integrity of the immune response to stop or attenuate intense of infection, especially among vulnerable populations (Laddu et al., 2020).

With the survival of infection over the period of a week, humoral and cell-mediated immune response specific to the virus can be developed offering a long lasting immune protection from re-infection and or having mild symptom of disease in case of recovered population. Usually a complex variety of fighting cells has a well-organized army on hand to protect the human body. The mechanisms available to the human organism for defense against viral infection can be classified in two groups as the specific (adaptive) and nonspecific (innate) immune defense, which is affected by exercise when come into play to eliminate an infecting virus. At the early stage, exercise acts on the nonspecific immune defenses with the action of interferon production, inflammation, fever, phagocytosis, and macrophage and NK cell activity. Moderate intensity regular exercise may increase in neutrophil and natural killer (NK) cell counts and enhanced immunoglobulin concentrations (Martin et al., 2009; Harris, 2011).

The prime role of nonspecific immune response is to limit the virus multiplication during the acute phase of viral infection. The specific immune defenses begin at the later stage almost immediately after exposure in 3–10 days which include both the humoral system (antibody-producing B cells) and the cell-mediated or cellular system (T helper cells and cytotoxic T lymphocytes) which play a protective role that function to help eliminating virus and to maintain specific resistance to infection through physical exercise.

Exercise induced humoral defense mechanism blocks the binding of virus, fusion of viral envelope to host cells plasma membrane and enhances phagocytosis of viral particles, thus preventing infection or re-infection. Nonspecific and specific immunity prevent infection and or obstruct the chronic infection whereas the cell-mediated immunity concentrates on finishing the source of infection by eliminating infected cells and make clear of the cellular reservoir of the virus (McCloskey & Heymann, 2020). In humoral immunity, virus-infected cells can stimulate B-lymphocyte to produce antibody that is specific for viral antigen. Antibodies can only attack viruses outside of their host cells, which mean that once an infection is established within an organ it can hardly be further influenced by antibodies, since the viruses spread directly from cell to cell. In principle, the humoral immune system is thus only capable of preventing a generalized infection, but only if the antibodies are present at an early stage (e.g., induced by a vaccination). Class IgG and IgM antibodies exhibit antiviral activity are active in the bloodstream (serum) and class IgA is active on the mucosal surface. The immunity conferred by infection, apparently IgA-dependent, is short lived. Re-infections are therefore frequent, whereby the antigenic variability of the virus may be a contributing factor. The function of antibody is to neutralize virus by recognizing viral antigen and blocking virus-host cell interactions that can lead to antibody-dependent-cytotoxic-cells (ADCC).

In most cases, cellular immunity is more important than humoral immunity. The cellular system is capable of recognizing and destroying virus-infected cells on the surfaces of which viral antigens are expressed. The humoral system can eliminate only extracellular viruses.

2.5   *Exercise and Immune Defense*

Exercise modulates the cell-mediated immune defense, which has a direct antiviral activity that kills virus-infected cells/ self-cells by antibody dependent cell-mediated cytotoxicity. Exercise can help to boost up immunity against the viral infections via a change in Th1/Th2 cell responses (Martin et al., 2009). Exercise has a positive effect on specific (humoral and cellular) and nonspecific immune defense against the viral infection for the development and regulation of immune function as shown in Fig. 2.

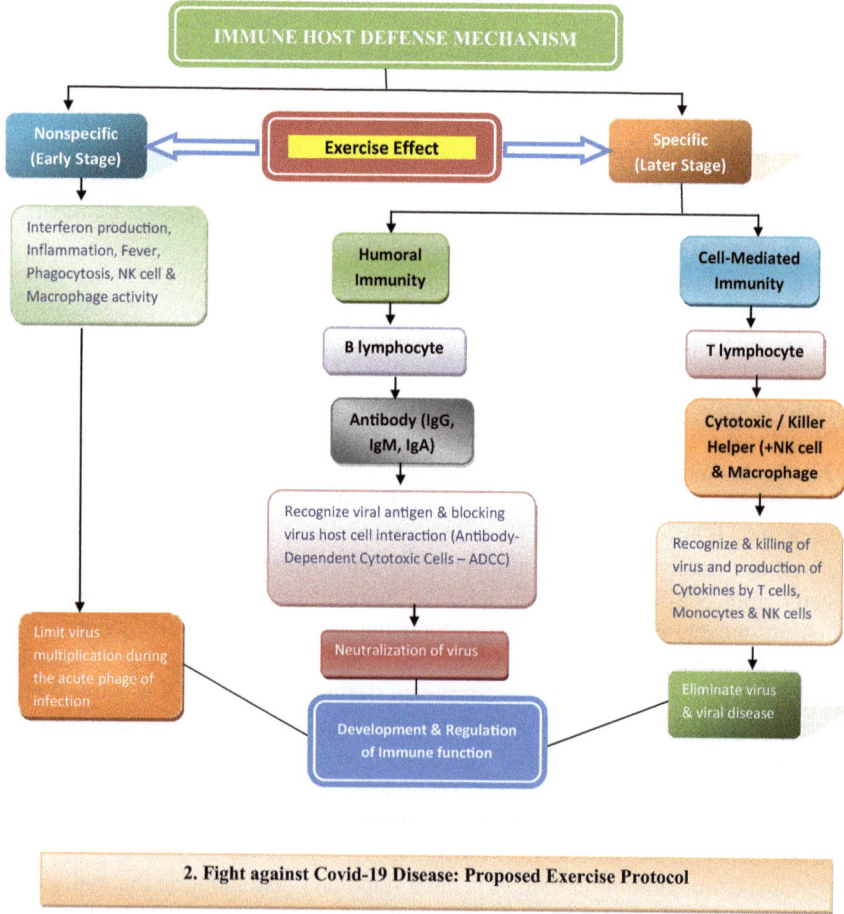

Fig. 2 Exercise effect on virus induced immune host defense mechanism

Cell-mediated immunity is the type of immune defense, which is far more important when it comes to fighting viral infections. It regulates the action of recognizing and killing of virus and production of cytokines. T lymphocyte is activated by antigen that consisting of Killer and Helper T cells and it combined with NK cells and macrophages which is essential in recovery from and control of viral infection. Killer cells recognize virus-infected cells by the viral antigens on their surfaces and destroy them. Finally, cytokines production by monocytes (monokines), T cells, and NK cells (lymphokines) help to encounter the viral agent by developing and regulating antiviral immune function during exercise. The observation that patients with defective humoral immunity generally fare better with virus infections than those with a defective cellular response underlines the fact that the cellular immune defense system is the more important of the two. (Klimpel & Baron, 1996; Fritz et al., 2005).

3 Fight Against Covid-19 Disease: Proposed Exercise Protocol

3.1 Exercise and Immune Function

Exercise is one of the key factors to safe and healthy living and making a healthy habit and with proper amount of daily exercise can provide an overall health benefit to the immune function of an individual. It has been observed that there are mainly four types of exercise such as endurance, strength, flexibility, and balance which can be improved the immune health and reduce the risk of developing disease (ACSM, 2018). The exercise can be practiced in both ways either acute or chronic type based on intensity, volume, and frequency. Acute exercise is expressed with the immediate effect of a single bout of exercise and when a single bout of exercise is performed repeatedly over time it becomes a chronic exercise as defined exercise training (Budde et al., 2016; Walsh et al. 2011). The exercise intensity indicates as low, moderate, or intense load of exercise, which determines the overall physiological response to exercise. The intensity of exercise literally expressed as a percentage of the maximum oxygen uptake capacity (VO_2max) of the individual, which represents the maximum aerobic capacity and high-energy use during exercise. It denotes as the combined capacities of the pulmonary and cardiovascular system to deliver oxygen to muscles of the body (Garber et al., 2011; Hawley et al., 2014). The physiological adaptation of acute and chronic exercise to innate and adaptive immune response is dependent on exercise mode and intensity of exercise during effort and recovery. Acute or chronic exercise have physiological and biochemical responses, which induce alteration in the homeostasis of the body by stimulating the immune cells (Budde et al., 2016; Peake et al., 2017; Kakanis et al., 2010). The physiological and biochemical adaptation for alteration in the homeostasis of the body takes place only in case of moderate intensity exercise and low and intense (below minimum and

Table 1 Moderate intensity acute and chronic exercise response to alteration of immune components for functional improvement of overall health

Moderate exercise and immune components alteration	
Innate (nonspecific)	Adaptive (specific)
↑granulocyte ↑neutrophils ↑NK cells (CD16, CD56) ↑dendritic cells ↑MHC II ↑macrophages ↑ monocytes (CD14, CD16)	↑T cells: CD4+ (Th1), CD8 + (Tc), memory CD45RO+ ↑B cells (CD19, native, memory) ↑IgA ↑↓cytokines (↑IL10, ↑IL1Ra, ↓IL1β, ↓IL6, ↓IL37, ↓IL38, ↓TNF-α, ↓IFN-γ, ↓CRP: Systemic inflammation)
Cellular, molecular and hormonal regulatory mediators of inflammation: Cortisol, catecholamine, growth hormone, prolactin, melatonin, leptin, adiponectin, Resistin, Visfatin, glutamine, specific mRNA, DAMPS	
Functional immune actions: Proliferation of T & B cells, cell activation, differentiation, production of pro-inflammatory cytokines, antibody production, migration properties & trafficking	
Improvement of immune and overall health function	

↑—Higher response/↓—Lower response

above maximum load) exercise intensity does not stimulate adaptation to change in the body homeostasis favoring a healthier lifestyle (Garber et al., 2011). Physical exercises especially regular moderate exercise is known to have a deep impact on the normal functioning of the immune system. Regular moderate exercise provides a gentle, natural means of supporting the immune system on a day-to-day basis and conditioning the lungs and respiratory tract, stimulating the lymphatic system to out toxins from the body to ensure optimal function. Moderate exercise is shown to boost up the functional activity of natural immune components of the human defense system like NK cells, circulating T- and B-lymphocytes, and cells of the monocyte–macrophage system for which probably minimizing the occurrence of some infections (Keast et al., 1988; Pedersen & Ullum, 1994; Woods & Davis, 1994).

Both acute and chronic exercise in short- or long-term period of time results to change the number and function of immune components of innate and adaptive immunity during, immediately after exercise and after recovery of exercise as presented in Table 1.

3.2 Regular Moderate Exercise and Immune Response

The number of studies have focused on how exercise is associated with improved immunity and reduced susceptibility to infection risk compared with sedentary lifestyle. Physical exercise and training may be well established to act as "immunotherapy" favorably representing high level cost-effective measures which can considerably improve the quality of human life (de Araújo et al., 2013). The infection of SARS CoV-2 effects on human body by causing immune health problems where

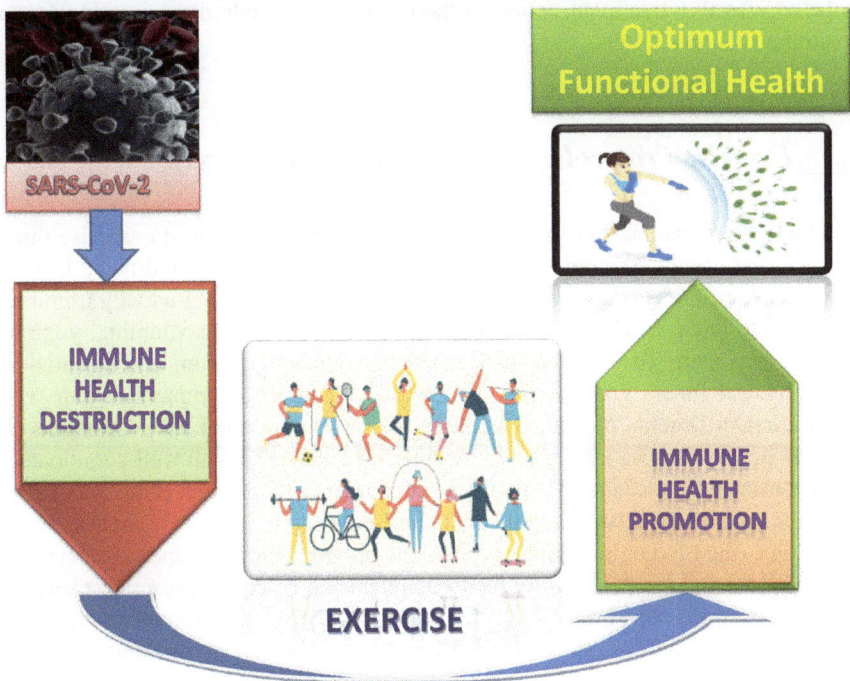

SARS-CoV-2

Optimum
Functional Health

IMMUNE
HEALTH
DESTRUCTION

IMMUNE
HEALTH
PROMOTION

EXERCISE

Fig. 3 Exercise Impact on Immunity for Optimum Functional Health to SARS-CoV-2 infection

exercise can help to promote immune health and achieve the optimum functional health as presented in Fig. 3.

It is well documented that moderate physical exercise can control and manage of some diseases such as diabetes, hypertension, lung disease, cardiac disease, obesity, cancer, etc. which are co morbid conditions in patients with COVID-19. Research data stated that regular exercise causes higher level of physical fitness that improves immune reactions to vaccination, reduces chronic low-grade inflammation, and increase many immune markers in various disease states including CVD, metabolic disorders, cancer, AIDS, and neuropsychological problems (Balchin et al., 2016).

Regular physical activity may boost immune systems and reducing some of the co-morbidities such as obesity, diabetes, hypertension, and heart conditions that make more vulnerable to severe COVID-19 illness (Siordia Jr., 2020). Exercise is useful in vulnerable population such as children, elderly, immune-compromised and sedentary persons to those who causes to suppress immunity that have resulted increasing the rate of morbidity and mortality. Patients who are low/moderate exercise group reduced risk of mortality than who "never/ seldom" exercised (Wong et al., 2008). Exercise reduced the risk of disease-specific and all-cause mortality (Ekelund et al., 2019). Physical exercise is not an alternative to medicines but rather a precautionary measure to keep diseases away. Exercise has been shown to be an effective treatment option in the control of lifestyle disease in long-term condition (Kujala, 2009).

3.3 Exercise Protocol and Specific Model of Exercise

Generally there can be followed the "333" rules which means to do exercise three times a week for a maximum of 30 mins a time and heart rate would be 130 beats/min or higher. It is best to extend the amount of time of exercise gradually from low to moderate intensity of load with jogging, walking, cycling, swimming, yoga or exercise in a gym. To make an ideal moderate exercise program, it is essential to meet physical fitness goals within the context of improving immune health. The general health benefits of regular exercise are widely accepted. Good amounts of daily moderate exercise can be safely maintained / increased with well-coordinated, more consistent, and centralized public health guidance (Constandt et al., 2020). The scientific basis of exercise technique includes frequency (how often), intensity (how hard), time (duration or how long), type (mode or what kind), volume (amount), and progression (advancement). In order to maintain the safety of prescribe exercise, clinicians can use the "frequency, intensity, time, type, volume, and progression" (FIIT-VP) principle (ACSM, 2018). The recommendation of moderate exercise prescription potentially consist of aerobic, resistance, flexibility, and neuro-motor components which to be done as per individual's own capacity to achieve and maintain health and fitness as expressed in Table 2.

At the beginning of the program one must do exercise with less number of repetitions and sets. In case of having any physical or medical problem such as heart disease, arthritis, diabetes, pregnancy, etc. individual must consult an expert or doctor before exercising. Asneurological (dementia), metabolic (obesity), cardiovascular (hypertension), pulmonary (asthma), and musculoskeletal disorders (osteoporosis) and psychiatric (depression) conditions and also for epilepsy, regular exercise is medicine (Arida et al., 2012, 2008; Pedersen & Saltin, 2015; Vancini et al., 2016).

3.4 Fitness Exercise Program During COVID-19 Pandemic

In the present times with the changing world to shift in the "New Normal" of living, it is compulsion to move towards the comprehensive primary health care which needs to encompass the preventive, promotive, curative, rehabilitative, and palliative healthcare with a strong emphasis on fitness and wellness. Regular physical exercise is like a living medicine for the prevention and treatment of chronic diseases and for the overall maintenance of health and well-being (Fletcher et al., 2018). The comprehensive aspects of healthcare have numerous benefits, especially in the times of COVID-19 pandemic when people are advised to stay at home and take steps to remain physically and mentally fit. Maintaining exercise routines during this

Table 2 FITT-VP recommendation of moderate exercise prescription to achieve and maintain health and fitness

1. Aerobic (cardio-respiratory endurance) exercise

Frequency	Intensity	Time	Type	Volume	Pattern	Progression
At least 5 days/week	Moderate (40–59% of HRR or VO2 max) *Light to moderate may be beneficial in de-conditioned individuals	30–60 min/day * <20 min/day can be beneficial for sedentary individuals	Regular, continuous, rhythmic and purposeful exercise involving major muscle groups	≥150 min/week. or pedometer counts of ≥5400–7900 steps/day	One continuous session in one interval or in multiple sessions in bouts of ≥10 min over the course of a day	A gradual progression of volume by adjusting duration, frequency & intensity *Increase exercise duration/session of 5–10 min every 1–2 week. over the first 4–6 week. of an exercise program

Example: Walking, jogging, running, cycling (leg and arm), cross-country, slow dancing, aerobics, rowing, stepping, hiking, downhill skiing, racquet sports, basketball, soccer, etc.

2. Resistance(strength) exercise

Frequency	Intensity	Time	Type	Repetitions/sets	Pattern	Progression
2–3 days/week (alternate days)	Light to moderate <50% 1 RM *20–50% 1 RM for older adults	No specific duration	Major muscle groups (single-joint and multijoint exercises targeting agonist and antagonist group)	8–12 repetitions of 2–4 sets *≥2 sets are effective to improve muscular endurance	2–3 min rest intervals between each set of repetitions *Rest of ≥48 h between session for any single muscle group is recommended	A gradual progression of greater resistance & or more repetitions/ set & or increasing frequency is recommended

(continued)

Table 2 (continued)

Example: Exercise with free weights, machines, stacked weights, resistance bands: Single joint: Biceps/leg curl, triceps/quadriceps extension, calf raises and multijoint: Chest/shoulder/ leg press, pull-down, push-ups, rows, squats, deadlifts, etc. targeting agonist and antagonist group. For core muscles-planks and bridges

3. Flexibility (stretching) exercise

Frequency	Intensity	Time	Type	Volume	Pattern	Progression
≥2– 3 days/ week *daily exercise is most effective	Stretch to the point of feeling tightness or slight discomfort *Seated exercises can be used to start	10–30s hold static stretch and 3–6 s PNF stretch (20–75% of max voluntary contraction) followed by 10–30s assisted stretch	Major muscle tendon unit (static: Active & passive), dynamic, ballistic & PNF stretching	60s of total stretching time for each flexibility exercise (two 30 s or four 15 s stretches)	2–4 repetitions for each exercise *Exercise is effective when muscles are warmed through aerobic activity or passive warm up method	A gradual improvement of range of motion around joints after about 3–4 week of regular stretching at least 2–3 times/ week is recommended

Example: Pilates, yoga, tai chi etc. (ballistic, slow stretch, Slow Stretch & Hold and post Iso-metric stretch method)

4. Neuromotor (coordination, Balance and Agility) exercise

Frequency	Intensity	Time	Type	Volume	Pattern	Progression
≥2– 3 days/ week	Not known	20–30 min/day	Exercise involving motor skills (coordination, balance, agility &gait)	≥60 min/ week	Not known	Not known

Example: Tai chi, yoga, gymnastics, qigong& multi-directional forms of running, jumping & skipping

Components of the exercise: *Warm-up-* 5–10 min; *conditioning-* 20–60 min/day of aerobic, resistance, neuromotor &/or sports activities; *cool-down-* 5–10 min of light aerobic & muscular endurance activities; *stretching-* at least 10 min of stretching exercise

1RM = one repetition maximum, the maximum amount of weight that patient can possibly lift for one repetition
Source: ACSM's Guidelines for Exercise testing and prescription (Tenth Edition: 2018)

pandemic period (Chen et al., 2020; Zhu, 2020), with simple exercise modalities were recommended to counteract the deleterious effects of lockdown and social distancing (Chen et al., 2020).

The fitness exercise program for children, youths, adults, and older adults in both genders need to be focused on developing components of endurance, strength, flexibility, agility, and balance with appropriate manner as presented in Table 3. Daily exercise enhances the immune function in individuals of all ages, and keeps safe them from infections (Campbell & Turner, 2018).

Exercise in any form is recommended during the COVID-19 pandemic due to its multiple benefits on psycho-physiological health. Personalized fitness training according to age, clinical conditions, specific suggestions to address home-based training during this situation are highly appropriated (Dwyer et al., 2020).

3.5 Recommended Exercises for Various Health Conditions During Covid-19

The COVID-19 pandemic has taken the globe in the most horrible situation where increased focus on the need to adopt sustainable practices and get back to the active lifestyle. This is the right time to save the humanity by involving physical exercise as fundamental pre-requisites to develop a strong immune system with healthy body and mind. Doing regular exercise during the COVID-19 outbreak will have significant health benefits to all patients. Exercise should be used to "precondition" patients prior to infection (Ahmed, 2020).The moment developing any symptoms of infection, bring to discontinue exercising and get back to do exercise slowly following recovery for returning to pre-infection fitness as described in Table 4.

A question has been raised to continuing exercise during or following an URTI over the neck including cough, sneezing and sore throat, individual is instructed to jog for 10 mins. Physical exercise must be discontinued until full recovery if the general condition and signs are declined. In case of no change of physical condition following the 10 mins to return, low to moderate physical activity below 80% of VO2 max is recommended. Following an URTI below the neck with the various chronic symptoms, exercise should be prohibited until full recovery (Harris, 2011; Eichner, 1993). Before infection, exercise is recommended to those asymptomatic individuals who tested negative with COVID-19 even once infected during the period of mild symptoms, permitted to resume moderate exercise practice. But those who asymptomatic with positive or active infection should abstain from exercise for at least 2 weeks from the date of positive test. It is reported that patients having COVID-19 symptoms should rest for$\geq$10 days from the beginning of symptoms plus 7 days from symptom resolution' (Hull et al., 2020). To strengthen cardiovascular health aerobic exercise is recommended before infection. After being infected, regular moderate aerobic exercise can help to improve lung ventilation during the period of mild symptoms. Such exercise may benefit immune function as

Table 3 Suggested age appropriate fitness exercise guidelines for active and healthy lifestyle

Age group	Aerobic/ endurance	Strength/ resistance	Flexibility/ stretching	Agility & balance
5–8 years (fundamental movement skills)	Running and jumping, jumping jacks, shuttle run (endurance and speed)	Animal walk: Crab walk, bear walk, Snake crawl, donkey kick	Cat and cow pose, overhead arm stretch, shoulder stretch, butterfly stretch	Walking on heels, crouch forward and backward,Standing like a stork, skipping, throwing & catching, Ziggag running

Fundamental movement skills: Walking, jumping, hopping, skipping, galloping, leaping, sliding, rolling, rotating, crawling, catching, bouncing, trapping, volleying, striking, curling, twisting, pushing, pulling, docking, swinging, balancing.
Minor games: Tag game, Hide and Seek, chain chain, stones, relay race: 3 legged race, lemon race, sack race, ball carry

Age group	Aerobic/ endurance	Strength/ resistance	Flexibility/ stretching	Agility & balance
9–14 years	Spot running (endurance, speed & core), climbing stairs, walking on toes, swimming, jumping jacks, March & Swing of arms	Straight leg raises-Forward & Sideward, push-ups on the wall, long jump, goal keeping	Calf stretch, Child's pose, knee to chest, bend down	Single leg stance, leg swings, walking on lines of different shapes
15–18 years	800 m race, brisk walking, quick air punches, 4 × 100/ 200/400 m relay race, swimming, walking lunges	Curl up, plank, push up, squat, heal raises, mountain climber, Forward & Sideward lunges, superman	Forward bend, calf stretch, Child's pose, knee to chest	Walking on lines of different shapes, single leg stance, leg swings
18–65 years	Spot running (endurance, speed & core), climbing stairs, jumping jacks, March & Swing of arms, brisk walking, quick air punches, swimming, walking lunges	*Lower body:* Straight leg raises-Forward & Sideward, squat; *upper body:* Push-ups on the wall, push up; *Core:* Curl up, plank	Forward bend, calf stretch, Child's pose, knee to chest, bend down	Single leg stance, leg swings, walking on lines of different shapes

(continued)

Table 3 (continued)

Age group	Aerobic/ endurance	Strength/ resistance	Flexibility/ stretching	Agility & balance
65 years and above	Spot running (endurance, speed and core), climbing stairs, walking on toes, jumping jacks, March & Swing of arms, 800 m or longer distance walking/running, brisk walking, swimming	Straight leg raises-Forward &Sideward, push-ups on the wall, squat, hip bridge, Back raises	Shoulder & Upper arm stretch, wall upper body stretch, chest stretch, cross arm stretch (shoulder), doorway stretch (chest & shoul-der), Child's pose (side), forward bend, calf stretch, knee to chest, bend down	Calf raises, seated sit-ups, shifting side to side, zigzag exercise, single leg stance, leg swings, walking on lines of dif-ferent shapes

Warm up exercise: Cat & cow combo, rope skipping, sport jumping, neck rotation, arm rotation, cross body toe touch

Cold down exercise: Neck flexion & extension, neck face to face, neck lateral flexion, sit & reach, quadriceps stretch, hamstring stretch, groin stretch, lumber stretch

Source: Age Appropriate Fitness Protocols and Guidelines, Fit India Mission, 2020, Ministry of Youth Affairs & Sports & Ministry of Health & Family Welfare, MHRD, Govt. of India

well (Medline plus, Exercise and immunity, 2020). To be physically active while social distance is maintained and feels well, exercise can be done during and following the COVID-19 pandemic. Suggestion with 2-m diameter "bubble" for safety movement as found dirty air produced by running or biking needs 5–20 m of spacing for an athlete following directly behind an infected person to stay in clean air (Blocken et al., 2020). Exercisers are advised to wear facemasks during low to moderate level exercise at the time of outside workout like brisk walking, jogging, etc. with maintaining a safe physical distance (2 m away), but use of facemasks may compromise breathing and oxygen uptake in high intensity activity, which are not recommended during these period (Lim & Pranata, 2020).

3.6   Exercise Load, Immune Response, and Infection Risk

Experts suggested a J-shaped connection between physical exercise and respiratory tract viral infection. Moderate intensity exercise boosts immune system and the risk of respiratory track viral infections is reduced whereas high intensity exercise with long periods weakens the immune system and the risk of inspiratory track infection is increased (Ahmadinejad et al., 2014). Moderate intensity exercise is beneficial for the immune system however single bouts of prolonged exercise may results to immune suppression (e.g., impairment of type I and II cytokine balance) in the hours and days following exercise, which can lead to higher infection risk (Simpson et al., 2015). There is a significant relation of exercise load between immune

Table 4 Exercise recommendations in people with various health conditions during COVID-19

Exercise protocol	Health condition		
	Asymptomatic (Healthy)	Mild symptomatic (URTI with limited symptoms)	Severe symptomatic (URTI with high risk)
Exercise intensity	Low to moderate	Low to moderate/ discontinue	Unpermitted
	High intensity exercise should be avoided		
Planning of exercise	Home based exercise or exercise in private environment with good ventilation; use of personal equipment	10-min jog test: If the general condition and signs are deteriorated: Prohibit physical activity until full recovery; if not changed: Allow low to moderate physical activity ($<$ 80% of VO2 max)	Exercise should be prohibited until full recovery
Exercise recommendations to get back in the normal & active lifestyle during COVID 19	Mixture of aerobic and strength exercise adjusting the intensity, volume, frequency and modality based on individual capability	Maintain active lifestyle. Do aerobic mode of exercise if capable (gradually carry on exercise)	Try to perform daily life activities independently if possible (return to exercise after testing)
	Perform multi-component exercise involving endurance, strength, balance & coordination after recovery		

Abbreviations: *URTI* Upper respiratory tract infection
Source: Halabchi et al. (2020) and Lim and Pranata (2020)

response and infection risk. Implementation of an appropriate physical exercise will eventually boost immunity and immune activation (Luzi & Radaelli, 2020). It has been found that inactivity results higher infection risk with suppression of immune response. Similarly, low and heavy exercise load increases the infection risk and causes suppress & impair immune response respectively. However, moderate load of exercise may lower the infection risk and can enhance immune response as given in Fig. 4.

4 Exercise Induced Immune Defense Mechanism of Covid −19

4.1 Exercise and Signaling Pathways of Immune System

Exercise as physical stress may influence in neuroendocrine signals of immune system. There are mainly two pathways that regulate immune function such as the hypothalamic-pituitary-adrenal (HPA) axis and the sympathetic nervous system

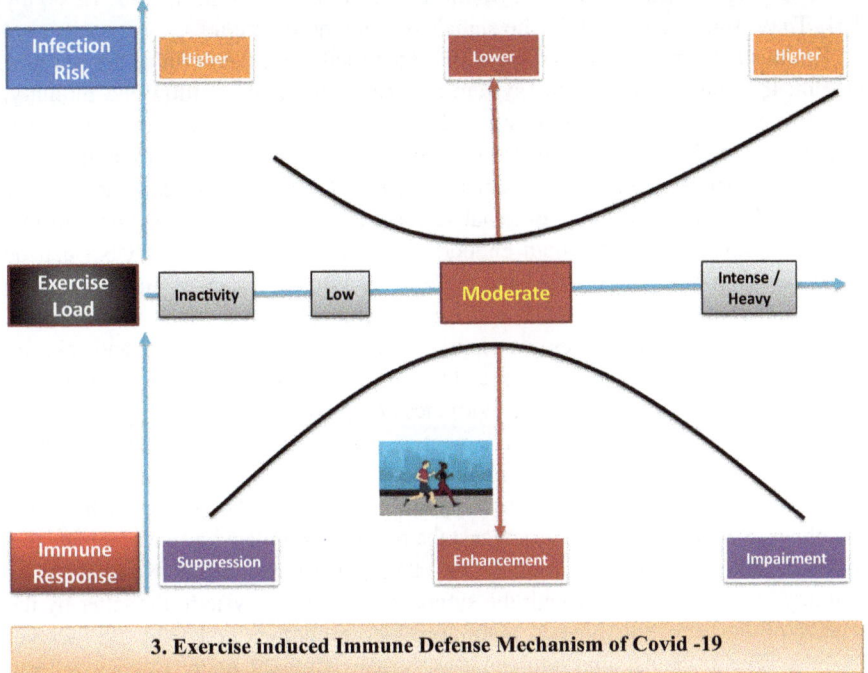

3. Exercise induced Immune Defense Mechanism of Covid -19

Fig. 4 Relationship of exercise load between immune response and infection risk

(SNS). Exercise stress can affect the brain that controls the immune system with the activation of HPA axis. When the HPA axis is stimulated, it helps to secret corticotrophin releasing hormone (CRH) from the hypothalamus. Due to activation of hypothalamus, it stimulates the anterior pituitary gland that results to secret ACTH into the systemic circulation. As secretion of ACTH, adrenal gland is activated for which glucocorticoids or cortisol is being released. Increased level of cortisol has an immunemodulatory effect that lead to regulate immune function. On the other way, stimulation of SNS may raise the secretion of adrenalin and nor-adrenalin (epinephrine and nor-epinephrine) as called catecholamine that can also regulates the immune function. This regulation of immune function acts as anti-inflammatory manner or an inhibition of the signaling pathways that enhance the immune response to viral infection. Furthermore, some other neuroendocrine factors, which induced by exercise stress also, modulate the immune function including prolactin, growth hormone, nerve growth factor (NGF), melatonin, leptin, etc. NGF, which is a neurotrophic hormone, can be increased by stress to regulate the immune response. The functional activity of NGF is elevated with the activation of hypothalamus to the HPA axis that leads to promote proliferation and differentiation of T and B cells. Exercise stress affects on some hormones (prolactin, melatonin, leptin, etc.) (Radogna et al., 2010; La Cava & Matarese, 2004). It acts as pro-inflammatory signals that help to promote immune cell activation, proliferation, and differentiation by producing pro-inflammatory cytokines such as IL-1, IL-12, TNF-α, IFN-γ by Th1. These support to enhance the signals of immune function.

The mechanisms of two pathways link identified from the brain to the immune system: the autonomic nervous system and neuroendocrine outflow via pituitary. Immune response alters neural and endocrine functions, and in turn, neural and endocrinal activity modifying immunologic functions. It has been shown that communication between the central nervous system (CNS) and the cellular immune system is bidirectional that endocrinal factors can alter immune function and that immune response can alter both endocrine and CNS response. The CNS can be involved in immune reactions arising from within the brain or in response to peripheral immune stimuli. Activated immunocompetent cells such as monocyte, lymphocyte and macrophages etc. can cross the blood-brain barrier and take up residence in the brain where they secrete their full repertoire of cytokine and other inflammatory mediators such as leukotrienes and prostaglanbius.

All aspects of immune and complement cascades can occur in the brain because of these nerves – macrophage communication. The CNS modulates immune cells by direct synaptic – like contacts in the brain and at peripheral site such as the lymphocyte organs. Researchers suggest the brain can regulate immunocompetence. Much of this neuroimmuno modulation takes place through the hypothalamic – pituitary system but also through the sympathetic nervous system, the latter by the release of the catecholamine at autonomic nerve endings and from the adrenal medulla. The principal immunoregulatory organs (lympth node, thymas, spleen, and intestinal payers patches) are abundantly supplied by autonomic nerve fibers. Sensory neuron contains a variety of neurotransmitters and neuropeptides that can influence cellular-immunological function. Mechanism underlying the alterations in immunity with exercise related to the activation of the sympathetic nervous system (SNS) which linked to altered activity of the hypothalamic-pituitary-adrenal (HPA) axis that result restores optimal antibody responses including antigen specific cell mediated delayed type hypersensitivity responses.

Moderate intensity exercise can increase the stress hormones, which leads to decrease in excessive inflammation (Harris, 2011). To explore the defense mechanism of immune health, exercise activates the motor cortex, which modulates the HPA axis triggering to release cortisol that can influence the activity of SNS. Activation of SNS increases the secretion of NE (norepinephrine) and by together cortisol with the help of WBC and NE (catecholamine) initiates the production of cytokines.

When the cytokine production takes place, it promotes to enhance the production of antibody and regulating the T cells response that results the development of immune health as presented in Fig. 5.

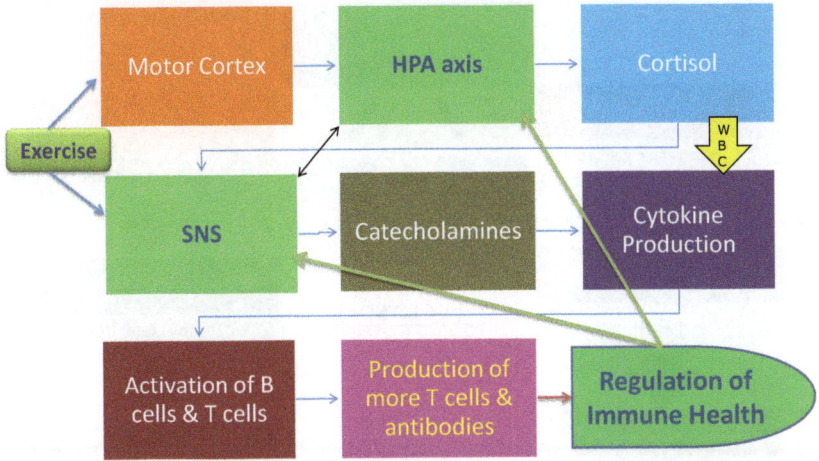

Fig. 5 Exercise induced immune defense mechanism

4.2   Exercise and Modulation of Immune Function

Moderate intensity exercise stimulates in the antipathogen activity of immune system macrophages in connection with temporary rises in the recirculation of key immune system cells, immunoglobins, and anti-inflammatory cytokines in the blood, collectively resulting in a reduced influx of inflammatory cells into the lungs and decreased pathogen load. Regular physical exercise minimizes the risk of systemic inflammation that is a main contributor to lung damage caused by COVID-19 (Sallis et al., 2020). Simultaneously, delicate elevations in stress hormones released from skeletal muscle, notably interleukin-6 (IL-6), is detected during acute bouts of moderate intensity exercise; even though, the pleiotropic nature of IL-6 seems to provide safeguard (versus harm) to immunity via directly suppressing potent inflammatory cytokines [e.g., tumor necrosis factor alpha (TNF-a)] in the lungs, making an anti-inflammatory milieu for several hours' post-exercise. Over time, these transient changes in cell-mediated immunity that happen after each bout of moderate intensity exercise are proposed to contribute to enhancing immune-surveillance against infectious pathogens and defend or diminish symptomatology of infectious diseases (Nieman & Wentz, 2019; Davison et al., 2016).

Exercise strengthens and stimulates hues number of immune cells that are able to recognize and kill virus-infected cells. To ensure a positive response and alteration of immune function against the reaction to the viral infection of SARS-Cov-2, exercise response plays an important role by stimulating the immune cells for which pro-inflammatory cytokines are released. These cytokines act to activates both CNS and SNS, which acts bidirectional to each other. Exercises can effect on CNS that can elevate the activity of the hypothalamus, which may increase the action of corticotrophin releasing factor (CRF). The CRF helps to activate the pituitary

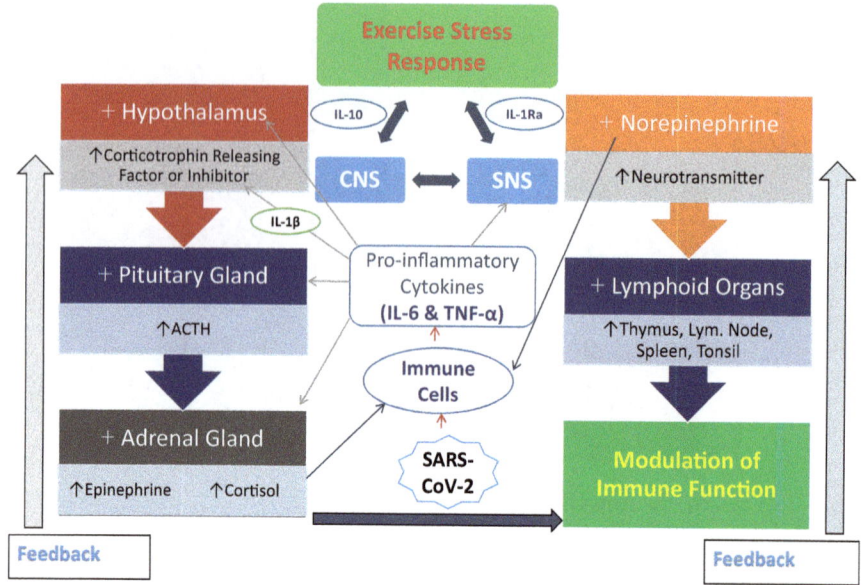

Fig. 6 Potential signaling pathways of exercise stress response to immune function against SARS-CoV-2

gland, which increases the secretion of ACTH. Secretion of ACTH results to trigger the activity of adrenal gland. Activated adrenal gland can raise the secretion of epinephrine and cortisol that lead to promote the alteration of immune function as shown in Fig. 6.

By rising level of cortisol it can enhance the T cell ability that lead to reduce the infection as recognized by NK and CD8 cells and killed by CD4 cells. On the other pathway SNS also can be stimulated by exercise response for which the release of NE is increased. Once the NE release is increased, it appears to produce more number of neurotransmitters, which help to lift up the activity of lymphoid organs by increasing thymus, lymphatic node, spleen, tonsil, etc. These lymphoid organs manage to alter the immune function.

These bidirectional communications of CNS and SNS has a feedback mechanism between exercise responses to immune health alteration. Exercise has important modulatory effects on immune function which are mediated by exercise induced release of pro-inflammatory cytokines, classical stress hormones (glucocorticoids and catecholamines), and hemodynamic effects leading to cell distribution, modification in expression of cell adhesion molecular, recruitment of mature lymphocytes, alteration in apoptosis and mitotic potential. HPA axis increases the efficiency of minaralo corticoid receptors that lead to lower the cortisol level followed by inhibition of cortisol synthesis. This synthesis is the result of efficient negative feedback mechanism, which increases the vasopressin / CRH ratio leading to positive effect on negative feedback by the reduction in pituitary stimulation. Exercise manages to decrease CRH mRNA transcription in the paraventricular muscles of the hypothalamus that diminish the activity in the anterior pituitary which proceed to improve immune response.

To prevent virus by increasing the resistance to infection, moderate exercise may help to increase the number amd percentage of T cells expressing IL-2, INF-gamma, TNF-alpha, IL-6, IL-4, and IL-10 in both stimulated and non-stimulated cells. Exercise also boost immune response through induction of IL-1, IL-8, and TNF-alpha stimulated by E and N and IL-6 responsible for elevated secretion of cortisol during exercise. IL-6 is important to stimulate the production of protective antibodies and T cells that can fight infection. Exercise induced catecholamine can increase the generation of ROS (reactive oxygen species) that stimulate IL-6 which produce antibody. Exercise helps to accumulates several proteins that can help preserve immunity, particularly muscle-derived cytokines like IL-6, IL-7, and IL-15. This has been observed that IL-6 appears to "direct" immune cell trafficking toward areas of infection, whereas IL-7 can improve the production of new T cells from the thymus and IL-15 helps to control the peripheral T cell and NK cell compartments, performing all together can increase the resistance to infection. (Simpson, 2020).

5   Conclusion

1. Resistance to Coronavirus infection is strongly associated with effectiveness of immune system and moderate exercise may enhance the immune efficiency to protect the host against COVID-19 virus.
2. Regular moderate exercise have a significant impact on the normal functioning of immune system and there is a great relation of exercise load between immune response and infection risk of COVID-19.
3. Exercise stress can influence the brain that controls and regulates the immune function to fight against the COVID-19 virus.

Within the limitation of the study, it may be concluded that the influence of regular moderate exercise strategy possibly has the potential to enhance the host immune health defense, which is suggested as a favorable pathophysiological mechanism to foster the normal homeostatic response to defense against corona virus disease (COVID-19) on human health.

6   Recommendations of Future Research

Further experimental research needs to be done to determine the effects of different specialized acute and chronic exercise variation of aerobic, resistance, stretching, and balance exercise based on intensity, volume, and frequency on various immune components of human body to make the targeted exercise prescription that could be used to improve and maintain immune health in the context of COVID-19 disease.

Acknowledgments The authors are thankful to Jadavpur University RUSA 2.0 Major Research Support and University Grant Commission (UGC), Govt. of India. The authors thankfully acknowledge cooperation and help extended by the Office Staff of the Director of Physical Instruction, members of Sports Board, Physical Education department and Jadavpur University authority. The authors are grateful to Prof. Sadhan Kumar Ghosh, Professor & Former Head, Mechanical Engineering Department and Ex-Dean, Faculty of Engineering and Technology, Jadavpur University, Kolkata, India for his expertise and valuable suggestions for smooth conduction of this study.

References

Age Appropriate Fitness Protocols and Guidelines, Fit India Mission. (2020). *Ministry of Youth Affairs & sports & Ministry of Health & family welfare*. MHRD, Govt. of India.

Ahmadinejad, Z., Alijani, N., Mansori, S., & Ziaee, V. (2014). Common sportsrelated infections: A review on clinical pictures, management and time to return to sports. *Asian Journal of Sports Medicine, 5*(1), 1–9. https://doi.org/10.5812/asjsm.34174. [PubMed: 24868426].

Ahmed, I. (2020). COVID-19 – does exercise prescription and maximal oxygen uptake (VO2 max) have a role in risk-stratifying patients? *Clinical Medicine, 20*(3), 28–24.

American College of Sports Medicine. (2018). *ACSM's guidelines for exercise testing and prescription* (10th ed.). Wolters Kluwer.

Arida, R. M., Cavalheiro, E. A., da Silva, A. C., & Scorza, F. A. (2008). Physical activity and epilepsy:Proven and predicted benefits. *Sports Medicine, 38*, 607–615.

Arida, R. M., Cavalheiro, E. A., & Scorza, F. A. (2012). From depressive symptoms to depression inpeople with epilepsy: Contribution of physical exercise to improve this picture. *EpilepsyRes., 99*, 1–13.

Balchin, R., Linde, J., Blackhurst, D., et al. (2016). Sweating away depression? The impact ofintensive exercise on depression. *Journal of Affective Disorders, 200*, 218–221.

Blocken B, Malizia F, van Druenen T, et al. (2020). Towards aerodynamically equivalent COVID19 1.5 m social distancing for walking and running.

Budde, H., Schwarz, R., Velasques, B., Ribeiro, P., Holzweg, M., Machado, S., Brazaitis, M., Staack, F., & Wegner, M. (2016). The need for differentiating between exercise, physical activity, and training. *Autoimmun Rev*, 110–111. https://doi.org/10.1016/j.autrev.2015.09.004

Campbell, J. P., & Turner, J. E. (2018). Debunking the myth of exercise induced immune suppression: Redefining the impact of exercise on immunological health across the lifespan. *Frontiers in Immunology, 9*, 648.

Chen, P., Mao, L., Nassis, G. P., Harmer, P., Ainsworth, B. E., & Li, F. (2020). Wuhan coronavirus (2019-nCoV): The need to maintain regular physical activity while taking precautions. *J. Sport Heal. Sci., 9*, 103–104. https://doi.org/10.1016/j.jshs.2020.02.001

Constandt, B., Thibaut, T., De Bosscher, V., Scheerder, J., Ricour, M., & Willem, A. (2020). Exercising in times of lockdown: An analysis of impact of COVID-19 on levels and patterns of exercise among adults in Belgium. *International Journal of Environmental Research and Public Health, 17*, 4144.

Davis, J. M., Kohut, M. L., Colbert, L. H., et al. (1997). Exercise, alveolar macrophage function, and susceptibility to respiratory infection. *Journal of Applied Physiology, 83*, 1461–1466.

Davison, G., Kehaya, C., & Wyn, J. A. (2016). Nutritional and physical activity interventions toimprove immunity. *American Journal of Lifestyle Medicine, 10*, 152–169.

de Araújo, A. L., Silva, L. C. R., Fernandes, J. R., & Benard, G. (2013). Preventing or reversing immunosenescence: Can exercise be an immunotherapy? *Immunotherapy, 5*, 879–893. https://doi.org/10.2217/imt.13.77

Dwyer, M. J., De Dominicis, S., & Righi, E. (2020). Physical activity: Benefits and challenges during the COVID-19 pandemic. *Scandinavian Journal of Medicine & Science in Sports, 30*, 1291–1294.

Eichner, E. R. (1993). Infection, immunity, and exercise. *The Physician and Sportsmedicine, 21*(1), 125–135. https://doi.org/10.1080/00913847.1993.11710319. [PubMed: 27414832].

Ekelund, U., Tarp, J., Steene-Johannessen, J., et al. (2019). Dose-response associations between accelerometry measured physical activity and sedentary time and all cause mortality: Systematic review and harmonised meta-analysis. *BMJ, 366*, l4570.

Emery, J.C.; Russell, T.W.; Liu, Y.; Hellewell, J.; Pearson, C.A.B.; Knight, G.M.; Eggo, R.M.; Kucharski, A.J.; Funk, S., CMMID 2019-nCoV Working Group; et al. (2020). The Contribution of Asymptomatic Sars-Cov-2 Infections to Transmission—A Model-Based Analysis of the Diamond Princess Outbreak. CMMID Repository. Retrieved August 7, 2020.

Fletcher, G. F., Landolfo, C., Niebauer, J., Ozemek, C., Arena, R., & Lavie, C. J. (2018). Promoting physicalactivity and exercise: JACC health promotion series. *J AmColl Cardiol, 72*, 1622–1639.

Fritz, M. D., Bienz, K. A., Johannes Eckert, D. V. M., Rolf, M., & Zinkernagel, M. D. (2005). *Medical microbiology.* Thieme.

Garber, C. E., Blissmer, B., Deschenes, M. R., Franklin, B. A., Lamonte, M. J., Lee, I. M., Nieman, D. C., & Swain, D. P. (2011). Quantity and quality of exercise for developing and maintaining cardiorespiratory, musculoskeletal, and neuromotor fitness in apparently healthy adults: Guidance for prescribing exercise. *Medicine and Science in Sports and Exercise*, 1334–1359. https://doi.org/10.1249/MSS.0b013e318213fefb

Halabchi, F., Ahmadinejad, Z., & Selk-Ghaffari, M. (2020 March). COVID-19 epidemic: Exercise or not to exercise; that is the question! Asian. *The Journal of Sports Medicine, 11*(1), e102630. https://doi.org/10.5812/asjsm.102630

Harris, M. D. (2011). Infectious disease in athletes. *Current Sports Medicine Reports, 10*(2), 84–89. https://doi.org/10.1249/JSR.0b013e3182142381. [PubMed: 21623289].

Hawley, J. A., Hargreaves, M., Joyner, M. J., & Zierath, J. R. (2014). Integrative biology of exercise. *Cell*, 738–749. https://doi.org/10.1016/j.cell.2014.10.029

Hull, J., Loosemore, M., & Schwellnus, M. (2020). Respiratory health in athletes: Facing the COVID-19 challenge. *Lancet Resp Med.* https://doi.org/10.1016/S2213-2600(20)30175-2

Kakanis, M. W., Peake, J., Brenu, E. W., Simmonds, M., Gray, B., Hooper, S. L., & Marshall-Gradisnik, S. M. (2010). The open window of susceptibility to infection after acute exercise in healthy young male elite athletes. *Exercise Immunology Review, 16*, 119–137. https://doi.org/10.1016/j.jsams.2010.10.642

Keast, D., Cameron, K., & Morton, A. R. (1988). Exercise and the immune response. *Sports Medicine, 5*, 248–267.

Klimpel, G. R., & Baron, S. (Eds.). (1996). *Immune Defenses, medical microbiology* (4th ed.). University of Texas Medical Branch at Galveston.

Kujala, U. (2009). Evidence on the effects of exercise therapy in the treatment of chronic disease. *British Journal of Sports Medicine, 43*, 550–555.

La Cava, A., & Matarese, G. (2004). The weight ofleptin in immunity. *Nature Reviews Immunology, 4*, 371–379.

Laddu, D. R., Lavie, C. J., & Phillips, S. A. (2020). Physical activity for immunity protection: Inoculating populations withhealthy living medicine in preparation for the next pandemic. *Progress in Cardiovascular Diseases*, 3. https://doi.org/10.1016/j.pcad.2020.04.006

Lim, M. A., & Pranata, R. (July, 2020). Sports activities during any pandemic lockdown. *Irish Journal of Medical Science (1971 -).* https://doi.org/10.1007/s11845-020-02300-9

Luzi, L., & Radaelli, M. G. (2020). Influenza and obesity: Its odd relationship and the lessons for COVID-19 pandemic. *Acta Diabetologica, 57*, 759–764. https://doi.org/10.1007/s00592-020-01522-8

Martin, S. A., Pence, B. D., & Woods, J. A. (2009). Exercise and respiratory tract viral infections. *Exerc SportSci Rev., 37*(4), 157–164. https://doi.org/10.1097/JES.0b013e3181b7b57b

McCloskey, B., & Heymann, D. L. (2020). SARS to novel coronavirus - old lessons and new lessons. *Epidemiology and Infection, 148*, e22.

Medline Plus, Exercise and immunity. (2020). Retrieved March 15, 2020, from https://medlineplus.gov/ency/article/007165.htm

Mousavizadeh, L., & Ghasemi, S. (2020). Genotype and phenotype of COVID-19: Their roles in pathogenesis. *Journal of Microbiology, Immunology, and Infection.* https://doi.org/10.1016/j.jmii.2020.03.022. Published online.

Nieman, D. C., & Wentz, L. M. (2019). The compelling link between physical activity and the body'sdefense system. *Journal of Sport and Health Science, 8*, 201–217.

Peake, J. M., Neubauer, O., Walsh, N. P., & Simpson, R. J. (2017). Recovery of the immune system after exercise. *J. Appl. Physiol., 122*, 1077–1087. https://doi.org/10.1152/japplphysiol.00622.2016

Pedersen, B. K., & Saltin, B. (2015). Exercise as medicine - evidence for prescribing exercise astherapy in 26 different chronic diseases. *Scandinavian Journal of Medicine & Science in Sports, 3*, 1–72.

Pedersen, B. K., & Ullum, H. (1994). NK cell response to physical activity: Possible mechanisms of action. *Medicine and Science in Sports and Exercise, 26*, 140–146.

Philippe, M., Gatterer, H., Burtscher, M., Weinberger, B., Keller, M., Grubeck-Loebenstein, B., et al. (2019). Concentric and eccentric endurance exercise reverse hallmarks of T-cell senescence in pre-diabetic subjects. *Frontiers in Physiology, 10*, 684. https://doi.org/10.3389/fphys.2019.00684

Radogna, E., Diederich, M., & Ghibelli, L. (2010). Melatonin: a pleiotropic molecule regulating inflammation. *Biochemical Pharmacology, 80*, 1844–1852.

Sallis, J. F., Adlakha, D., Oyeyemi, A., et al. (2020). An international physical activity and public health research agenda to inform coronavirus disease-2019 policies and practices. *Journal of Sport and Health Science, 9*, 328–334.

Simpson, R., Kunz, H., Agha, N., & Graff, R. (2015). Exercise and the regulation of immune functions 2015. *Progress in Molecular Biology and Translational Science, 135*, 355–380.

Simpson, R. J. (30 March, 2020). *Exercise, immunity and the COVID-19 pandemic.* Accessed Mar 3, 2022, from https://www.acsm.org/blog-detail/acsm-blog/2020/03/30/exercise-immunity-covid-19-pandemic

Siordia, J. A., Jr. (2020). Epidemiology and clinical features of COVID-19: A review of current literature. *Journal of Clinical Virology, 127*, 104357. https://doi.org/10.1016/j.jcv.2020.104357

Vancini, R. L., Andrade, M. S., & de Lira, C. A. (2016). Exercise as medicine for people with epilepsy. *Scandinavian Journal of Medicine & Science in Sports, 26*, 856–857.

Vina, J., Sanchis-Gomar, F., Martinez-Bello, V., et al. (2012). Exercise acts as a drug; thepharmacological benefits of exercise. *British Journal of Pharmacology, 167*(1), 1–12.

Walsh, N. P., Gleeson, M., Shephard, R. J., Gleeson, M., Woods, J. A., Bishop, N. C., Fleshner, M., Green, C., Pedersen, B. K., Hoffman-Goetz, L., Rogers, C. J., Northoff, H., Abbasi, A., & Simon, P. (2011). Position statement part one: immune function and exercise. *Exercise Immunology Review, 17*, 6–63.

Wong, C.-M., Lai, H.-K., Ou, C.-Q., et al. (2008). Is exercise protective against influenza-associated mortality? *PLoS One, 3*, e2018.

Woods, J. A., & Davis, J. M. (1994). Exercise, monocyte/macrophage function, and cancer. *Medicine and Science in Sports and Exercise, 26*, 147–157.

Zhu, W. (2020). Should, and how can, exercise be done during a coronavirus outbreak? An interview withDr. Jeffrey A. Woods. *Journal of Sport and Health Science, 9*, 105–107. https://doi.org/10.1016/j.jshs.2020.01.005

Zou, L., Luan, F., Huang, M., Hong, Z., Yu, J., Kang, M., Yu, J., Kang, M., Song, Y., Xia, J., et al. (2020). SARS-CoV-2 viral load in upper respiratory specimens of infected patients. *The New England Journal of Medicine, 382*, 12.

SARS-CoV-2 Variants and Third Wave: Impact on Social, Economic, Health, Educational, and Waste Management Aspects

Sadhan Kumar Ghosh, Sutripta Sarkar, Sannidhya Kumar Ghosh, Kaniska Sarkar, Aida Ben Hassen Trabelsi, Arisman, S. M. Tariqul Islam, Shafiul Islam, Mst. Farzana Rahman Zuthi, Tusar Kanti Roy, Natela Dzebisashvili, Nugzari Buachidze, Maruful Hasan Mazumder, Sharmin Sultana, and Vladimir Maryev

1 Introduction

The onslaught of COVID-19 shows no signs of relenting. The pandemic of COVID-19, with the primary variant "Alpha" of the virus (WHO classified), causing the pandemic globally, has so far evolved significantly in genetic mutations since

S. K. Ghosh (✉) · K. Sarkar
Jadavpur University, Kolkata, India

S. Sarkar
Barrackpore Rastraguru Surendranath College (affiliated to West Bengal State University), Kolkata, India

S. K. Ghosh
University of Colorado, Boulder, CO, USA

A. B. H. Trabelsi
Centre de Recherches et des Technologies de l'Energie (CRTEn), Hammam Lif-Tunisi, Tunisia

Arisman
Centre for South East Asian Studies, Banten, Indonesia

S. M. T. Islam · T. K. Roy
Khulna University of Engineering and Technology (KUET), Khulna, Bangladesh

S. Islam · S. Sultana
Department of Public Administration, Rajshahi University, Rajshahi, Bangladesh

M. F. R. Zuthi · M. H. Mazumder
Chittagong University of Engineering and Technology (CUET), Chattogram, Bangladesh

N. Dzebisashvili · N. Buachidze
Georgian Technological University, Tbilisi, Georgia

V. Maryev
Russian Environmental Operator, Government of Russian Federation, Moscow, Russia

December 2019 initiated at Wuhan, China. A destructive initiation ever experienced in the last 100 years in the globe. It continues to rage throughout the world and the whole world is affected by the economy, health of people, business, administration, and many other aspects causing loss of thousands of people. Most of the countries, including India are currently witnessing the third wave of COVID-19 mutated variants. According to World Health Organization, the current wave of infections is driven by two variants of the SARS-CoV-2, Delta and Omicron. Omicron (B.1.1.529) was first detected in Botswana, South Africa in November 2021 and quickly spread across the globe as it is fast spreading. The world is witnessing a huge wave of infection by the Omicron variant of SARSCoV2. Estimates based on the Institute for Health Metrics and Evaluation (IHME) model show that as of January 17, 2022, there were 125 million Omicron infections per day worldwide, with a peak delta wave in April 2021. It is equivalent to ten times or more. The waves of Omicron have mercilessly reached all continents, and in Eastern Europe, North Africa, Southeast Asia and Oceania, very few countries have not yet started the waves of this SARSCoV2 variant.

The global growth is expected to decelerate markedly to 4.1% in 2022 after rebounding to an estimated 5.5% in 2021. These are due to the continued COVID-19 flare-ups, diminished fiscal support, and lingering supply bottlenecks. The advanced economies in part of Europe, Japan, Rep. of Korea, USA and a few others are projected to return to pre-pandemic trends next year, in emerging market and developing economies (EMDEs), specifically in small states and fragile and conflict-afflicted countries. These countries will remain markedly below, as the vaccination rates are low, due to tighter fiscal and monetary policies, and more persistent scarring from the pandemic as per the world bank flagship report January 2022.

The first two waves of COVID-19 have numerous losses of lives and creating immense pressure on the healthcare systems. In addition, prolonged downtime in economic activities, increased rate of unemployment, irrecoverable disruption in the educational system, and environment through inadequate waste management practice are some of the major challenges. The people have lived through series of lockdowns when their normal business activities were largely restricted.

The world has thought to have survived the disasters of the first two waves of COVID-19 with its vaccine becoming available in 2021 and also being effective at protecting severe illness caused by the Delta and previous other variants of concern. However, the third wave of the pandemic has already begun in many countries around the world due to the ominous faster rate of COVID-19 transmission likely due to the emergence of latest Omicron variant. Preliminary assessment data shows that the transmissibility of Omicron is about two to three times faster than the Delta variant. The Omicron variant has also raised major concern among the researchers for evaluating vaccine effectiveness against it since the Omicron variant has significantly mutated in crucial parts of its original genome.

Although the nations around the world have taken initiatives to ease lockdowns to resume normal operations in international trade, business, and socio-economic activities, the future trends and challenges of COVID-19 pandemic seem to be very uncertain especially due to the outbreak caused by Omicron. It will be challenging to further impose preventive measures such as lockdowns risking global economic recession. In these backdrops, this article presents an overview of the current situation of COVID-19 and challenges posed by the pandemic in social, economic, health, educational, and waste management aspects.

Like the previous two waves, the third wave has affected all aspects of public life and governance. Governments have refrained from complete lockdowns due to the severe inconvenience and economic losses to general public and countries as a whole. Medical and health workers having maximum exposure are the most affected. Financial as well as educational institutions have partially or wholly shut down due to the high rates of infections. The pandemic has severely affected the marginalized people who are struggling for survival (UNICEF Report, 2021). This may be noted that the waves have been defined differently in different countries. While India, Bangladesh, ASEAN Countries, Georgia, USA, UK, Europe, Russia, and many other countries are referring the present status as third wave, Tunisia and a few other African countries referring to this as fifth wave. Since the outbreak of COVID 19 in December 2019, Hong Kong, China, experienced four waves. Hong Kong had recorded 10,453 reverse transcription PCR (RT-PCR) confirmed cases by February 1, 2021. Most of those occurred during the last second waves. The third wave occurred during June to early September 2020 and was caused by a single introduction of GISAID (https://platform.gisaid.org) clade GR virus. In early November 2020 the fourth wave began caused by a newly introduced GISAID clade GH SARS-CoV-2 (Siu et al., 2021) (Daniel et al., 2021).

Southeast Asian countries have weathered the first year of the pandemic relatively unscathed, but due to variants of SARSCoV2, application of public health patches, and delays in vaccine deployment, in ASEAN countries, namely, Vietnam, Malaysia, Myanmar, Thailand, and Indonesia, a large outbreak has occurred in many countries (Fig. 1).

Testing using PCR method to detect COVID-19 in Georgia started on January 30, 2020. Coronavirus-infected person was first identified in Georgia on February 26, 2020[1]. After the first case, the Minister of Health of Georgia held a special briefing for the public. The disease was diagnosed by a study conducted in Lugar's laboratory. On March 21, 2020, the President of Georgia declared a state of emergency in the country (see Footnote 1). On March 23, 2020, the Prime Minister of Georgia confirmed the case of the first internal transfer in Marneuli Municipality, which was followed by the announcement of a strict local quarantine regime[2]. The crisis caused by the new coronavirus (COVID-19) has posed new challenges to Georgia, like the rest of the world. The Government of Georgia is actively working

---

[1] https://ka.wikipedia.org/wiki/%E1%83%99%E1%83%9D%E1%83%A0%E1%83%9D%E1%83%9C%E1%83%90%E1%83%95%E1%83%98%E1%83%A0%E1%83%A3%E1%83%A1%E1%83%98%E1%1

[2] https://ka.wikipedia.org/wiki/%E1%83%99%E1%83%9D%E1%83%A0%E1%83%9D%E1%83%9C%E1%83%90%E1%83%95%E1%83%98%E1%83%A0%E1%83%A3%E1%83%A1%E1%83%98%E1%4

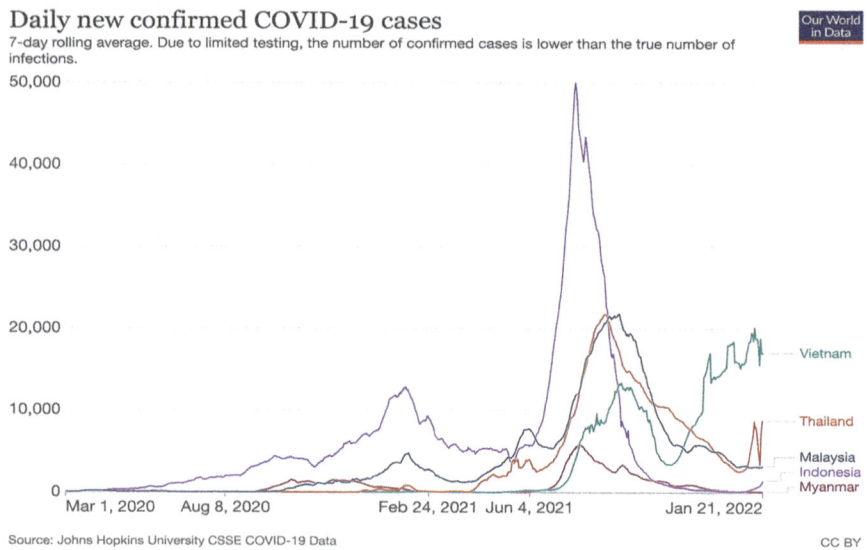

Daily new confirmed COVID-19 cases

7-day rolling average. Due to limited testing, the number of confirmed cases is lower than the true number of infections.

Source: Johns Hopkins University CSSE COVID-19 Data

Fig. 1 Confirmed COVID-19 cases in ASEAN countries during March 2021 to January 2022

to prevent the spread of the new coronavirus in Georgia. The first government decree "Measures to prevent the possible spread of the new coronavirus in Georgia and the plan for operational response to cases caused by the new coronavirus" was approved on January 28, 2020, after which the relevant resolutions and regulations were adopted[3], various services, call centres, and online portals have been set up.

In order to make a data and to prevent the spreading of the infection in Georgia, free antigen-based testing is carried out by Ag-RDTs test systems, at the time of application of any citizen of Georgia, in all medical institutions of the country. Testing using the PCR1 method is free of charge in case of complaints and doctor referral. However, in the case of infection both during the pandemic and currently in the weak detection of each patient's disease, he is supervised by a family doctor, with the help of further necessary studies. In addition, the state offers such patients the use of coveted hotels for quarantined medical care throughout the simplified procedure. And patients with moderate to severe symptoms are served inpatient.

As per the New York Times dated 25th January 2022, the New York City schools will shorten isolation to 5 days for most students who test positive. In Europe, some countries are lifting restrictions and others are adding them. Researchers identify biological factors that may increase a person's chances of having long COVID. With record virus numbers and Omicron on the rise, South Korea tries a new testing plan. Omicron's spread underscores the potential consequences of the global vaccine gap, experts say. New York State's mask policy is back in effect after a judge grants a stay. In the UK, the prime minister announces "No Mask" policy recently (https://

---

[3] http://matsne.gov.ge

www.nytimes.com/interactive/2021/us/covid-cases.html). The prime minister of India announced that there will be no lockdown in the country. These are a few encouraging initiatives in many countries.

2 SARS-CoV-2 Variants: An Overview

Of all viruses, SARS-CoV-2 virus causes COVID-19 and over times the virus is transformed into different variants. WHO experts and scientists group have named SARS-CoV-2 variants under Variants of Concern (VoC) as Alpha, Beta, Gamma, Delta, and Omicron; and under Variants of Interest (VoI) as Lamda and Mu (World Health Organization, 2022a, 2022b, 2022c, 2022d, 2022e).

Viruses keep evolving and change in their genome are incorporated during replication. A single change in the virus genetic material is known as mutation and a variant may contain one or several of these mutations (https://www.cdc.gov/coronavirus/2019-ncov/variants/variant-classifications.html). All variants share a common lineage, i.e., they have a common ancestor. Centre for Disease Control and Prevention (USA) has classified the variants in accordance with their impact on public health, viz. Variants Being monitored (VBM), Variants of Interest (VOI), Variants of Concern (VOC), and Variants of High Consequence (VOHC).

A Virus Evolution Working Group of the World Health Organization has devised a naming scheme for a uniform and clear communication regarding Variants of Concern (VOC) and Variants of Interest (VOI) (Konings et al., 2021). WHO label, area, and dates with earliest documented samples to designated dates of the variants are given below (World Health Organization, 2022a, 2022b, 2022c, 2022d, 2022e).

- Alpha (B.1.1.7 and Q lineages) (UK, September-2020 to 18-Dec-2020),
- Beta (B.1.351 and descendent lineages) (South Africa, May-2020 to 18-Dec-2020),
- Gamma (P.1 and descendent lineages) (Brazil, Nov-2020 to 11-Jan-2021),
- Epsilon (B.1.427 and B.1.429)
- Eta (B.1.525)
- Iota (B.1.526)
- Kappa (B.1.617.1)
- 1.617.3
- Lamda (Peru, Dec-2020 to 14-Jun-2021)
- Mu (B.1.621, B.1.621.1) Mu (Colombia, Jan-2021 to 30-Aug-2021)
- Zeta (P.2)
- Delta (B.1.617.2 and AY lineages) (India, Oct-2020 to VoI: 4-Apr-2021 and VoC: 11-May-2021),
- Omicron (B.1.1.529 and BA lineages) (Multiple countries, Nov-2021 to VUM: 24-Nov-2021 and VoC: 26-Nov-2021).

While the second wave was mostly driven by the Delta variant, experts are of the opinion that the third wave of infections has been caused by Omicron (https://www.

thehindu.com/sci-tech/science/omicron-a-deviant-from-covid-19-pandemic-progres sion-script-virologist-t-jacob-john/article38279889.ece dated 17/1/2022).

Bangladesh reported its first COVID-19 case on eighth March 2020 and confirmed first death from COVID-19 on 18th March (IEDCR, 2020). As of 22 January 2022, Bangladesh reported 1,664,616 confirmed cases of COVID-19 across 64 districts, including 28,192 confirmed deaths (IEDCR, 2020). In Bangladesh, COVID first wave was started in June–July 2020. COVID patients increased at 4% in the peak period of the week from 26 June to 2 July 2020, where maximum 4019 COVID patients were detected on 2 July. The second wave of COVID started in March 2021 by Delta variant. It took more than a month to rise the detection from 1000 to 7000, where 7600 patients were detected on 7 April 2021. COVID detection was found highest at 22% in the week from 19 July to 28 July 2021, where maximum detection was 16, 230 on July 28. With the emergence of the Omicron variant, governments in Southeast Asian countries have returned to a trial-and-error strategy to handle it in more effective ways. They doubled the border restrictions. Even in the third year of the COVID-19 pandemic, the borders of the region remain severely restricted. However, many experts are wondering if it is an effective strategy to stay close to contain COVID-19 to tourists, students, and business travellers now that Omicron has taken over. The COVID-19 strategy of ASEAN countries (border closure, quarantine requirements, popular mask-wearing culture coupled with vaccination campaigns) worked brilliantly.

At the end of May 2021, a Delta Strain, the so-called Indian B.1.617.2 was identified in the country, strains with the ability to propagate even faster through a series of new mutations in the S gene sequence (L452R, D614G, P681R, $\pm$ (E484Q, Q107H, T19R, del 157/158, T478K, D950N))[4]. Figure 2 shows the trends of COVID-19 Cumulative and daily number of confirmed cases in Georgia from 26.02.2020 to 1.10.2021. The first case of Omicron strain was detected in Georgia on December 2020. On the fourth day after the confirmation of two cases of Omicron new strain of coronavirus in Georgia, 24 people have been infected with this strain in the country.

According to healthcare experts, the Omicron option will become dominant in Georgia in the next 2 months, January and February 2022 (https://www. radiotavisupleba.ge/a/31652296.html). As per the information published by the Georgian National Centre for Disease Control, as of January 13, 1009 cases of coronavirus infection with the "Omicron" variant have been laboratory confirmed in the country. To date, laboratory tests have identified 951. The director of the National Centre for Disease Control said on January 11 that the circulation rate of Omicron strains in Georgia is around 45% in all positive cases. "This figure is 60% in the capital and 30% in the regions (https://euronewsgeorgia.com/2022/01/11/ pandemiis-meekvse-talga/).

---

[4]http://test.ncdc.ge/Handlers/GetFile.ashx?ID=07ad44ba-95c0-4a9b-9682-f19ded67d51a

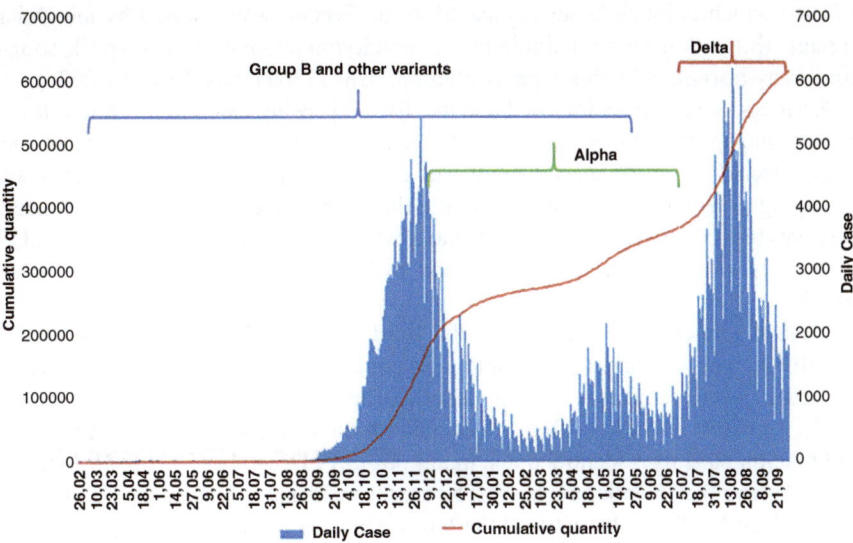

Fig. 2 COVID-19 Cumulative and daily number of confirmed cases Georgia, 26.02.2020–1.10.2021 (https://euronewsgeorgia.com/2021/12/22/omicron-sakartveloshi/)

3 The Appearance of Third Wave

The pandemic of COVID-19, with the primary variant "Alpha" and all other variants of the viruses causing the pandemic globally, has so far evolved 352,796,704 confirmed cases, including 5,600,434 deaths, reported to WHO with a total of confirmed cases of 40 million cases in India. African countries have also been affected so much in the pandemic COVID-19 with 7,908,943 confirmed cases and 161,875 deaths as up to January 23, 2022 as per WHO data. As per the WHO Corona Virus dashboard, the present situation as on January 25, 2022, the confirmed cases are 132.8 million in Europe, 128.9 million in the USA, 50.5 million in South East Asia, 18.4 million in Eastern Mediterranean, 14.5 in Western Pacific, and 8.0 million in African Countries.

India currently ranks second among all the countries in the number of COVID cases being reported daily (https://covid19.who.int/table). From the end of December 2021 and January 2022 there has been an upsurge in the number of COVID cases in most of the Indian states. Experts believe that India is already into the third wave and up to 90% of infections are driven by the Omicron variant, highly contagious one. This has also been relieved by the genome sequencing data of the samples collected mostly from the metropolitan cities in India (https:// economictimes.indiatimes.com/news/india/omicron-now-accounts-for-over-90-of-cases-in-indian-metros/articleshow/89029077.cms dated 21/1/2022). According to the Chairman of the COVID-19 working group of the National Technical Advisory Group on Immunization, the overall hospitalization rate is around 1–2% of those infected which is much lesser compared to the Second wave caused by the Delta variant (https://timesofindia.indiatimes.com/city/pune/omicron-wave-peak-soon-but-it-may-also-quickly-ebb-experts/articleshow/88783694.cms dated 9/1/2022).

Restrictions have been imposed and the disease is being controlled at micro-level by introducing containment zones in affected areas. Most of the Indian States and other affected countries around the world are refraining from complete lockdown and trying to keep the economic activities going. Government of India has mandated the states to increase testing and isolate those who are infected. Instructions have also been given to avoid mass gatherings and to Work from home (WFH) as far as possible.

Though Omicron is dominant in the third wave of COVID-19, most of the COVID patients are affected by Delta variant. It implies that Delta and Omicron are active at the same time in Bangladesh (The Daily Prothom Alo, 20 January 2022). Omicron is detected first in Bangladesh on 19 December 2021. A total of 55 Omicron patients is identified in a month, from 19 December 2021 to 19 January 2022. Among the 55 omicron patients, 52 are the residents of Dhaka, the Capital city of Bangladesh. Dhaka district is at the top of the COVID risk, where the detection rate is 28.11% and Rangamati district is at second position, where the detection rate is 10.71%. The third wave of COVID-19 starts in Bangladesh mainly in the second week of January 2022. Detection is recorded at 183% during the week from 13 January to 19 January 2022. The rate of COVID-19 detection in mid-December 2021 was only 1%, which has raised at 25% on 19 January 2022, just after a month. Bangladesh reported 232% (24,011 vs 7234 and 14.6 new cases per 100,000) increase of COVID-19 new cases in the week from 10 to 16 January 2022 (World Health Organization, 2022a, 2022b, 2022c, 2022d, 2022e). Omicron variant takes the place of Delta variant of COVID-19. About 73% Omicron patients have runny nose, 68% have headache, 64% have fatigue, 60% have sneezing, 60% have sore throat, and 44% have cough (The Daily Star, January 23, 2022). The number of patients at hospitals over the country increases significantly, and this trend marks uprising.

Indonesia is the most populous country among ASEAN members with 272 million people. Despite rising cases Indonesia has started new normal and put less restriction in public at the end of June 2020 by allowing mall, workplaces to open. In the middle of September Jakarta as one of COVID hotspot in Indonesia starts to go back to adopt large social restriction as the cases continue to rise, while other local regions remain in new normal situation.

When the first case was confirmed on January 25, 2020, Malaysia with a population of approximately 31 million was added to the list of coronavirus-infected countries. Malaysia may have kept the cases identified prior to the sudden outbreak low by a religious group rally attended by 16,000 people at the end of February 2020. To flatten the pandemic. The main measures are as follows. On March 13, 2020, the Government banned all rallies, including international conferences, sporting events, social and religious rallies, until April 30, 2020. On March 18, 2020, the government issued an Exercise Management Order (MCO) by March 31 to address the outbreak of COVID-19 under the Infectious Disease Prevention and

Control Act of 1988 and the Police Act of 1967. The Malaysian government has launched a series of stimulus measures to reduce the impact of COVID-19 on sectors and communities.

Myanmar's population is currently around 54,336,457. The first confirmed case in Myanmar was March 24, 2020. Although the number of confirmed cases is relatively small compared to other ASEAN countries, there are concerns about more cases due to the slow spread of tests in Japan. The United Nations has announced plans to donate 50,000 test kits to Myanmar, complementing the previous 3000 donations from Singapore and 5000 from South Korea. The country is fragile and Myanmar's public health system is terribly inadequate to accommodate the scale of the pandemic. In addition, Myanmar does not have a safety net, making the poor the most vulnerable group during a pandemic-induced health and economic crisis. The blockage will affect the country's livelihood and food security.

Thailand, with a population of about 70 million, was the first country to be infected with COVID-19 on January 13, 2020, outside of ASEAN and China. The Thai government has announced a partial blockade to curb the spread of the virus. The situation in Thailand has improved as the number of new cases has decreased and new cases have not been imported due to the almost total ban on flights from early April. There are criticisms of the lack of testing and the suspiciously small number of cases. Thailand confirmed about 2000 cases.

Vietnam shares a long and busy border with China, but the pandemic is still under government control and employs a variety of swift responses. With a total population of 97,338,597, the first COVID-19 cases were identified on January 23, 2020. As of September 27, the Vietnamese government has confirmed a total of 1069 cases. Of these, 999 recovered and 35 died from a pandemic. The Vietnamese government has shown a swift and proactive response in the fight against such an unprecedented illness. At the beginning of February, Vietnam became the first country after China to limit the negative effects of the COVID-19 pandemic, expelling large residential areas to quarantine zones. By isolating the infected and tracking their contact, you can completely block the communities and villages that have been at risk of this pandemic due to their close relationship with the infected.

The appearance of the third wave in the world is due to the spread of the omicron. The prevalence of the strain in Georgia, as already mentioned, is fixed after December 20, 2021. The peak of Omicron strain infection is expected to be in early February. According to the forecast of the National Centre for Disease Control and Public Health, it is possible to have up to 60 thousand cases per day (15 cases per 1000 population), of which 20–30% will be detected in the laboratory, i.e., somewhere between 10 and 15 thousand cases will be laboratory confirmed daily in Georgia (https://www.matsne.gov.ge/ka/document/view/5084798?publication=1). Given this forecast from the National Centre for Disease Control, this wave of pandemics in Georgia, which has already begun, will be the most difficult. Internal distribution of "Omicron" has already started in Georgia. By mid-January 2022, 45% of COVID-19 positive citizens across the country were infected with the Omicron strain. In the capital alone, the figure is 60%, and in the regions—30% (https://www.matsne.gov.ge/ka/document/view/5084798?publication=1).

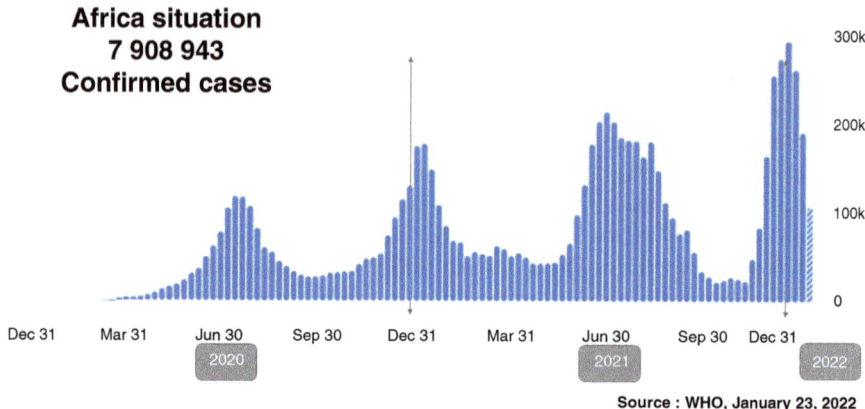

Fig. 3 Status of Confirmed cases of COVID 19 affected people in African countries

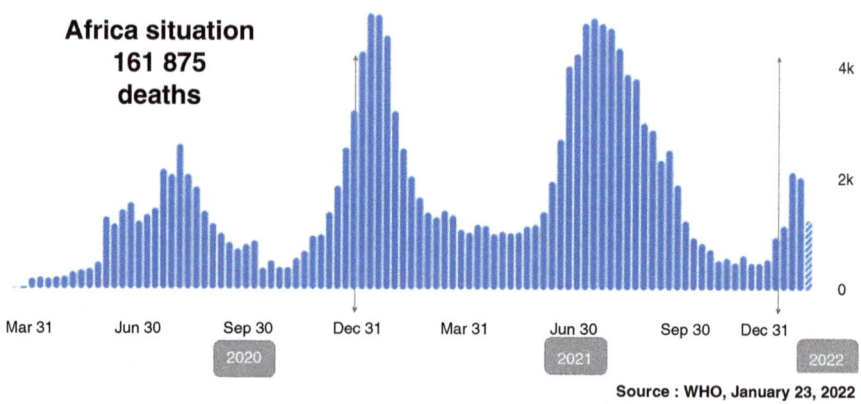

Fig. 4 Status of death of COVID 19 affected people in African countries

African countries have also been affected so much in the pandemic COVID-19 with 7,908,943 confirmed cases and 161,875 deaths as up to January 23, 2022 as per WHO data (Figs. 3 and 4). As for most countries, the global COVID-19 pandemic has deeply affected Tunisian economy, social activities, health systems, and young people education. There were five waves evolved in Tunisia. Since January 2020, date for the declaration of the COVID-19 as a public health emergency by the World Health Organization (WHO) and until the third week of January 2022, more than 300 million cases of COVID-19 have been reported in the world with more than five million of confirmed death (WHO, January 22, 2022). By the end of 2021 and in the first weeks of 2022, a new wave of infections COVID-19 took place with millions of confirmed cases of COVID-19 every day. The African countries accounted for the lowest confirmed cases (around eight million, close to 2% of the total cases reported) by the WHO on January 22, 2022, with 161,875 confirmed deaths (around 3% of the total death in the world). In Tunisia, from 3 January 2020 to 21 January 2022, there have been 817,051 confirmed cases of COVID-19 with 25,881 deaths, as published by WHO in January 22, 2022. In Tunisia, until January 15, 2022, a total of 14,097,156 vaccine doses have been administered with 8,647,792 persons vaccinated with at least one dose and 6,107,788 persons fully vaccinated and with an average of 119,28 total dose administrated per 100 population (WHO, January 22, 2022). The first vaccination start date in Tunisia was 13 March 2021.

The COVID-19 disease reached Tunisia on 2 March 2020 (reference https://news.trust.org/item/20200302144359-otpd5/, accessed January 25, 2022). Thus, the first wave in Tunisia started in March 2020. The second wave in Tunisia started in July 24, 2020 (Reference COVID-19: 9 nouveaux cas testés positifs dont 4 cas de contamination locale". *lapresse.tn* (in French). 17 July 2020. Retrieved 17 July 2020, accessed January 25, 2022). The third wave in Tunisia started in March 2021. On 2 March, the first case of lineage B.1.1.7 (the "UK variant") was reported in Tunisia (Tunisia records first cases of UK variant—as it happened 2 March 2021 *www.theguardian.com*, accessed 3 March 2021, accessed January 25, 2022). The fourth wave in Tunisia started July 2021 Variant Delta (reference Fourth wave of Covid-19 hits Middle East, WHO sounds alarm over delta variant, https://www.hindustantimes.com/world-news/fourth-wave-of-covid-19-hits-middle-east-who-sounds-alarm-over-delta-variant-101627612069509.html, Jul 30, 2021, accessed January 25, 2022).

The fifth wave in Tunisia started December 2021–January 2022 Variant Omicron. Tunisia's first case of the Omicron variant was reported on 3 December 2021 (reference Coronavirus Cinquième vague: Les institutions sanitaires sur le qui-vive" *lapresse.tn* (in French). 10 January 2022, accessed January 25, 2022).

As reported to WHO, the data of confirmed cases of COVID-19 from 3 January 2020 to 21 January 2022 in Libya and Senegal are 401,444 and 83,752, respectively, while the number of deaths that occurred in Libya and Senegal are 5889 and 1917, respectively. A total of 2929 961 and 1,923,773 vaccine doses have been administered in Libya and Senegal, respectively, as of 6 January 2022. In South Africa, from 3 January 2020 to 21 January 2022, there have been 3,572,860 confirmed cases of COVID-19 with 93,846 deaths, reported to WHO. As of 16 January 2022, a total of 28,903,010 vaccine doses have been administered. In Nigeria, from 3 January 2020 to 21 January 2022, there have been 251,694 confirmed cases of COVID-19 with 3123 deaths, reported to WHO. As of 18 January 2022, a total of 18,567,498 vaccine doses have been administered.

In 2020, with the beginning of the COVID-19 pandemic, Tunisia, as most countries, adopted good practices in order to limit the human-to-human spread such as general lockdown, partial lockdown, non-essential travel restrictions, and many additional restrictions for specific sectors, partial or full school closures, isolation in dedicated isolation centres, self-isolation for 14 days for all travellers arriving in Tunisia, etc. Besides, during the pandemic, Tunisia follows the guidelines and recommendations of the WHO to reduce the spread of COVID-19 such as maintaining social distancing, wearing medical masks, frequent handwashing, and some good preventative measures like coughing and sneezing into a flexed elbow, etc. Moreover, as recommended by the WHO, and since the beginning of the COVID-19 pandemic, Tunisia has recommended the use of personal protective equipment (PPE) such as face masks and other specific equipment like single use gloves and aprons for workers in COVID services and other agents who handle and who are in direct contact with patients infected with COVID-19. The COVID-19 new wave linked to the Omicron variant—called the fifth wave in Tunisia—started in the beginning of December 2021 with the confirmation of the first Omicron variant infection by the National Ministry of Health on third of December 2022. Compared to the previous wave, this new Omicron variant wave is characterized by very high transmissibility of the virus and less severe clinical forms. The key indicator in this new wave is the low hospital admission and the slow death rate compared to previous waves. This could be explained by the acquisition of great immunity, thanks to vaccination and the spread of contaminations, the appearance of new chains of less virulent viruses, and the use of immunity boosters.

4 Impact and Effectiveness of Vaccination

As of 23 January 2022, a total of nearly 10 billion vaccine doses have been administered in the globe, while nearly 52.6% of population (4.70 billion persons) have been vaccinated with full doses. India has vaccinated nearly 49.6%, 0.685 billion citizens are fully vaccinated as on January 25, 2022. India achieved a landmark in its vaccination drive against COVID-19 on 21st December 2021 as the cumulative doses administered in the country surpassed the 100-crore mark (1.0 billion) starting from 16th January 2021 targeting 30 crore beneficiaries based on priority groups (Fig. 5).

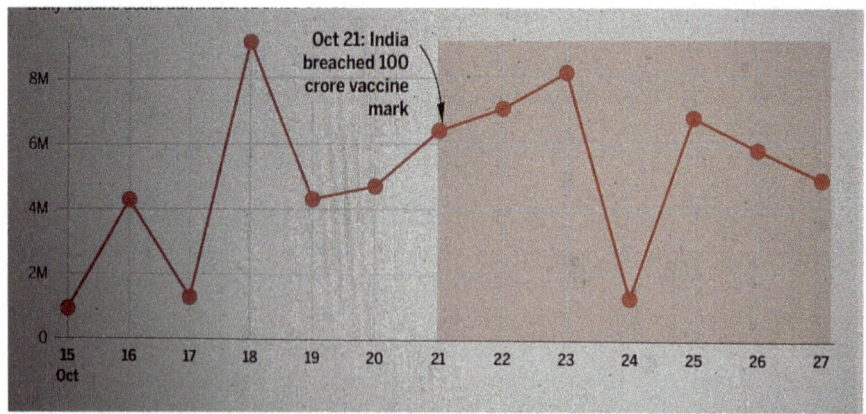

Fig. 5 India's COVID vaccination progress—Daily vaccination administered from Oct 15, 2021 (source: CoWIN)

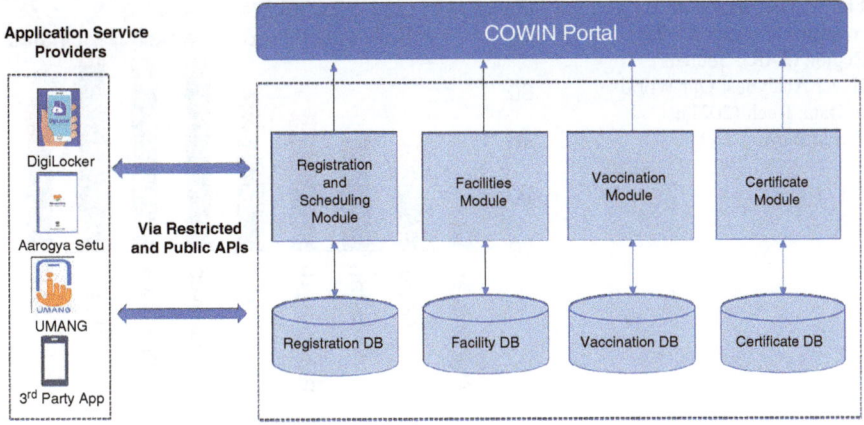

Fig. 6 Current technical architecture of CoWIN Platform in India for vaccination

The CoWIN platform for management of registration, appointment scheduling, managing vaccination, and certification has been rolled out by the Ministry of Health and Family Welfare (MoHFW) and it is being used by all participating facilities in India's National Covid-19 Vaccination Programme. The CoWIN Platform offers several functionalities (Fig. 1) as per the Guidelines for Integration of CoWIN with Third-Party Applications Developed by Ecosystem Partners and Guidance note for CoWIN 2.0 CoWIN is deployed on a cloud account of the MoHFW and can be accessed across the country through the internet. The progress of vaccination showing the reaching 100-crore milestones is shown in Fig. 6.

Soon after the outbreak of SARS-COV-2 scientists and researchers started to search for remedies to combat the virus. Within a year into the pandemic, researcher successfully came up with vaccines which were approved by WHO for emergency use. Till date at least ten vaccines have been approved and are being administered in different countries.

In India vaccination started in January 2021 but by the time the vaccination rate picked up, the second wave of infections was already playing havoc in the country. Currently India is one of the top countries in the vaccination list and has managed to fully vaccinate (both the doses) 49% of its population. Concerns have been expressed about the efficacy of vaccines since Omicron has affected large parts of the world. While some studies suggest that vaccines reduce severe illness and hospitalization (Preprint by Moghadas et al., 2021), other studies have reported significant reduction in antibodies after 6 months of vaccination (Alagoz et al., 2021). Studies conducted by vaccine manufacturers have shown that a booster dose helps in increasing neutralizing antibodies against the virus (https://directorsblog.nih.gov/tag/covid-19-vaccine). WHO has recently mandated to administer an additional booster dose to the vulnerable population (who.int). Throughout the world so far hospitalizations have been less in the third wave but as Omicron spreads, at some point hospitalization might also increase and large number of

Fig. 7 Vaccination in countries in South Asian Region (SAR). Source: Haver Analytics; Our World in Data; Ruch (2021); World Bank

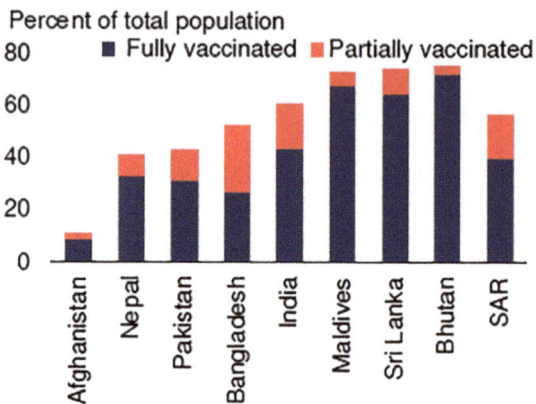

people might get severe disease (https://www.euro.who.int/en/health-topics/health-emergencies/pages/news/news/2022/01/the-omicron-variant-sorting-fact-from-myth). It has been clearly stated by WHO that despite vaccination, people have to wear mask and maintain social distance.

Before vaccination, WHO provided a list of guidelines, including handwashing, wearing mask, maintaining social distance strictly, to prevent transmission of coronavirus. But there was an urgent need of vaccine to save human lives. As WHO says that "*vaccines save millions of lives each year. Vaccines work by training and preparing the body's natural defences—the immune system*" (WHO, 2022b). However, vaccination has a great impact on human lives. Study reveals that "*about 67% of the people must be immunized against the coronavirus to stop transmission*" (Islam et al., 2021). Therefore, effective vaccination is an urgent requirement to get rid of the pandemic COVID-19. India and countries with small populations achieved higher vaccination rates. Inflation has risen above targets, partly reflecting supply problems and increasing food and energy prices. Domestic financial conditions have remained accommodative in India, with evidence of only limited increases in expectations of inflation or policy rates, or in term premia (World Bank, 2022). Figure 7 demonstrates the% of total population vaccinated in the countries in SAR.

Initially, people of Bangladesh did show interest to get vaccinated but fearing adverse impacts on life. A study reveals that fear of adverse consequences such as side effect was the most common reason for hesitation to take vaccines among people of Bangladesh and that was around 86.67%. The study also reveals that they had no sufficient information about fear and it was around 73.85% (Rahman et al., 2021). However, the mass awareness programs conducted globally by WHO and locally by the government agencies, people became aware of such vaccination. Moreover, the government asked the government employees to take vaccine and it was compulsory for them. It created confusion among the mass people in the country. Later, the Government of Bangladesh set a road map for vaccination program in the country and set a target to provide vaccines to 119,221,953 people, covering 70% of its total population in the country. As of 22 January 2022, the government of Bangladesh administered the first dose of vaccine to 92,426,233 people (about 54.27% against target population), second dose to 58,005,259 (around 34.06% against population), and third dose to 958,552 (around 0.56% against population) in the country (GoB, 2022a).

Moreover, in the first phase, frontline fighters of COVID-19 such as doctors, nurses, personnel of different cadre services were provided COVID-19 vaccine on the priority basis. Later, people of 60 years and above were considered priority group for the vaccines. Then the government reduces the age limit to 40 years old and later students of university, college, school were also considered for COVID-19 vaccine. The nationwide vaccination campaign started on February 7, 2021 (Islam et al., 2021). Later, the vaccination program started as "mass campaign" on different occasion all over the country to make wider coverage. Later, on November 11, 2021, the government started vaccination for school children aged between 12 and 17 years old for the first time in the country. The government set a target to provide vaccines to three million students in the first phase (Kamruzzaman, 2021) and all students will be covered in different phases. Now, booster doses for 50 years and above are being administered considering the third wave of COVID-19 known as Omicron variant.

To achieve the target, initially, a collection of COVID-19 vaccines was depended only on Serum Institute of India, for the people of Bangladesh. (Hossain et al., 2021). Subsequently, the government of Bangladesh collected vaccines from Russia, China, and USA to cater to the increasing needs. The relaxed mode of vaccine collection and management hampered to implement an effective vaccine management program in Bangladesh. Conversely, cold-chain management was another challenge to implement the vaccine program in the country. UK Research and Innovation (UKRI) observed that *"Bangladesh has one of the largest pharmaceuticals and vaccine industries in the world, but, like many countries, doesn't have the capacity to deliver fast-track mass vaccination"* (UKRI, 2022). However, WHO asserts that *"safe and effective vaccines are a game-changing tool: but for the foreseeable future we must continue wearing masks, cleaning our hands, ensuring good ventilation indoors, physically distancing and avoiding crowds"* (WHO, 2022a).

Vaccination rates across Southeast Asia remain low, with 30.3% of Malaysia's population, 15.4% of Thailand, 15.2% of Indonesia, 4% of Vietnam, and 3.3% of Myanmar's population being vaccinated. Figure 8 demonstrates the COVID-19 Vaccination trends in ASEAN countries during Jan 2021 to Jan 2022. Vaccines are arguably one of the most cost-effective public health measures available, but they are undervalued and underutilized worldwide, including in ASEAN countries. Vaccination does not mean that public regulations can be relaxed quickly.

The national plan for the introduction of COVID-19 vaccine in Georgia was developed on 21.01.2021. The vaccination process against COVID-19 is carried out in accordance with the National Plan for the Introduction of the COVID-19 Vaccine, which describes in detail the vaccination process in the country, taking into account the relevant legislation[5].

---

[5] https://www.matsne.gov.ge/ka/document/view/5084798?publication=1

Daily share of the population receiving a first COVID-19 vaccine dose
7-day rolling average

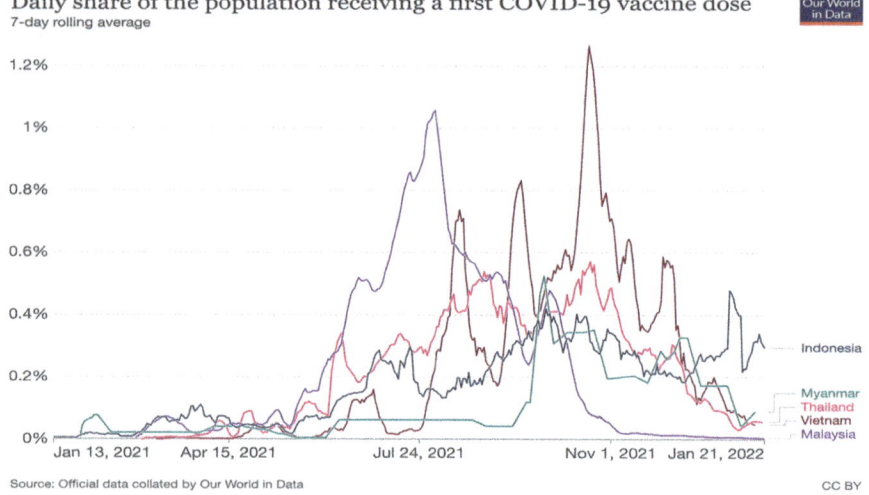

Fig. 8 COVID-19 Vaccine administered in ASEAN countries during Jan 2021 to Jan 2022

On March 13, 2021, the first 43,200 doses of AstraZeneca vaccine arrived in Georgia. A 73-year-old employee of Lugar Laboratory became the first person to register for the vaccine[6]. The head of the National Centre for Disease Control became the first person to be vaccinated in Georgia[7].

According to the data of the electronic module of immunization management on January 22

- The total number of vaccinations carried out in the country: 2,651,265;
- At least one dose: 1,317,488;
- Vaccinated with two doses: 1,205,638 (coverage rate per 21,000 population— 21,816), accounting for 28.7% of the adult population. Approximately 85 doses of vaccine are given per 100,000 population per day.

According to a study by the National Centre for Disease Control and Public Health, the proportion of infected people in fully vaccinated individuals across the country is 0.19% (207/106084). According to the vaccines, the calculation of specific vulnerability rates showed that during the reporting period, the rate of COVID-19 vulnerability in fully vaccinated individuals, regardless of the vaccine

---

[6]https://ka.wikipedia.org/wiki/%E1%83%99%E1%83%9D%E1%83%A0%E1%83%9D%E1%83%9C%E1%83%90%E1%83%95%E1%83%98%E1%83%A0%E1%83%A3%E1%83%A1%E1%83%98%E1%6

[7]https://ka.wikipedia.org/wiki/%E1%83%99%E1%83%9D%E1%83%A0%E1%83%9D%E1%83%9C%E1%83%90%E1%83%95%E1%83%98%E1%83%A0%E1%83%A3%E1%83%A1%E1%83%98%E1%7

administered, was significantly lower than the overall COVID-19 vulnerability rate of 2.16% (1723/79680)[8]. The Interagency Coordination Council of Georgia calls on citizens to be actively involved in the vaccination process, including the booster vaccination process, which is important for the easy transmission of the virus. According to the decision of the Government of Georgia, the 200-GEL incentive measure will last until the end of January, and the age of those who will receive the money after being vaccinated with the COVID vaccine has been reduced to 50 years. According to the plan of the National Centre for Disease Control, 70% of the population of Georgia should be vaccinated by July 2022[9].

In Tunisia, this new wave linked to Omicron variant comes when a high vaccination coverage is achieved 52% of fully vaccinated population compared to the global rate around 52.4% of fully vaccinated population (Our world in Data, January 21, 2022). Besides, the comparison of the number of COVID-19 confirmed cases and the deaths cases in this new wave to previous waves (when the population was not vaccinated, the vaccination started on March 13, 2021, in Tunisia), we can conclude that vaccination provided protection against severe outcomes.

5   Impact of Third Wave on Economy and Lockdown

The COVID-19 pandemic has economically impacted both the rich and the poor nations; however, the developed world is already showing signs of recovery. Vaccination rate has been quite high in these countries hence hospitalization has been low though in the third wave infection has been very high. According to the International Monetary Fund (IMF) report in October 2021 (World Economic Outlook Update, 2021), the Global economic growth rate fell to -3.2% in 2020 but consolidated to 5.9% in 2021 and has been predicted to be around 4.9% in 2022. The IMF has also warned that the developed countries will continue to face economic difficulties as long as the lesser developed nations continue to reel under the effects of pandemic and due to shortage of supply and disparity in distribution of vaccines.

Since the end of the year 2019 at the completion of 2 years of the COVID-19 crisis, the progress in economy has been both encouraging and troubling, clouded by many risks and considerable uncertainty while in 2021 after a sharp decline in 2020 the outputs are being revived in many countries. Advanced economies and many middle-income countries have reached encouragingly significant vaccination rates. USA and a few other countries have covered their significant number of citizens by the booster doses which have been started recently in India. International trade has picked up, and high commodity prices are benefiting many developing countries.

---

[8] http://test.ncdc.ge/Handlers/GetFile.ashx?ID=07ad44ba-95c0-4a9b-9682-f19ded67d51a

[9] https://www.ncdc.ge/

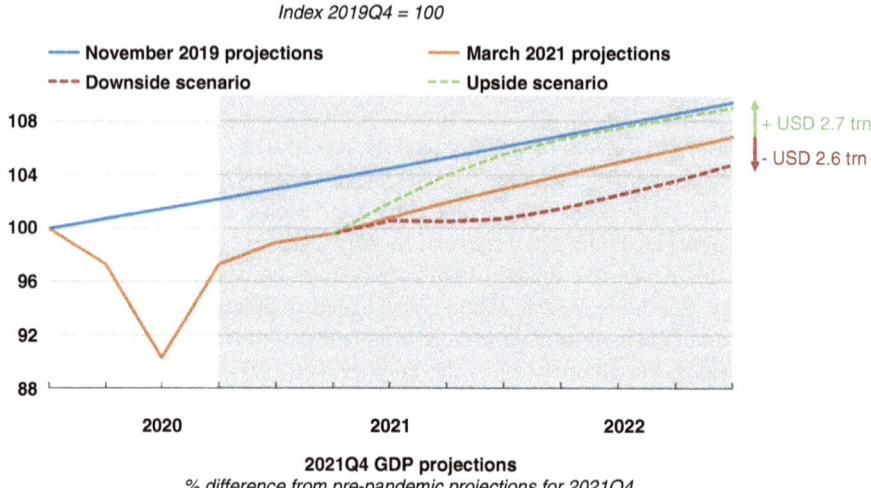

Fig. 9 Global GDP Projections and difference from Pre-pandemic Projections. Source: OECD, 2021

Domestic financial crises and foreign debt restructurings have been less frequent than might have been expected in a time of severe global shocks.

Output in South Asia is projected to expand by 7.6% in 2022, accelerating from 7.0% the previous year, as COVID-19 vaccination progresses and contact-intensive sectors recover. According to a report by OECD (Tackling Coronavirus COVID-19, 2021) unemployment had drastically increased in some OECD regions in 2020 compared to the same period of the previous year. A massive decline was observed in GDP of most of the nations during the pandemic. Figure 9 shows the Projected global GDP growth pre-pandemic and the projection adjustment made post-pandemic. Though there are signs of recovery of the economy in 2022 yet it is likely to fall short of the previous projections.

Following the major setback to health and economic activity caused by the mid-2021 second wave of COVID-19 in South Asia (SAR), as demonstrated in Fig. 10, the economic activity has recovered. New cases of COVID-19 stabilized at lower levels last year but are again accelerating in parts of the region as the Omicron variant spreads rapidly in early 2022. The economic damage caused by the second wave in India has already been unwound with output effectively back to levels reached prior to the pandemic (2019Q4) as COVID-19 cases and restrictions subsided.

The surge in infections in 2021 related to the Delta variant sapped consumer demand, but to a much more limited degree than previous waves. Persistent supply bottlenecks have weighed on global production and trade. In advanced economies, high vaccination rates and sizable fiscal support have helped cushion some of the adverse economic impacts of the pandemic. In sectors particularly sensitive to the

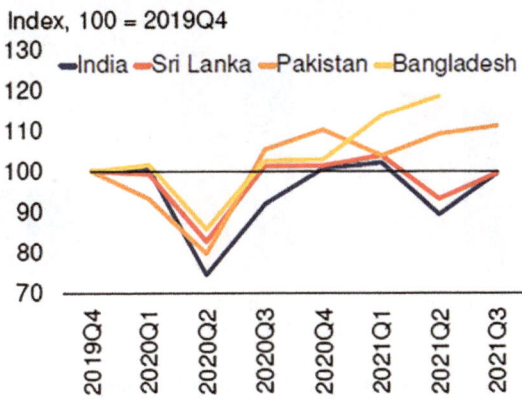

Index, 100 = 2019Q4

Fig. 10 Output in the SAR rebounded following a mid-2021 sharp COVID 19 waves. Source: Haver Analytics; Our World in Data; Ruch (2021); World Bank

pandemic, including trade, tourism, and hotels, damage has lingered, however, and remains well below pre-pandemic levels. Sri Lanka also saw a rebound in activity despite a resurgent pandemic, with new cases peaking in late 2021. In Bhutan, growth has been revised down because of the effects of strict COVID-19 protocols, setbacks in infrastructure projects due to limited migrant labour, and the standstill in tourism. In Bangladesh, strong export growth, supported by returning readymade garment demand from abroad, and a rebound in domestic demand—with improving labour income and remittance inflows—supported the recovery.

India's economy is expected to expand by 8.3% in fiscal year 2021/22 (ending March 2022). In FY2022/23 and FY2023/24 growth has been upgraded, to 8.7 and 6.8%, respectively, to reflect an improving investment outlook with private investment, particularly manufacturing, benefiting from the Production-Linked Incentive (PLI) Scheme, and increases in infrastructure investment. Growth prospects have improved in the region since June 2021, reflecting forecast upgrades for Bangladesh, India, and Pakistan (World Bank (2022)).

Compared to the first two waves of the COVID-19, a faster rise in cases as well as a higher rate of infection positivity is being reported in many countries around the world since the onset of the third wave of infections reported in South Africa in December 2021. This has led economists of many countries to be more cautious to revise their near-term fiscal growth to a lower estimated value. According to the World Bank's recent report of "Global Economic Prospects" in countries around the world, global economic growth may decelerate markedly to as low as 3.4% due to Omicron related further economic disruptions and lingering supply chain snarls entangle the world's economy (World Bank, 2022).

However, the major economies of South Asian countries are generally expected to grow in the fiscal years 2021–2022 and 2022–2023 (World Bank, 2022). The estimated fiscal growth of economy of Bangladesh and her neighbouring countries in the fiscal year 2021–2022 and 2022–2023, according to the predictions of

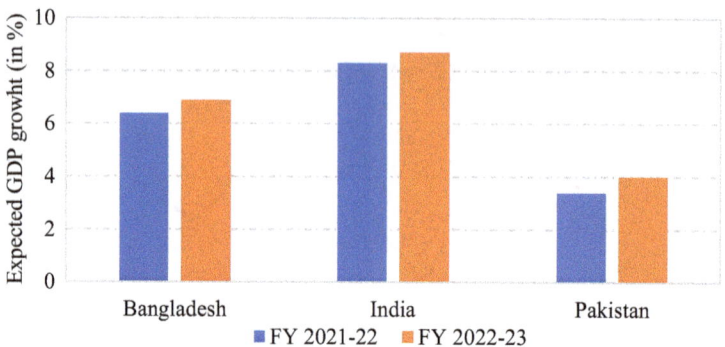

Fig. 11 Country forecasts of expected real GDP growth of major economies of South Asia (Source: The World Bank, 2022)

World Bank's recent report as shown in Fig. 11, is perhaps indicative of the fact of resumption in their normal business operations at the state of the pandemic observed in the second half of 2021.

Although the economy of Bangladesh has suffered from serious damages due to series of lockdowns and restrictions in business operations imposed in 2020 and 2021, it is expected to grow around 6.4% in the fiscal year 2021–2022. With several measures taken to combat the outbreak of COVID-19, the country has so far been successful to reduce the rate of mortality due to COVID-19 and enable almost full-scale resumption of her major economic activities. There have been no further lockdowns imposed on normal business operations in Bangladesh since the restrictions largely eased on August 10, 2021. The country's economy is expected to reach a further growth of around 6.9% in the fiscal year 2022–2023, subject to the condition of sustaining favourable conditions of normal economic activities.

However, the global economy is facing new challenges of COVID-19 due to the onset of the Omicron variant with no exceptions observed in South Asian countries including Bangladesh. The economists still remain optimistic to suffer far less economic damage from the outbreak of third wave of the pandemic as preliminary data suggests lowered cases of severe illness or hospitalizations of fully vaccinated people. Sustaining the expected momentum of future economic growth is based on highly optimistic predictions of full-dose vaccination coverage of about 80% by the end of 2022 and also on reliance on current vaccines' effectiveness at protecting severe illness or deaths due to emerging new mutants of COVID-19 virus. As of real time government data released on 21st January on DGHS, *Bangladesh* website, there has been 11,434 confirmed cases of COVID-19 infections with 28% rate of infection positivity (DGHS, 2022a, 2022b). According to the most recent official data collated on January 21, 2022 by *Our World in Data*, Bangladesh has achieved only about 35% of full-dose vaccination coverage and 21% share of people only partly vaccinated against COVID-19. Among the South Asian countries, Bhutan, Maldives, and Sri Lanka have reached above 65% full-dose vaccination coverage of their people against COVID-19, while the three major economies in South Asia (World Bank, 2021), as mentioned in Fig. 11, are lagging behind to reach full-dose vaccination coverage to 50% of their population (World Bank, 2022; Our World in Data).

Scientists and researchers are still uncertain whether the current vaccines are effective against protecting the spread of the virus through new variants of concern such as Omicron. The faster rise of infections in the third wave of COVID-19 coupled with a higher rate of infection has also led the policymakers uncertain about further imposing restrictive measures such as nationwide lockdown which is obvious to decelerate the momentum of expected economic growth in future. Therefore, the expected future economic growth of Bangladesh and her neighbouring countries remains highly volatile due to the vulnerable economic systems likely to suffer major disruptions caused by the emerging new variants of COVID-19.

In ASEAN, pandemics have so far caused immediate economic turmoil evolved from forced blockade of offices, academic institutions, markets, factories, community quarantine, stay-at-home order, temporary business closure and travel restrictions, prohibition of virus containment, and many such processes. Especially because the government requires community measures to contain the pandemic. The vulnerable and marginal groups are the most affected by "Blockage" or other forms of movement restrictions, such as informal or daily wages workers, workers in affected industries, etc. The elderly, especially, when switching to a digitally supported office, "Work from Home (WFH), is another problematic system which has been implemented in office, schools, colleges, etc. Small and medium-sized enterprises (MSMEs) are also at risk given the limited resources to wait for a pandemic. In addition to ensuring the effectiveness of policy responses to pandemics, the government has a public responsibility to support it to those who are most affected, especially in the absence of social welfare protection. Management Communication with the people is indispensable in order to promptly provide information and reassure the people.

The Indonesian government has budgeted an additional fiscal stimulus package of Rs 126.2 trillion to accelerate the National Economic Recovery (PEN) program. These incentives include more cash grants to healthcare workers and non-healthcare workers, which will be available with a budget of 23.3 trillion rupees by December 2020. In addition, Rs 81.1 trillion is allocated to the use of sectoral programs by various government sectors, institutions, and local governments to provide productive support to SMEs at a face value of Rs 2.4 million annually. The Malaysian government has announced a MYR 20 billion (US $ 4.56 billion) stimulus to help the country's tourism and other industries address the effects of the coronavirus pandemic. The wall will also exacerbate the outbreak. As for stimulus, the first 100 billion baht (US $ 3.2 billion) package was aimed at supporting businesses in the form of soft loans, tax withholding deductions and VAT refunds (ASEAN Briefing, 2020).

On March 30, the Prime Minister of Georgia, Giorgi Gakharia, declared the country a "de facto universal quarantine" and a "de facto curfew"[10]. There is no lockdown in the country at this point and no plans are made. Exceptions are point locks in educational and pre-school institutions. At the same time, from the summer of 2021, COVID passports came into force in Georgia, which are issued to citizens who have been vaccinated twice or transferred to COVID. Regulations were issued in the autumn of the same year, according to which citizens who do not have COVID passports are also not allowed to attend numerous public events: restaurants, restaurants, cafes, and bars in open and closed spaces, cinemas, theatres, opera, museums, concert halls, casinos and gambling establishments, spas, gyms, hotels, and also the ski resorts [11].

The COVID-19 pandemic is first a health and humanitarian crisis, and businesses are rapidly adjusting after nearly 2 years of pandemic. While the crisis unfolds, the industry leaders, researchers, academicians, and political leaders should prepare for what is coming next. But not the next "normal". Normal may not be available to us anymore in near future, and "business as usual" even less. The present reality taking shape is made of complexity, uncertainty, and opportunities, while the steps are to be taken analysing the threats and weaknesses. To adapt and thrive, organizations must accelerate and become resilient and agile. The leaders and organizations must address the deep changes needed in this new environment, from the employees, interested parties, operations, and technology perspective, at speed and at scale with resilience.

6   Impact of Third Wave on Social Aspects

Bangladesh is yet not so much into the adverse impacts of the latest variant of COVID-19 virus, which is named as Omicron, but it is creating worry among the people in the country specifically the health care providers. Though the rate of death due to Omicron is low, the global death graphs bring psychological concern on mental health in the country.

Along with the economic impact of the outbreak of COVID-19, it has had an unimaginable impact on healthcare system creating immense fear and anxiety among health professionals and mass people. The sufferings due to the COVID-19 have already made our social system fragile over the duration of the crises in the past 2 years. The unique societal structure of togetherness in Bangladesh has been destroyed due to the impacts of the first two waves of the pandemic which has had tremendous impacts on other associated socio-economic aspects. Among many such

---

[10] https://ka.wikipedia.org/wiki/%E1%83%99%E1%83%9D%E1%83%A0%E1%83%9D%E1%83%9C%E1%83%90%E1%83%95%E1%83%98%E1%83%A0%E1%83%A3%E1%83%A1%E1%83%98%E1%5

[11] https://www.interpressnews.ge/ka/article/691633-kartuli-presis-mimoxilva-13012022/

impacts highly notable are major disruptions in economic activities and systems of trade, loss in agricultural economy, worsening social violence and abuse due to psychological trauma of COVID-19, crisis in public health services for treating mental health conditions or diseases other than COVID-19. About 20 million workers lost their jobs in Bangladesh in the informal sector due to the lockdown and shutdown of garments industry. Losses in privately owned businesses, especially small business industries, are still uncounted. Moreover, 66% farmers of which 53% engaged in crop and vegetables' farming and 99% of fish farmers have earned less money during the first two waves of the pandemic than they would have earned during normal period (Gautam et al., 2022). The pandemic of COVID-19 has already put the agricultural systems and small businesses into great risk and the crisis is expected to worsen further due to the unknown impacts of the third wave of the pandemic. This obviously would damage the supply and communication chains among different societal groups creating major income disparity among different social groups. The end results of the impacts of the COVID-19 pandemic may be observed through massive social transformation negatively affecting our economic systems, health, and well-being.

Women and girls have gone through increased domestic violence, counts of early marriages have increased, of dropouts from school have increased among rural girls. The healthcare workers who have served COVID-19 patients have been socially condemned due to the fear of infection. Among many causes that have created major psychological impacts or disorders in Bangladesh major causes are sudden lockdown and enforced quarantine limiting physical movement, stress due to the forced conditions of work from home, online education instead of conventional schooling, etc. (Mamun et al., 2021; Tasnim et al., 2021). Tasnim et al. (2021) estimated moderate to severe depression and anxiety (38.9% and 35.2%, respectively) among the people living in Bangladesh. Gautam et al. (2022) reported that women faced more depression and anxiety than men and 43% of children had subthreshold mental disturbances. Digitalization is a paradigm shift towards virtual communication in this pandemic situation and telehealth is certainly a forward-looking social impact of COVID-19. Rahman et al. (2021) indicated gender disparity in telehealth. They found 14% of the female patients opted for telehealth visits only with 57% cases of missed doses of medication whereas 20% of males received telehealth visits and only 29% of them missed their doses of medication.

The third wave of COVID-19 has just been initiated in Bangladesh and in other South Asian countries. It is an optimistic assumption that the country might not face social impact to a greater extent than it has suffered during the first two waves of COVID-19. Bangladesh has already had severe impacts of numerous social challenges that have negatively transformed the social structure of communication, business, and trade. Further sufferings due to the unknown impacts of third of subsequent waves of COVID-19 will put the country to face unknown major social challenges due to the complex crosslinks among different social, public health, and human welfare aspects. It is high time to take appropriate and effective policies for the quick social reclamation against the likely impacts of the third and subsequent waves of COVID-19.

7   Impact of Third Wave on Waste Management

The novel coronavirus diseases (COVID-19) pandemic tremendously affected the waste sector, with different types of wastes like domestic waste, solid waste, medical waste, etc. are subjected to uncontrolled dumping and open burning, leading to public health risks. The use of Personal Protective Equipment (PPE), including medical masks, goggles or face shields, gowns, and other respiratory protective equipment and their disposal evolve a challenge in the waste management systems. These generated increased amounts of hazardous waste during the pandemic will certainly impact the country's waste management and overall environmental performances. Pandemic COVID-19 has exposed the weakness of the waste management in cities while the plastic waste has been increased by 15% to 40% as well the biomedical waste both hazardous and non-hazardous in many of the cities worldwide.

The biomedical waste management as well as the management of COVID waste in India has been tackled effectively from the very beginning as the Biomedical Waste Management Rule (2016) in India has been very effectively implemented in most of the states and Union Territories. All the health care units including the COVID quarantine facilities have their own segregated waste storing facilities and either have own treatment or tagged with the authorized service providers who are paid per bed rate system fixed by the government. UNIDO has been working to develop and implement environmentally sound practices and techniques for medical waste with more than 160 hospitals across Karnataka, Punjab, Odisha, Maharashtra, and Gujarat (Express News Service, Pune, October 24, 2020). Each municipality is responsible to tackle the COVID waste and domestic Hazardous wastes stored at domestic levels as per SWM Rules 2016.

A study was conducted during the second wave of COVID-19 and during the start of third wave in a temporary camp shelter in Cox's Bazar, Bangladesh (Fig. 12 and Table 1). A comparison of waste management system is presented in the following table that shows before COVID-19 practice and during second and third wave of it.

To address the COVID-19 response, about 654 government and 5055 non-government hospitals along with 9061 testing centres are generating COVID-

Fig. 12 Field Survey at waste management plant in a temporary camp shelter in Bangladesh

Table 1 Comparison of waste management system before COVID-19 and during second and third wave in the temporary camp shelter No. 1 (East), 1 (West), 2 (East), 3, 4 (Ext.), 15, 16 and 21

Category	Before COVID-19	During second wave	During third wave
Type of generated SW	– Organic – Inorganic	– Organic – Inorganic	– Organic and Inorganic – Infectious waste, i.e., PPE
Types of bin/system used for storing SW	– Two types of bin (organic-blue, inorganic-red)	– Two types of bin (organic-blue, inorganic-red) – Communal bin – Paddle bin	– Two types of bin (organic-blue, inorganic-red) – Communal bin – Paddle bin
SW collection method	– House to house – Sub-block wise – Drainage – Communal bin – Mini-dumping chamber	– House to house – Communal bin – Mini-dumping chamber SW bin and paddle bin	– House to house – Communal bin – Mini-dumping chamber SW bin and paddle bin
Types of initiatives taken and strategies followed for SWM	Collection →Segregation →Composting →Filling	Collection →Segregation →Composting →Screening →Packing	Collection →Segregation →Composting →Screening →Packing
Transportation methods	– Man power – Trolley and Van	– Man power – Trolley and Van	– Man power – Trolley and Van
SW generation trend	As usual	Decreasing	As usual
Types of problems/challenges in SWM system	– They are not agreeing to use SW bin	– Behaviour habit – Not used to paddle bin, communal bin	– Behaviour habit – Not used to paddle bin, communal bin
Types of measures were taken during COVID-19 if SW collection was temporarily shut down	– Continuing as usual process like boots, gloves, aprons	– Reduce collection manpower – Continuing as usual process; using PPE, mask, and maintaining social distance	– Continuing as usual process – Using PPE, mask, sanitizer – Maintaining social distance
Monitoring	– In the field	– Remote monitoring	– Remote monitoring

Source: Field Survey, 2022

19 medical waste—many of them without an appropriate disposal and resource recovery process. The Daily Star last year had estimated that at least 14,500 tons of medical waste were generated in Bangladesh during April 2020—the initial month of the COVID-19 pandemic, and this may have further increased.

In addition to medical waste, household waste is another issue in COVID-19 pandemic. Adopted in many countries during the pandemic, the Stay-at-Home Order, (Cahya, 2020) encourages 4444 people to work, study, worship, and engage in other activities at home (Cahya, 2020). People's behaviour is changing because people have to stay home all day. In pandemic as the stay-at-home and social distancing are governed by orders in many countries, people have to rely more on food delivery (Brunell, 2020). As the most populous country in ASEAN, Indonesia must make special efforts to dispose of its waste, especially the currently explosive amount of medical waste. Moreover, as an archipelagic state, Indonesia tends to dump medical waste into all waterways. The amount of landfilled waste has decreased by up to 40% due to reduced waste generation from offices, restaurants, and industry, but the increase in medical waste including huge plastics wastes from hospitals and quarantine centres is now an urgent concern. During COVID-19's rapid growth, especially in ASEAN countries, rapid adaptation to current waste management policies is important. Pandemics can be a positive opportunity for better waste management.

On the one hand, the pandemic has ecologically improved a number of issues related to ecosystems, on the other hand, along with a number of positive effects, it has also had some negative effects, such as new contaminants in the environment, including medical gloves and COVID-19 face masks, which are found both in surface waters and in landfills. In Georgia, this type of waste is abundant in the country and they are often found in illegal landfills, which unfortunately are many today in Georgia, especially in regions where these types of problems are not given due attention by local governments. First of all, it should be noted that the Georgian population is less informed about the dangers of such medical waste treatment and how dangerous these types of contaminants can be in the process of spreading the virus.

Due to the adopted preventive measures to fight against COVID-19, a great number of facemasks, gloves, and medical suits are daily manufactured and used during the pandemic, creating huge amounts of solid waste. These COVID-19 wastes pose environmental and health threats since the coronavirus can survive on material surfaces (e.g., metals, glass, and plastics) for up to 9 days according to Kampf et al., 2020. Thus, guaranteeing the suitable collection and management of these occasional specific waste is a growing challenge for several countries especially for developing ones that have poor waste management systems. In Tunisia, since the waste management strategy is based on landfilling, the improper management of contaminated COVID-19 waste may increase the spread of viral disease. Besides, solid wastes manipulation by waste pickers without wearing proper PPE can also increase the COVID-19 transmission.

According to Nzediegwu & Chang, 2020, in Africa, the number of used PPE (e.g., facemasks) is estimated to reach seven hundred million per day. In Tunisia, with confirmed COVID-19 cases increase, wearing face masks is compulsory in Tunisia for all citizens. Statistical data regarding COVID-19 specific waste are not readily available for Tunisia, but overall medical waste generation has increased by 40% due to COVID. Tunisia is taking active measures to contain and reduce the spread of COVID-19, strategies to manage solid wastes, including used PPE, during the pandemic.

8   Impact of Third Wave on Health and Health Care System

The COVID-19 pandemic requires a resilient healthcare system to respond appropriately, both in mobilizing resources to address the COVID-19 pandemic and in addressing the decline in essential public health services. The lessons learned from this pandemic include the need to reform the health care system and build a resilient health care system in anticipation of future outbreaks and pandemics. The field of health care system reform is education and utilization of human resources in line with future health problems. Strengthening primary health care, modernization of medical facilities in remote areas, borders and geographically isolated areas, and building autonomy for drugs and medical devices are required. Health safety management must be improved by improving prevention and management of infectious diseases, including administering vaccination. Promotion of the use of health funds and information technology and strengthen community empowerment to practice a healthier lifestyle are very much necessary to prevent or reduce the effects of third wave. We have to be very careful and preventive actions are to be initiated well in advance to protect us from potential evolution of fourth and/or fifth waves of COVID-19 variants. The learning points and experience of handling previous variants will definitely be helpful. We have to be alert that there may be evolution of different variants in any part of the globe that may bring forward the 4th and/or subsequent waves. Hence the basic preventive measures, like, social distancing, hand washing, use of face masks and restricted gatherings are to be practised and monitored.

It was reported that during the first two waves of novel coronavirus, the normal health care services seriously affected in the country as the doctors, nurses, and other health workers were not available at the hospitals and clinics as well as diagnostic centres. Consequently, telemedicine appeared as an alternative option to get health care services in the country. But it was not sufficient as per requirements and demands. Besides, people of the country were not accustomed with such services. So, it creates dissatisfaction among the service seekers. In the meantime, among many other initiatives, the government of Bangladesh finalized a national digital health strategy. However, there are 524 health providing specialized institutes, medical college hospitals, district hospitals, and Upazila health complex in Bangladesh (GoB, 2022b). Besides, there are more than 13,000 community clinics in the country (Riaz et al., 2020). "To face the challenges and address the populations' needs, digital technology has been used in various aspects like mobile devices for laboratory report, commodity management, telemedicine for COVID and non-COVID healthcare, targeted patient communications, community awareness, and contact tracing. Moreover, digital health information systems have allowed us to track and analyse the trends in service utilization and make evidence-based policy interventions and strategies during the pandemic (WHO, 2021)".

According to the Interagency Coordination Council in Georgia, due to the high rate of spread of the Omicron strain, the number of cases of infection is increasing in the country, but the workload of the hospital sector has not increased. Specialists estimate that the infection "Omicron" as well as the "Delta" strain is going smoothly in vaccinated persons and there is no need to hospitalize the infected persons.

At the same time, in January, the terms of isolation and quarantine for vaccinated, unvaccinated, and covariated persons were changed in Georgia. The quarantine period for asymptomatic and mild patients was 8 days, and for moderate and severe cases—10 days. In both cases, it is mandatory to wear a veil for the next 5 days after the isolation is completed. At this stage, 2050 citizens are in clinical hotels, and 4465 people are in hospital under supervision. 281 patients remain in resuscitation, while 1083 patients are treated in the intensive care unit. The vast majority of these cases are caused by the coronavirus Delta strain. In case of Omicron infection, only individual citizens (5–10%) needed hospitalization. So far, none of them did not need intensive care or resuscitation—they were discharged from the clinic very soon[12].

9   Impact on Education System

Another area that has been hit hard during this pandemic is education. With increasing uncertainty, many governments are considering whether to keep schools open. One thing we certainly know is that another wave of widespread school closures will have a devastating impact on children. The evidence is clear that there were limited resources for students, teachers, and parents during this pandemic. The lack of access to distance learning has restrained decades of progress in education and a blurred childhood. Two years of this pandemic is black holes for the students that retards their metal age. It is a significant loss to the mankind. During the COVID-19 pandemic, schools and academic institutions including universities worldwide have been closed for many months depending on the online teaching and learning which created a lot of problems adversely impacting the development of human resources in the country.

Due to the corona infection situation, the educational institution has been closed for 2 weeks now. The school reopened after being closed for about a year and a half

---

[12] https://www.radiotavisupleba.ge/a/31648960.htm

due to coronavirus. In the meantime, due to the rising tide of infection, the country's educational institutions are back to normal after 4 months. For the time being, from January 21 to February 6 of 2022, all the equivalent educational institutions in the country are being closed. Only universities will take action in their respective fields. It has been decided to stop physical class activities in the universities till February 6. If the direct class activities are stopped, the online class-examination will continue and the residential halls of the university will be open in compliance with the hygiene rules. A limited range of university offices have been declared to open from 9 a.m. to 1:00 p.m. and some cases till 5:00 p.m. The university has advised the students to stay at the place of residence following proper hygiene rules. Concerned parties are also requested not to hold public meetings on campus.

Impacts of COVID-19 on education system of Bangladesh:

1. Dropout rates at the primary grades have been increased (ADB, 2021).
2. Low attendance rate as the access to digital devices for remote learning and e-learning is inadequate (ADB, 2021).
3. Learning loss like decline in student's morale to study (ADB, 2021).
4. Spending less time on education especially female students as they got involved in household chores (WB, 2021).
5. Parents were disable to provide continuous educational support to children in lockdown (UNESCO, 2021).
6. Increased mental distress, increased child labour, and increased child marriage due to longtime school closure (UNESCO, 2021).
7. The future of 37 million children in Bangladesh is at risk with their education severely affected by the COVID-19 pandemic (UNESCO, 2021).
1. Inadequate evaluation (Auto-Pass) at a very important stage of learning (HSC) makes the future uncertain (UNESCO, 2021).

Steps taken by the Government of Bangladesh to ensure smooth education system:

2. The government shut all education institutions for 543 days from March 2020 to Sept 12, 2021 (The Financial Express, 2022).
3. Reopened school has to follow the rules of first to fourth grades, and sixth to ninth grades will attend classes once a week and end of fifth, tenth and 12th grades will attend classes every day (Aljazeera, 2021).
4. The government makes COVID-19 vaccine mandatory for school participation.
5. The government shut all education institutions again from January 21 to February 6 (Dhaka Tribune, 2022).
6. Coaching centres across the country have been closed.
7. Government has allowed education institutions to take virtual classes.
8. The government took decision to give auto-pass to candidates of Higher Secondary Certificate and equivalent examinations of the year 2021 based on their secondary school certificate and junior schools certificate results (New Age Bangladesh, 2020).

9. Government provides a short syllabus that could be completed in 3 months for SSC and equivalent exams and 4 months for HSC and equivalent exams.
10. Health guideline such as ensure to hygiene and maintain distance has been provided to be followed by the students (New Age Bangladesh, 2020).

Corona infection situation has not been brought under control, and in 2020 and 2021, vacations at educational institutions have been gradually increased. Due to the closure for a long time, no one, including the government wanted the educational institution to be closed again. But the infection is on the rise again and even many teachers and students are getting infected. Again, it is normal for about 4 crore students of the country to lose their education due to the closure of educational institutions. But public health experts say the decision was made because of health risks. They need to be proactive in keeping alternative topics like online class assignments effective in closing their counselling.

Distance (online) learning has never been used in Georgia's education system before the pandemic, and the country has faced a new challenge unprepared. By 2020, 38% of households did not have a computer and 16% did not have access to the internet. 11% of teachers and students did not use the distance learning program TEAMS at all[13]. The distance learning process is more complex in villages and regions compared to large cities. The share of households in rural areas without computers and internet was 68% and 26%, respectively. However, during the same period, the share of households in cities with similar challenges was 24% and 9%, respectively[14].

Up to date, the secondary education system has returned to normal, although in some cases (identifying cases at will and in institutions) there is a point-to-point and individual shift to online learning. Part of the vocational and higher education institutions has also switched to hybrid learning, which involves combining online and conventional learning.

The impact of the pandemic on student development has not yet been assessed. Given the duration of distance learning, its quality, and the challenges involved in enrolling in it, a significant lag in students' academic skills development is expected[15].

Due to partial or full lockdown and school closures, the COVID-19 pandemic caused many disruptions in the education of young people. According to a recent study conducted by UNESCO on the education in a post-COVID world, in 2020, nearly 1.6 billion students worldwide were affected by school closures (UNESCO, 2021).

In many countries, the vaccination of school children has already started so that the schools and colleges can start offline classes. In the first week of January 2022, India has reached the historic milestone of administering 1.50 billion vaccinations to

---

[13] https://osgf.ge/ganatlebis-seqtoris-mimokhilva-covid-19-pandemiis-pirobebshi/

[14] https://idfi.ge/public/upload/Analysis/covid_19_and_the_georgian_eductation_sector.pdf

[15] https://www.kutaisipost.ge/ka/akhali-ambebi/article/22758-2022-01-19-13-27-52

its citizen while the vaccination for the school children and booster doses for the seniors have started in early January 2022. Schools, colleges and universities have started offline classes and examinations for the students. The citizens started enjoying their normal daily life slowly. However, the bad news is that there is a trace of fourth wave with new variety of variants in China, South Korea and a few other countries.

10   Discussion and Data Analysis

The total caseload found up to 23 January 2022 is 28,223, according to the latest data of the Government of Bangladesh, whereas the total number of recovered patients in the country stood at 1,554,268 (GoB, 2022a). The COVID-19 fatality rate in Bangladesh is 1.72% and the current recovery rate is 94.64% as per the Government data. Around 153 million doses of vaccine are given to Bangladeshi people, whereas 58.4 million people are fully vaccinated. Only 35.4% people are vaccinated for two doses in Bangladesh which is far less than world's score 52.5% (GoB, 2022a). Fig. 13 shows the Number of Confirmed Cases by Divisions in Bangladesh and the death rates as in Jan 2022.

It is expected that all sectors such as social, economic, environment, education, health, and so on so forth will face manifold challenges. Among many others, the public health will bear a huge cost in terms of physical and mental health in Bangladesh. It is feared that students will face a huge burden of eye-related diseases as they are pushed for online learning. Most of the students use mobile phone for this purpose and the experts especially the eye-specialists/physicians opine that mobile phone is not a teaching-learning friendly device. "It brings huge pressure on eye, causing negative impact on eye-vision", an eye specialist asserts. Consequently, policymakers should keep in mind to formulate effective and pragmatic policies to overcome the challenges appeared due to the pandemic COVID-19 and beyond. For example, the alternative teaching-learning mode—blended teaching—learning would the better option to continue the academic activities. Digital health care management system would be the alternative option to provide regular treatment. Work from home would be also another alternative mode of action to continue office management. As a result of such activities, waste management would be challenging especially for the developing countries like Bangladesh. So, dependency on ICT would be increased. Here, the instruments of the fourth industrial revolution (4iR) could be facilitators to such waste management in the country. For this, capacity building both of institutional and individuals would be one of the major tasks of the policymakers. The policymakers should adopt short-term and long-term strategic plans for capacity building. The short-term plans would be effective for immediate crisis management, while the long-term strategic plans would enhance enabling institutional and individual capacity for managing unforeseen crisis and disaster.

As the coronavirus Omicron or B.1.1.529 mutant spreads faster, ASEAN countries need to be more vigilant to prevent the mutant from invading their country. In

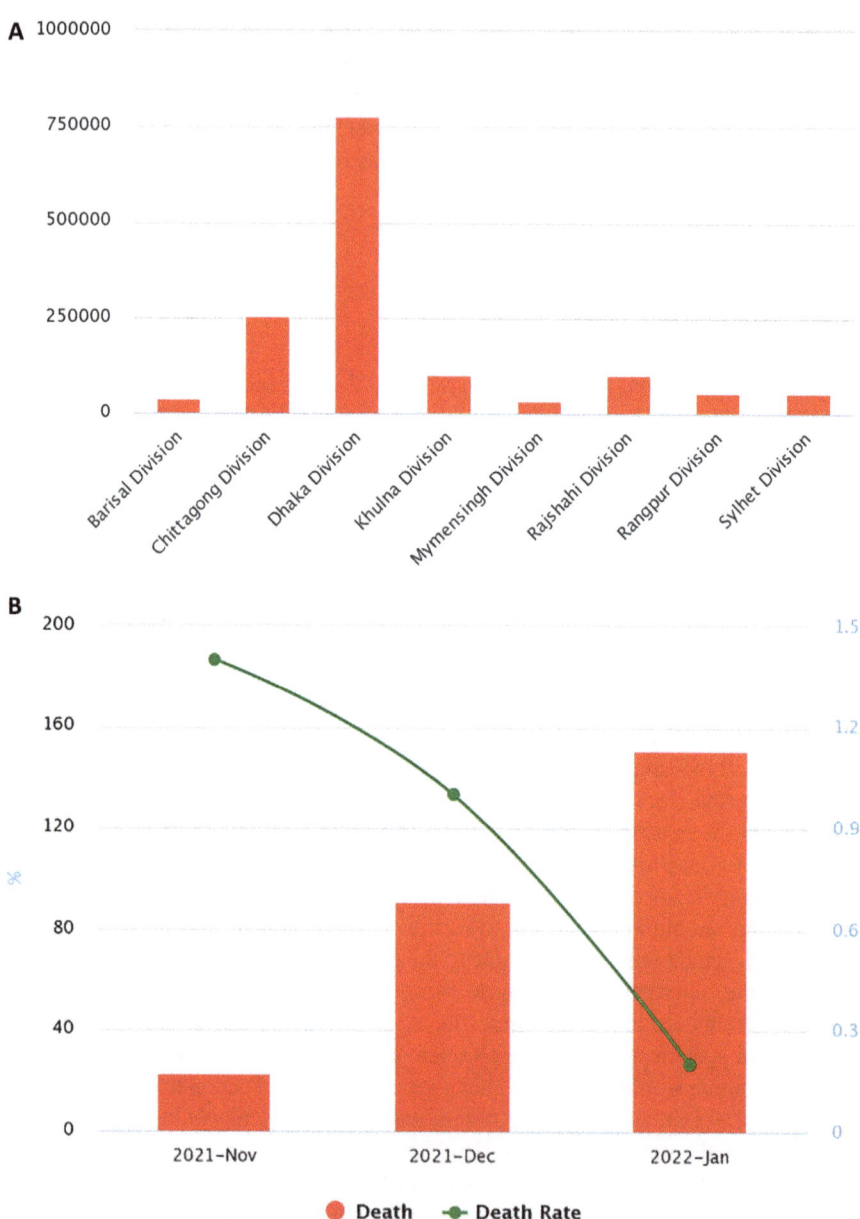

Fig. 13 A. Number of Confirmed Cases by Divisions of Bangladesh and the death rate as in Jany 2022.; (Source: DGHS, 23 January, 2022)

addition, coronavirus variants that fall into this category can have a serious impact on vaccine efficacy. Vaccination programs continue to be one of the most important strategies in ASEAN countries to combat pandemics, especially to reduce deaths and protect the economy. One of the reasons for the spontaneous spread of the virus, as well as in the world, the new strain of COVID19 in Georgia is called "Omicron". According to the National Centre for Disease Control on January 18–19, 50% of coronavirus-positive cases across the country come from "Omicron". This figure is much higher in the capital and it is around 65%. In addition, numerous cases of reinfection have been identified in the country. While Omicron is considered by some epidemiologists to be a relatively "simple" strain, however, there is a serious risk that the increased rate of infection will place a heavy burden on the hospital sector. Although the hospitalization rate has been reduced compared to the previous Delta strain in the case of Omicron, the high number of simultaneous infections will still cause the hospital sector to reset. According to the preliminary data of the National Centre for Disease Control, 15 patients were found in the clinic from those infected with Omicron, which was officially confirmed. None of them was vaccinated.

One of the main ways to stop the spread of "Omicron", epidemiologists and doctors still name vaccination and adherence to safety rules (proper use of the veil and keeping a distance). However, the vaccination rate in Georgia does not exceed 30–35%. Studies have also shown that the effectiveness of protection against infection in fully vaccinated individuals is 9.44 times lower than in non-vaccinated individuals, while the effectiveness of protection against fully vaccinated COVID-19 vaccine is 89.47%[16].

11 Conclusion

The current coronavirus (COVID-19) pandemic is having a profound multi-sectoral and multi-lateral impacts through all over the world. The most impacted sectors, among others, are economic, health, and education. This pandemic impacted another sector profoundly along with causing a major children's rights crisis, the most disadvantaged being disproportionately affected. Though the sectors must vary country to country. However, among all the negative impact of COVID-19 pandemic, there are some positive outcomes from it, such as the habit and good practice of hygiene, wearing mask, washing hands, keeping a minimum distance where possible and most importantly for the natural environment. The environment has got enough time to revive as a result of lockdown. This come back of the natural environment will help it to endure future pressure of human activities.

The catastrophic third wave of the coronavirus pandemic has hit several ASEAN countries as delta variants gain momentum in the region, leading to record numbers of infections and deaths. ASEAN countries such as Indonesia, Malaysia, Vietnam, Myanmar, and Thailand have so far avoided such large-scale outbreaks. Currently, they are struggling to contain new outbreaks, but Indonesia and Myanmar are

---

[16] Report of the National Center for Disease Control and Public Health, 7 reviews, as of 01.10.2021

suffering from low vaccination rates and overcrowded hospitals. The medical systems of both countries are on the verge of collapse.

It is impossible to stop the spread of the infection because "Omicron" has a high rate of delay, which is unimaginable even in the conditions of a strict "lockdown". In this situation, vaccination is paramount, which can reduce the number of both infections and severely ongoing disease. Against the background of the growing number of infected, the issue of complicated communication with family doctors is to be expected. This problem has been reported by patients before, but it seems that the first blow to themselves will now have to be the primary health care ring. Of course, "Omicron" kills again. It will again take patients to the clinic and once again, there is a very high risk of rapid infection of the same medical staff, putting us at risk of losing a working hand.

Experts comment that half or more of economies in East Asia and Pacific, Latin America and the Caribbean, and the Middle East and North Africa, and two-fifths of economies in Sub-Saharan Africa will still be below their 2019 per capita GDP levels by 2023. The support from the developed countries will be very much required for the revival of these countries in risk.

References

Alagoz, O., Sethi, A. K., Patterson, B. W., Churpek, M., Alhanaee, G., Scaria, E., et al. (2021). The impact of vaccination to control COVID-19 burden in the United States: A simulation modeling approach. *PLoS One, 16*(7), e0254456. https://doi.org/10.1371/journal.pone.0254456

Aljazeera. (2021). Bangladesh reopens schools after 18-month COVID shutdown. News Published on 13 September 2021.

ASEAN Briefing. (2020, November). EPRS | European Parliamentary Research Service Author: Martin Russell Members' Research Service PE 659.338.

Asian Development Bank. (2021). Impact of COVID-19 on primary school students in disadvantaged areas of Bangladesh, ADB Briefs, NO. 200.

Biomedical Waste Management Rule. (2016). https://dhr.gov.in/sites/default/files/Biomedical_Waste_Management_Rules_2016.pdf

Brunell, D. C. (2020). Medical waste piles up during outbreak. *The Spokesman-Review*. https://www.spokesman.com/stories/2020/apr/09/don-c-brunell-medical-waste-piles-up-during-outbre

Cahya, G. H. (2020). *Stay home, President Says*. The Jakarta Post. https://www.thejakartapost.com/news/2020/03/16/stay-home-president-says.html

Daniel, K. W. C., Kenrie, P. Y. H., Haogao, G., Ronald, L. W. K., Pavithra, K., Daisy, Y. M. N., Gigi, Y. Z. L., Carrie, K. C. W., Man-Chun, C., Ka-Chun, N., Nicholls, J. M., Dominic, N. C. T., Malik, P., Michael, C. W., & Chan, L. L. M. (2021, 2021). Introduction of ORF3a-Q57H SARS-CoV-2 variant causing fourth epidemic wave of COVID-19, Hong Kong, China. *Emerging Infectious Diseases, 27*(5). https://doi.org/10.3201/eid2705.210015

DGHS. (2022a). Director General Health Service, Press release on 23 January, 2022.

DGHS. (2022b). Directorate general of health services (DGHS), Government of the People's Republic of Bangladesh. COVID-19 Dynamic Dashboard for Bangladesh. Retrieved January 21, 2022, from http://dashboard.dghs.gov.bd/webportal/pages/covid19.php

Gautam, S., Setu, S., Khan, M. G. Q., & Khan, M. B. (2022). Analysis of the health, economic and environmental impacts of COVID-19: The Bangladesh perspective. *Geosysystems and Geoenvironment, 1*, 100011.

Government of Bangladesh. (2022a). COVID-19 Vaccination Dashboard for Bangladesh. Retrieved January 22, 2022, from 247.238.92/webportal/ pages/covid19-vaccination-update. php/

Government of Bangladesh. (2022b). Real time health information dashboard, directorate general of health services, Mohakhali, Dhaka, Bangladesh. Retrieved January 22, 2022, from http://103.247.238.81/webportal/pages/rtlhb_menu.php/

Hossain, M. J., Rahman, S. M. A., Emran, T. B., Mitra, S., Islam, M. R., & Dhama, K. (2021). Recommendation and roadmap of mass vaccination against COVID-19 recommendation and roadmap of mass vaccination against coronavirus disease 2019 pandemic in Bangladesh as a lower-middle-income country. *Archives of Razi Institute, 76*(6). https://doi.org/10.22092/ARI. 2021.356357.1824

IEDCR. (2020). Institute of Epidemiology Disease Control and Research 2020. "Covid-19 Status Bangladesh." Retrieved from https://www.iedcr.gov.bd

Islam, M. R., Hasan, M., Nasreen, W., Tushar, M. I., & Bhuiyan, M. A. (2021). The COVID-19 vaccination experience in Bangladesh: Findings from a cross-sectional study. *International Journal of Immunopathology and Pharmacology, 35*, 1–13. https://doi.org/10.1177/ 20587384211065628

Kampf, G., Todt, D., Pfaender, S., & Steinmann, E. (2020). Persistence of coronaviruses on inanimate surfaces and their inactivation with biocidal agents. *The Journal of Hospital Infection, 2020*(104), 246–251.

Kamruzzaman, M. (2021). Bangladesh kicks off COVID-19 vaccination drive for schoolchildren. Anadolu Agency. Retrieved January 23, 2022, from www.aa.com.tr/en/asia-pacific/bangladesh-kicks-off-covid-19-vaccination-drive-for-schoolchildren/2408956/

Konings, F., Perkins, M. D., Kuhn, J. H., Pallen, M. J., Alm, E. J., Archer, B. N., & Van Kerkhove, M. D. (2021). SARS-CoV-2 variants of interest and concern naming scheme conducive for global discourse. *Nature Microbiology, 6*, 1–3.

Moghadas, S. M., Vilches, T. N., Zhang, K., Wells, C. R., Shoukat, A., Singer, B. H., & Galvani, A. P. (2021). The impact of vaccination on COVID-19 outbreaks in the United States. medRxiv.

Mamun, M. A., Sakib, N., Gozal, D., et al. (2021). The COVID-19 pandemic and serious psychological consequences in Bangladesh: A population-based nationwide study. *Journal of Affective Disorders, 279*, 462–472,. ISSN 0165-0327. https://doi.org/10.1016/j.jad.2020.10.036

New Age Bangladesh. (2020). 'HSC auto pass will have long-run impact on students, edn system', News published on 24 October, 2020.

Nzediegwu, C., & Chang, S. X. (2020). Improper solid waste management increases potential for COVID-19 spread in developing countries. *Resources, conservation, & Recycling, 161*, 104947. https://doi.org/10.1016/j.resconrec.2020.104947

OECD. (May, 2021). *Tackling Coronavirus COVID-19: Contributing to a global effect*. OECD.

Rahman, M. M., Chisty, M. A., Sakib, M. S., Quader, M. A., Shobuj, I. A., Alam, M. A., Halim, M. A., & Rahman, F. (2021). Status and perception toward the COVID -19 vaccine: A cross-sectional online survey among adult population of Bangladesh. *Health Science Reports, 4*(4), 1–10. https://doi.org/10.1002/hsr2.451

Riaz, B. K., Ali, L., Ahmad, S. A., Islam, M. Z., Ahmed, K. R., & Hossain, S. (2020). Community clinics in Bangladesh: A unique example of public-private partnership. *Heliyon, 6*(5), e03950. https://doi.org/10.1016/j.heliyon.2020.e03950

Siu, G. K., Lee, L. K., Leung, K. S., Leung, J. S., Ng, T. T., Chan, C. T., et al. (2021). Will a new clade of SARS-CoV-2 imported into the community spark a fourth wave of the COVID-19 outbreak in Hong Kong? *Emerging Microbes and Infection, 9*, 2497–2500. https://doi.org/10. 1080/22221751.2020.1851146

SWM Rules. (2016). Govt of India. https://cpcb.nic.in/uploads/MSW/SWM_2016.pdf

Tasnim, R., Sujan, M. S. H., Islam, M. S., Ferdous, M. Z., Hasan, M. M., Koly, K. N., & Potenza, M. N. (2021). Depression and anxiety among individuals with medical conditions during the COVID-19 pandemic: Findings from a nationwide survey in Bangladesh. *Acta Psychologica, 220*, 103426.

The Financial Express. (2022). Petition filed with High Court seeking closure of educational institutions. News published on 19 January, 2022.

Tribune, D. (2022). Educational institutions closed for two weeks amid Covid surge. News published on 21 January, 2022

UKRI. (2022). The logistics of mass vaccination in Bangladesh and beyond. Retrieved January 22, 2022, from https://www.ukri.org/our-work/tackling-the-impact-of-covid-19/our-global-contribution/the-logistics-of-mass-vaccination-in-bangladesh-and-beyond/

UNESCO. (2021). Situation analysis on the effects of and responses to COVID-19 on the education sector in Asia, Bangladesh case Study, October, 2021.

UNICEF Report. (2021). https://www.unicef.org/india/press-releases/community-based-monitoring-gathers-voices-marginalized-families-india-assess-impact

World Bank. (2021). Shifting gears: Digitization and services-led development. South Asia economic focus, fall 2021. Washington, DC: World Bank. © World Bank. https://openknowledge.worldbank.org/handle/10986/36317

World Bank. (2022). Global Economic Prospects, January 2022. Washington, DC: World Bank. © World Bank. https://openknowledge.worldbank.org/handle/10986/36519

World Economic Outlook Update. (2021). *International Monetary Fund* (p. 6). International Monetary Fund.

World Health Organization. (2021). High-level Consultation for Finalization of National Digital Health Strategy. Retrieved January 22, 2022, from https://www.who.int/bangladesh/news/detail/22-11-2021-high-level-consultation-for-finalization-of-national-digital-health-strategy//

World Health Organization. (2022a). COVID-19 Weekly Epidemiological Update. https://www.who.int/docs/default-source/coronaviruse/situation-reports/20220118_weekly_epi_update_75.pdf?sfvrsn=3f8800dc_5 & download=true

World Health Organization. (2022b). Tracking SARS-CoV-2 variants. https://www.who.int/en/activities/tracking-SARS-CoV-2-variants/

World Health Organization. (2022c). Last consultation on January 22, 2022.

World Health Organization. (2022d). COVID_19 Vaccines. Retrieved January 22, 2022, from www.who.int/emergencies/diseases/novel-coronavirus-2019/covid-19-vaccines/

World Health Organization. (2022e). COVID-19 advice for the public: Getting vaccinated. Retrieved January 22, 2022, from https://www.who.int/emergencies/diseases/novel-coronavirus-2019/covid-19-vaccines/advice/